Dynamics and the Problem of Recognition in Biological Macromolecules

NATO ASI Series

Advanced Science Institutes Series

A series presenting the results of activities sponsored by the NATO Science Committee, which aims at the dissemination of advanced scientific and technological knowledge, with a view to strengthening links between scientific communities.

The series is published by an international board of publishers in conjunction with the NATO Scientific Affairs Division

A	**Life Sciences**	Plenum Publishing Corporation
B	**Physics**	New York and London
C	**Mathematical and Physical Sciences**	Kluwer Academic Publishers Dordrecht, Boston, and London
D	**Behavioral and Social Sciences**	
E	**Applied Sciences**	
F	**Computer and Systems Sciences**	Springer-Verlag
G	**Ecological Sciences**	Berlin, Heidelberg, New York, London,
H	**Cell Biology**	Paris, Tokyo, Hong Kong, and Barcelona
I	**Global Environmental Change**	

PARTNERSHIP SUB-SERIES

1. Disarmament Technologies	Kluwer Academic Publishers
2. Environment	Springer-Verlag
3. High Technology	Kluwer Academic Publishers
4. Science and Technology Policy	Kluwer Academic Publishers
5. Computer Networking	Kluwer Academic Publishers

The Partnership Sub-Series incorporates activities undertaken in collaboration with NATO's Cooperation Partners, the countries of the CIS and Central and Eastern Europe, in Priority Areas of concern to those countries.

Recent Volumes in this Series:

Volume 285 — Molecular, Cellular, and Clinical Aspects of Angiogenesis
edited by Michael E. Maragoudakis

Volume 286 — Analytical Use of Fluorescent Probes in Oncology
edited by Elli Kohen and Joseph G. Hirschberg

Volume 287 — Light as an Energy Source and Information Carrier in Plant Physiology
edited by Robert C. Jennings, Giuseppe Zucchelli, Francesco Ghetti, and Giuliano Colombetti

Volume 288 — Dynamics and the Problem of Recognition in Biological Macromolecules
edited by Oleg Jardetzky and Jean-François Lefèvre

Series A: Life Sciences

Dynamics and the Problem of Recognition in Biological Macromolecules

Edited by

Oleg Jardetzky

Stanford University
Stanford, California

and

Jean-François Lefèvre

ESBS, Université Louis Pasteur
Illkirch Graffenstaden, France

Springer Science+Business Media, LLC

Proceedings of a NATO Advanced Study Institute and of
the International School of Biological Magnetic Resonance's Second Course on
Dynamics and the Problem of Recognition in Biological Macromolecules,
held May 19 – 30, 1995,
in Erice, Sicily, Italy

NATO-PCO-DATA BASE

The electronic index to the NATO ASI Series provides full bibliographical references (with keywords and/or abstracts) to about 50,000 contributions from international scientists published in all sections of the NATO ASI Series. Access to the NATO-PCO-DATA BASE is possible in two ways:

—via online FILE 128 (NATO-PCO-DATA BASE) hosted by ESRIN, Via Galileo Galilei, I-00044 Frascati, Italy

—via CD-ROM "NATO Science and Technology Disk" with user-friendly retrieval software in English, French, and German (©WTV GmbH and DATAWARE Technologies, Inc. 1989). The CD-ROM also contains the AGARD Aerospace Database.

The CD-ROM can be ordered through any member of the Board of Publishers or through NATO-PCO, Overijse, Belgium.

Library of Congress Cataloging-in-Publication Data

Dynamics and the problem of recognition in biological macromolecules /
   edited by Oleg Jardetzky and Jean-François Lefèvre.
      p.   cm. -- (NATO ASI series. Series A, Life sciences ; v.
   288)
      "Published in cooperation with NATO Scientific Affairs Division."
      "Proceedings of a NATO Advanced Study Institute and of the
   International School of Biological Magnetic Resonance's Second
   Course on Dynamics and the Problem of Recognition in Biological
   Macromolecules"--CIP t.p. verso.
      Includes bibliographical references and index.
      ISBN 978-0-306-45388-5   ISBN 978-1-4615-5839-2 (eBook)
      DOI 10.1007/978-1-4615-5839-2
      1. Molecular recognition--Congresses.   2. Macromolecules-
   -Congresses.   I. Jardetzky, Oleg. II. Lefèvre, Jean-François.
   III. Series.
   QP517.M67D96   1996
   574.8'8--dc20   96-31701
                                                              CIP

ISBN 978-0-306-45388-5

© 1996 Springer Science+Business Media New York
Originally published by Plenum Press, New York in 1996

http://www.plenum.com

10 9 8 7 6 5 4 3 2 1

PREFACE

From within complex structures of organisms and cells down to the molecular level, biological processes all involve movement. Muscular fibers slide on each other to activate the muscle, as polymerases do along nucleic acids for replicating and transcribing the genetic material. Cells move and organize themselves into organs by recognizing each other through macromolecular surface-specific interactions. These recognition processes involve the mutual adaptation of structures that rely on their flexibility. All sorts of conformational changes occur in proteins involved in through-membrane signal transmission, showing another aspect of the flexibility of these macromolecules.

The movement and flexibility are inscribed in the polymeric nature of essential biological macromolecules such as proteins and nucleic acids. For instance, the well-defined structures formed by the long protein chain are held together by weak noncovalent interactions that design a complex potential well in which the protein floats, permanently fluctuating between several micro- or macroconformations in a wide range of frequencies and amplitudes.

The inherent mobility of biomolecular edifices may be crucial to the adaptation of their structures to particular functions. Progress in methods for investigating macromolecular structures and dynamics make this hypothesis not only attractive but more and more testable. In this field, X-ray crystallography and nuclear magnetic resonance are essential experimental methods, and more and more sophisticated approaches are being developed to integrate dynamic aspects into the detailed structural description of biological macromolecules. In this task, molecular dynamics simulation supplies the analysis with a wealth of tools and ideas. The complete understanding of the relationship among structure, dynamics, and function depends upon effective cooperation between these experimental and theoretical methods. Many aspects of the subject have to be considered, going from methodological developments to the study of various systems, each providing fragmental information on the dynamic behavior of biological macromolecules and its role in recognition processes. The second Advanced School in Biological NMR, held in Erice in the spring of 1995 and sponsored jointly by NATO and FEBS, was inspired by these principles.

CONTENTS

THE COOPERATIVE SUBSTRUCTURE OF PROTEIN MOLECULES

Yawen Bai and S. Walter Englander

The Johnson Research Foundation
Department of Biochemistry and Biophysics
University of Pennsylvania School of Medicine
Philadelphia, Pennsylvania 19104-6059

1. ABSTRACT

A new hydrogen exchange method makes it possible to study the unfolded state of a protein and its partially unfolded forms under native conditions. The hydrogen bonded groups that are exposed in each state, identified by NMR-detected hydrogen exchange measurements, define the structure of each intermediate. The free energy level of each state can be obtained from the measured hydrogen exchange rates. Initial results with cytochrome c depict four structural units which together account for the entire protein structure. The intermediate forms reveal the cooperative substructure of the protein and appear to represent the major kinetic intermediates in the protein folding pathway.

2. THE PROBLEM OF PROTEIN INTERMEDIATE STATES

To understand the principles of protein structure - their cooperativity, design, folding, evolution - one would like to be able to adopt the divide and conquer approach of the biochemist, to take proteins apart into their component elements, study them, and see how they fit back together. This kind of approach is defeated by the great cooperativity of protein structures. Separated parts of a protein molecule are not in general independently stable. In recent years approaches have been devised to study partially folded proteins, either as short-lived kinetic intermediates (Englander & Mayne, 1992; Baldwin, 1993) or as non-native forms at equilibrium under mildly destabilizing conditions (Ptitsyn, 1994; Kuwajima, 1989). Here we show how partially folded intermediates can be characterized by hydrogen exchange measurements under native conditions.

The results obtained identify four cooperative substructural units that compose the cytochrome c protein (Fig. 1). The units of substructure include two different omega loops (red and yellow), a helix plus an omega loop (green), and a unit that involves two docked helices (blue). The colors indicate the relative free energy levels of the different units (lowest

Dynamics and the Problem of Recognition in Biological Macromolecules, edited by Jardetzky and Lefèvre
Plenum Press, New York, 1996

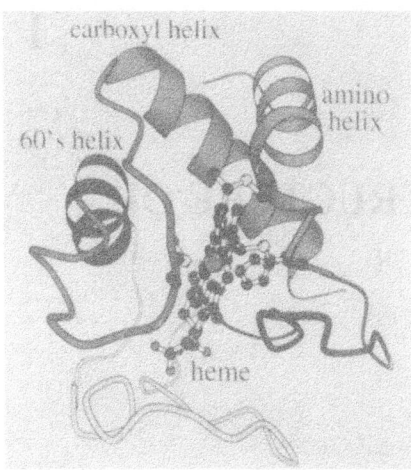

Figure 1. The cooperative substructural units that compose the cytochrome c protein, obtained from the native state hydrogen exchange experiment. A color representation of this figure can be found facing p. 112.

energy is red). In addition to showing the fundamental structural design of cyt c, these results appear to reveal the entire folding pathway that carries the protein from its unfolded to its folded native form.

3. EXPERIMENTAL AND THEORETICAL BASES

3.1. Hydrogen Exchange

The vast majority of slowly exchanging protein hydrogens represent peptide group NHs held in hydrogen bonded structure. For these hydrogens to exchange with hydrogens in the solvent, the protecting hydrogen bonds must be transiently broken. In the so-called EX2 (bimolecular exchange) limit, exchange is governed by Eq. 1.

$$k_{ex} = K_{op} \, k_{ch} \tag{1}$$

The exchange rate (k_{ex}) is the chemical exchange rate characteristic of the freely exposed NH (k_{ch}) (Bai et al, 1993; Connelly et al., 1993) multiplied by the fraction of time the protecting H-bond is broken, given by the equilibrium opening constant, K_{op}. From the measured exchange rate the free energy of the structural opening reaction can be obtained as in Eq. 2.

$$\Delta G_{op} = \text{-RT ln } K_{op} = \text{-RT ln}(k_{ex}/k_{op}) \tag{2}$$

For any given NH hydrogen, the opening that governs exchange may represent the breaking of a single H-bond, the global unfolding of the whole molecule, or some sub-global intermediate-sized opening. In general any NH can exchange through these various pathways, so that the exchange rate is given by Eq. 3.

$$k_{ex} = [K_{op}(\text{local}) + K_{op}(\text{subglobal}) + K_{op}(\text{global})] \, k_{ch} \tag{3}$$

Usually one kind of opening will dominate the exchange, but the dominant opening may be changed by adjusting ambient conditions.

3.2. Conformationally Excited States

Fig. 2 shows an energy diagram that includes the native state (N), the unfolded state (U), and some possible partially unfolded forms (PUFs) at intermediate energy levels. All possible PUFs must be populated at a level determined by the Boltzmann distribution, and over time all the protein molecules must cycle through all possible states. If the protein is extremely cooperative, PUFs will exist only at free energy levels higher than U, but some PUFs may exist at lower levels with free energy between N and U. These forms are invisible to most measurements, which are dominated by the abundant native state. Hydrogen exchange methods escape this limitation because N makes no contribution to the HX rate of protected NHs. Information on some partially unfolded forms can be obtained from HX measurements if the H-bond breakage reactions that produce a given intermediate dominate the HX rate of the proteins exposed in the unfolding. This condition can be met by the suitable manipulation of denaturants at low concentration.

3.3. Energy Level Manipulation by Denaturant

Denaturants like guanidinium chloride (GdmCl) exert their effects by binding weakly to solvent exposed protein surface (Pace, 1986; Schellman, 1987). Because U exposes additional GdmCl-binding surface, GdmCl shifts the N to U equilibrium toward U by changing the free energy gap (ΔGu) between U and N, according to Eq. 4.

$$\Delta G_u(GdmCl) = \Delta G_u(0) - m[GdmCl] \tag{4}$$

The protein chemist uses this relationship to determine the stabilization free energy of proteins, that is the energy gap, $\Delta G_u(0)$, between U and N at zero GdmCl concentration. To obtain ΔG_u, it is necessary to measure the equilibrium constant for the N to U equilibrium. At normal solution conditions, the equilibrium is far over toward N, so that one cannot quantify [U] and directly obtain [U]/[N]. One overcomes the problem by increasing GdmCl concentration to carry the protein through the melting transition. The free energy gap as a function of GdmCl [$\Delta G_u(GdmCl)$] can then be measured through the transition (from K = U/N) and extrapolated back to zero GdmCl, according to Eq. 4. The slope m proportions to the additional surface exposed in the equilibrium unfolding reaction measured.

Some results of this kind for cyt c are shown by the open symbols in Fig. 3. Till now, this kind of analysis has required a long extrapolation to obtain ΔG_u at zero GdmCl. It has

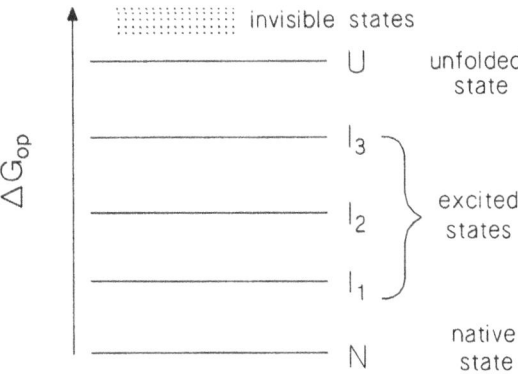

Figure 2. An energy level diagram that includes the lowest free energy native state (N), the unfolded state (U), and some possible partially unfolded forms (PUFs).

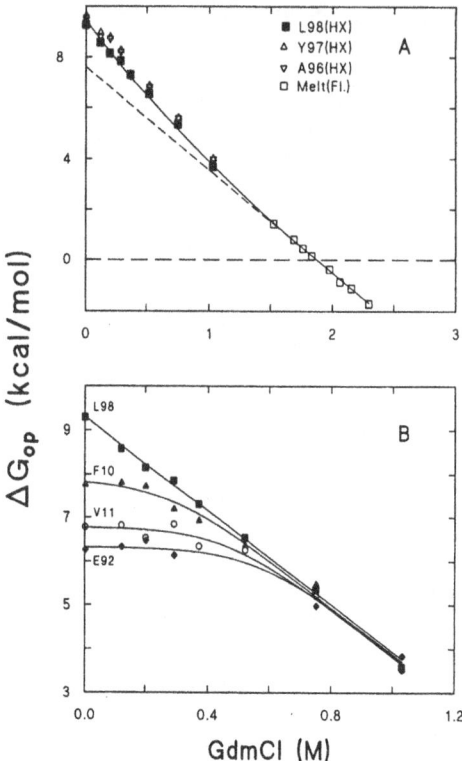

Figure 3. Free energy for structural opening reactions as a function of GdmCl concentration. A) The closed symbols are from HX rates for the 3 slowest NHs in cyt c. The open symbols are from a classical GdmCl melting experiment. B) Faster exchanging NHs in the amino- and carboxy-helices join the global unfolding isotherm. Only a few of the total of 16 NHs that join the global isotherm are shown.

not till now been possible to obtain data for the U/N equilibrium and therefore for ΔG_u at conditions below the melting transition.

4. THE CONFORMATIONALLY EXCITED STATES OF CYTOCHROME C

4.1. Global Unfolding by Hydrogen Exchange

Some NH groups in cyt c are so well protected that they can only exchange when cyt c experiences transient global unfolding. The direct measurement of the HX rate of these NHs will then yield K_{op}(global) (Eq. 1) and ΔG_{op} (Eq. 2). The values obtained from the HX rates of the 3 slowest NHs in cyt c are shown by the closed symbols in Fig. 3A. The HX curve merges smoothly with the ΔG_u values determined from classical melting experiments at high GdmCl and point to the correct value for the N to U free energy gap at zero GdmCl (Bai et al., 1994). Similar agreement was demonstrated for the slowest exchanging peptide group NH in ribonuclease A.

This experiment makes it possible to detect and quantify the cycling of proteins through the globally unfolded state even at essentially native conditions. One simply puts the protein into solution in D_2O and observes the time dependent decrease of the NMR resonances of the slowest exchanging NHs, using either 1D or 2D NMR methods.

4.2. Local and Global Unfolding

The HX behavior of some other NHs in cyt c is shown in Fig. 3B. At low GdmCl concentration, these NHs exhibit small m values (Eq. 4) because their exchange is dominated by small unfolding reactions that expose little new GdmCl sensitive surface (Eq. 4). As GdmCl is increased, the large scale global unfolding reaction is promoted, as indicated in Fig. 3B by the Leu98 NH, and rises to dominate the exchange (Eq. 3) of all the NHs in the amino-terminal and the carboxy-terminal helices of cyt c. All these NHs merge into a common HX curve (Fig. 3B) that we will refer to as an HX isotherm.

4.3. Subglobal Unfolding Reactions

It can be expected that all the exchanging hydrogens will ultimately be taken over by the large scale global unfolding equilibrium as GdmCl is increased. Fig. 4 shows that before this occurs, something more interesting happens. Different NHs merge into a series of lower lying HX isotherms that exhibit serially decreasing values of m and ΔG_{op}. Fig. 4A shows four such isotherms; for each isotherm only two hydrogens are shown to avoid confusion. The more complete behavior is shown for one of the isotherms in Fig. 4B, which includes a number of NHs in the 60's helix of cyt c.

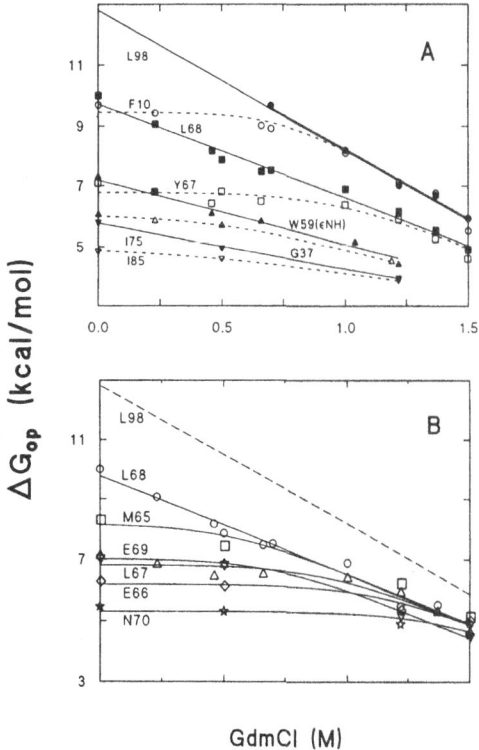

Figure 4. Subglobal unfolding isotherms. A) Two hydrogens are shown for each of the 4 isotherms found. B) The more complete behavior is shown for the isotherm that includes a number of NHs in the 60s helix.

5. THE COOPERATIVE SUBSTRUCTURE OF CYTOCHROME C

5.1. The Equilibrium Substructure

These HX isotherms represent large but still subglobal unfolding reactions, as indicated by their free energy and their slope. The identity of each unfolding unit can be inferred from the NH hydrogens seen to join the isotherm. The isotherms define the structural units shown in Fig. 1, which represent entire helices and/or entire omega loops, described before by Leczscinzki & Rose (1986).

The data reveal that cyt c is composed of four units that fold and unfold reversibly in a cooperative way. That the subglobal unfolding reactions are cooperative is indicated by the fact that openings intermediate between these units are not seen.

5.2. Implications for Protein Folding

The HX data obtained in this way do not give a complete description of the structure that unfolds to produce each overall partially unfolded form (PUF). For example, consider the yellow isotherm suggested in Fig. 4A. The more complete data show that the isotherm is joined by 5 NHs that appear to define the large bottom loop, shown in yellow in Fig. 1. It is possible that the yellow loop unfolds independently to produce the isothermal behavior seen. Another possibility is that when the yellow loop unfolds, the red loop is also unfolded. One cannot tell since the NHs that define the red loop have already exchanged and become invisible to the HX measurements that determine the yellow isotherm. The different models, for independent opening and for sequential opening, are represented by the upward pointing reaction arrows in Fig. 5.

The principle of microscopic reversibility requires that each up arrow in Fig. 5 must be matched by a down arrow, representing an equal and opposite flow of material in the refolding direction. This is so because the HX experiment was done at equilibrium conditions, in this case at pD 7 and 30°C, where the number of molecules that occupy any given energy state does not change in time. If the unfolding reactions detected by the native state HX experiment are properly represented by the sequential model in Fig. 5, then the sequence of unfolding reactions that carry cyt c reversibly from N to U must be matched by equivalent reactions that carry it back from U to N. These reactions (Eq. 5) would then describe the

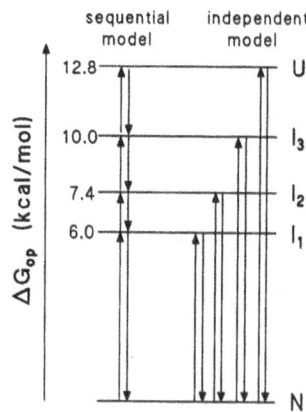

Figure 5. Illustration of the sequential opening and independent opening models. The free energy level measured for each unfolded state is shown.

major unfolding and refolding pathway of cyt c, and the PUFs would represent the intermediates in the cyt c folding pathway.

$$N_{RYGB} <=> P_{YGB} <=> P_{GB} <=> P_B <=> U \qquad (5)$$

A number of external observations support this most interesting possibility. The last unit to unfold in Eq. 5, the amino-terminal plus carboxy-terminal helices (blue), is the first element seen to fold in HX pulse labeling experiments (Roder et al. 1988). Other work has shown that the barrier that blocks folding at this stage is due to the misligation to the heme of either His26 or His33, both of which are in the green unit (Fig. 1). The sequence in Eq. 5 would predict this result since the misfold would allow the blue unit to form, but would block at the step of folding the green unit. Again, the blue unit appears to have independent stability (Kuroda, 1993; Wu et al., 1993), consistent with the suggestion that it forms first in folding and the other units, which are not independently stable, fold subsequently by docking with it to gain the necessary stability. All these observations support the last step in Eq. 5. The first step is a necessary component of either the independent or the sequential model. Thus the only intermediate not supported by external data is P_{GB} in Eq. 5. In the independent unfolding model, the unfolding of the yellow unit would generate P_{RGB} and not P_{GB}. A test (Bai et al., 1995) that compares the m value measured with the surface calculated to be exposed by these two different openings favors the form, P_{GB}, as in Eq. 5.

6. CONCLUSIONS

This work demonstrates a hydrogen exchange method that can reveal sizeable but still subglobal unfolding units in protein molecules. The experiment is done at native conditions, and takes advantage of the ability of denaturants at relatively low concentration to selectively promote larger unfolding reactions. Results so far available for cytochrome c reveal the fundamental cooperative substructural design of the protein. The results also appear to define the metastable intermediates that compose the major folding pathway.

7. REFERENCES

Bai, Y., Milne, J. S., Mayne, L., and Englander, S. W. (1994) *Proteins 20*, 4-14.
Bai, Y., Sosnick, T. R., Mayne, L. and Englander, S. W. (1995) *Science*, in press.
Bai, Y., Milne, J. S., Mayne, L., and Englander, S. W. (1993) *Proteins 17*, 75-86.
Baldwin, R. L. (1993) *Curr. Opin. Struct. Biol.* 3, 84-91.
Connelly, G. P., Bai, Y., Jeng, M-F., and Englander, S. W. (1993) *Proteins 17*, 87-92.
Englander, S. W. and Mayne, L. (1992) *Annu. Rev. Biophys. Biomol. Struct. 21*, 243-265.
Kuroda, Y. (1993) *Biochemistry 32*, 1219-1224.
Kuwajima, K. (1989) *Proteins 6*, 87-103.
Kraulis, P. J. (1991) *J. Appl. Crystallogr.* 24, 946-950.
Leszczynski, J. F. and Rose, G. D. (1986) *Science 234*, 849.
Pace, C. N. (1986) *Methods Enzymol. 131*, 266-280.
Ptitsyn, O. B. (1994) *Protein Eng.* 7, 593-596.
Roder, H., Elove, G. A. and Englander, S. W. (1988) *Nature 335*, 700-704.
Schellman, J. A. (1987) *Biopolymers 26*, 549-559.
Wu, L. C. et al. (1993) *Biochemistry 32*, 1027-76.

2

NMR STRUCTURES OF SERINE PROTEINASE INHIBITORS LDTI AND RBI, AND COMPARISON WITH X-RAY STRUCTURES

Tad A. Holak and Richard A. Engh

Max-Planck-Institute for Biochemistry
D-82152 Martinsried, F.R.G.

1. ABSTRACT

Serine protease inhibitors are extremely widespread and are found in bacteria as well as in eukaryotic cells. They are divided into at least 13 distinct classes (for reviews, see Laskowski and Kato, 1980; Laskowski, 1986; Laskowski et al., 1987a-b; Richardson, 1991). Representative three-dimensional structures have been determined for most of these families (Bode and Huber, 1992). Although their overall protein folds are different, their common feature is a structurally conserved binding loop, which binds in the active site of the corresponding serine protease in a substrate-like fashion. Three-dimensional structures of two serine protease inhibitors are described in this chapter: LDTI, leech derived tryptase inhibitor and RBI, the bifunctional α-amylase/trypsin inhibitor from seeds of *ragi*. The characteristic structural properties of these inhibitors are illuminated in terms of their interaction with target proteases.[*]

---

[*] Abbreviations: RBI, bifunctional α-amylase/trypsin inhibitor from seeds of ragi (*Eleusine coracana* Gaertneri, Indian Finger Millet); P1, P2, P3 and P1', P2', P3', designate inhibitor residues amino-terminal and carboxy-terminal of the scissile peptide bond, respectively; BPTI, bovine basic pancreatic trypsin inhibitor; CMTI, *cucurbita maxima* trypsin inhibitor; LDTI, leech derived tryptase inhibitor; WAI 0.28, wheat α-amylase inhibitor 0.28; WAI 0.53, wheat α-amylase inhibitor 0.53; 2D, two-dimensional; 3D, three-dimensional; CD, circular dichroism; CT, constant time; D_2O, deuterated water; DQF-COSY, double quantum filtered homonuclear correlated spectroscopy; NMR, nuclear magnetic resonance; NOE, nuclear Overhauser effect; NOESY, two-dimensional NOE spectroscopy; HMQC, heteronuclear multi-quantum coherence spectroscopy; HSQC, heteronuclear single quantum coherence spectroscopy; HNHA, 3D $^1H^N$ - $^{15}N^H$ - $^1H^\alpha$ correlation spectrum; TOCSY (HOHAHA), total correlation spectroscopy (homonuclear Hartmann-Hahn spectroscopy); TPPI, time proportional phase incrementation; RMSD, root mean square difference; rms, root mean square; SA, simulated annealing; {SA}, an ensemble of the SA structures; (SA)m, the structure obtained by constrained minimization of the mean structure, the mean structure is obtained by averaging the coordinates of the {SA} structures.

Dynamics and the Problem of Recognition in Biological Macromolecules, edited by Jardetzky and Lefèvre
Plenum Press, New York, 1996

2. STRUCTURES OF TRYPSIN-LIKE PROTEASE INHIBITORS

2.1. Structure of Leech Derived Tryptase Inhibitor (LDTI-C) in Solution

2.1.1. Introduction. The leech derived tryptase inhibitor (LDTI) is a small protein from *Hirudo medicinalis* which inhibits human tryptase (Sommerhoff et al., 1994). The actual physiological functions of the human tryptase, a tetrameric trypsin-like serine prote-inase from mast cells, remain to be clarified (for a review see Eklund and Stevens, 1993). *In vitro* studies suggest that the enzyme is involved in the catabolism of extracellular matrix proteins (Schwartz et al., 1987), in coagulation (Maier et al., 1983), and in neuropeptide turnover (Caughey et al., 1988). Furthermore, a pathogenetic role of tryptase in allergic reactions and destruction of cartilage as in asthma (Wenzel et al., 1988), rheumatoid arthritis (Stevens et al., 1992), fibrosis (Claman, 1989), artherosclerosis (Kokkonen and Kovanen, 1989), scleroderma (Hawkins et al., 1985), anaphylaxis, and mastocytosis (Schwartz et al., 1985), has been supposed.

LDTI exists in three isoforms, designated LDTI-A, LDTI-B, and LDTI-C (Sommer-hoff et al., 1994). Form A, the shortest one, is 42 amino acids long and could not be separated from form B, which contains an additional C-terminal Gly. Form C contains 46 amino acids and has a molecular mass of 4738 Da (Auerswald et al., 1994). In this chapter, we describe the three-dimensional structure of LDTI-C in solution as determined by 2D NMR.

2.1.2. Results and Discussion. Fig. 1 summarizes the sequential and medium-range NOEs between the NH protons and the NH, C^αH and C^βH protons, observed in the NOESY spectra of LDTI-C. These NOEs are the basis for the determination of the secondary structure of the protein. For LDTI-C, examination of the pattern of the NOEs suggests the presence of a helix like structure between residue 24 and 29. Strong sequential $H^N(i)$-$H^N(i+1)$ cross-peaks, typical for helices, were found in the spectra. Additionally, $H^N(i)$-$H^N(i+2)$ or

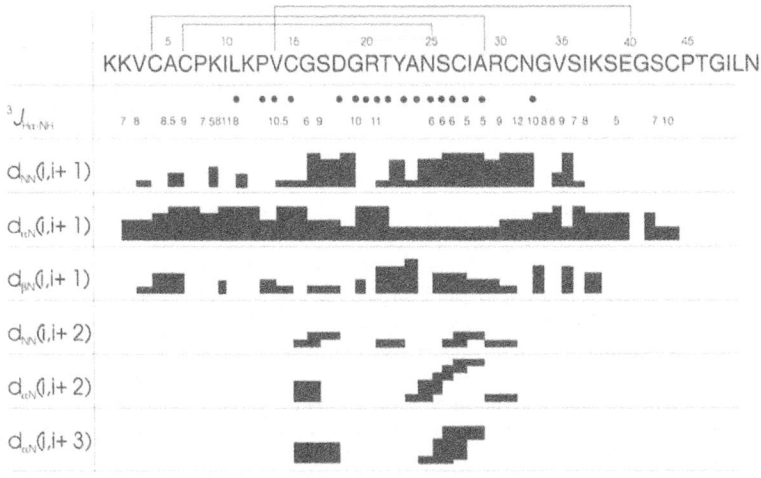

Figure 1. Amino acid sequence of LDTI-C and survey of NMR data used for identifying secondary structure. The NOEs ($|i$-$j|<5$), classified as weak, medium, strong, and very strong, are represented by the heights of the bars and were extracted from the NOESY spectra with mixing times of 100 ms. The C^αH(NH)(i)-C^δH(i+1) (Pro) NOE is shown along the same line as the C^αH(NH)(i)-NH(i+1) connectivities. Filled circles indicate NHs that did not exchange against D$_2$O after 6 hrs. The values of the $J_{C\alpha H\text{-}NH}$ coupling constants (in Hz) are indicated below the sequence. The disulfide pairing is C4-C29, C6-C25 an C14-C40.

$H^\alpha(i)$-$H^N(i+3)$ connectivities were observed. However, the $H^\alpha(i)$-$H^N(i+4)$ and the $H^\alpha(i)$-$H^\beta(i+3)$ cross-peaks, which are necessary to characterize this helix as a α-helix, were missing. On the other hand, the $H^\alpha(i)$-$H^N(i+2)$ cross-peaks, typical for a 3_{10}-helix, were present. For a regular 3_{10}-helix, however, one should also observe weak $H^\alpha(i)$-$H^\beta(i+3)$ cross-peaks, which as mentioned above are not observed. This segment of the protein can then be best described as a 3_{10}-helix-loop.

LDTI-C contains 6 cysteine residues and up to 3 disulfide bridges are possible. Based on the homology of LDTI-C to ovomucoid 3rd domain and other Kazal inhibitors, we predicted disulfide pairing between Cys 4-Cys 29, Cys 6-Cys 25, and Cys 14-Cys 40, confirmed by NOE contacts among the $C^\beta H$ and $C^\alpha H$ protons of corresponding cysteine residues in the disulfide bridges. Since, however, Cys 4 and Cys 6 are close to each other, the initial structures were calculated without the presence of the disulfide bridges i.e. no corresponding disulfide distance constraints nor S-S bonds were introduced in the calculations (Holak et al., 1989a). These structures also indicated the predicted disulfide pairing and in the final calculations the S-S bridges were treated as normal bonds.

The three-dimensional structure of LDTI-C was calculated from 262 approximate interresidue distance constraints (no intraresidue constraints were used). This number of constraints corresponds to the almost complete assignment of all cross-peaks in the NOESY spectra. A total of 20 structures were calculated by a hybrid distance geometry-simulated annealing method (Holak et al., 1989a) with the program XPLOR 3.1 (Brünger and Nilges, 1993). The global folding of the peptide chain in the core is uniquely defined due to a large number of NOEs; an average number of interresidue constraints per residue in the core of the protein was 18 (Fig. 2). All structures satisfy the experimental constraints (r.m.s.d.s for residual NOE and dihedral constraint violations were 0.0023 Å and 0.078°, respectively) with small deviations from idealized covalent geometry. The average atomic r.m.s difference for heavy atoms in the core of LTDI (residues 4-34) among the structures was 0.4±0.1 Å for

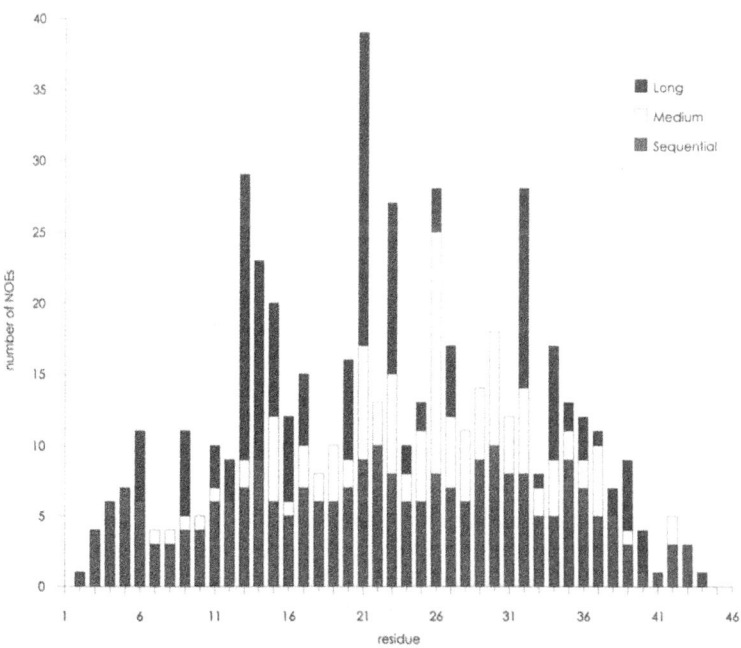

Figure 2. Plot of the number of NOE distance constraints per residue versus amino acid sequence of LDTI-C. All constraints appear twice, once for each interacting residue. No intra residue constraints were used.

A

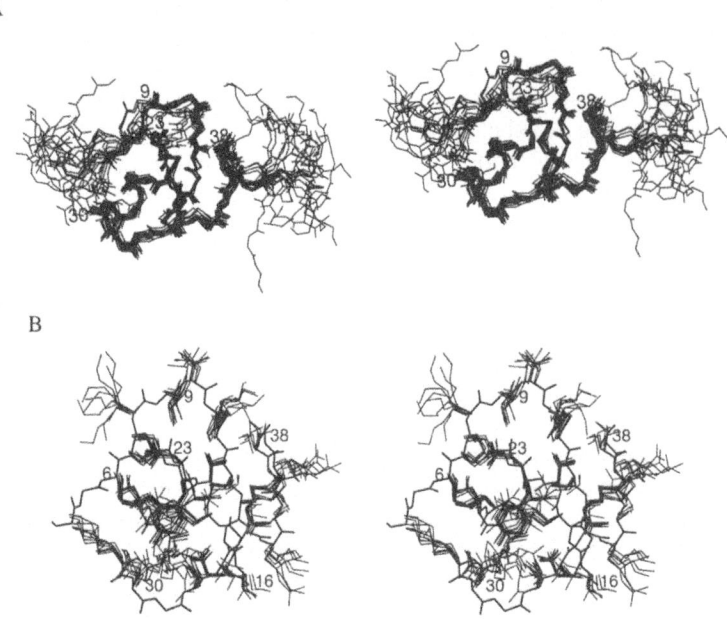

B

Figure 3. (A) Stereo view of the backbone atoms (N, Cα, C, and O) of LDTI-C structures best fitted to N, Cα, and C atoms of the residues 12-29. (B) Stereo picture of Cα tracing of LTDI with side chains of residues in the core.

the backbone atoms and 1.1±0.3 Å for all atoms. The average r.m.s difference between pairs of the structures (Holak et al., 1989a) in the φ, ψ angles was <25° and <40 °, respectively. Larger r.m.s. differences were observed for the C- and N-terminus (ca. 3 Å for the backbone atoms).

Fig. 3 indicates that the backbone of LDTI-C is well determined with the exception of residues 1-3 and the C-termini residues from 41-46. These two regions of the protein seem to be completely unstructured. There were no long-range NOEs observed for these regions, and the sequential NOEs were much weaker than NOEs of the residues in the core of the protein. Also, the TOCSY peaks were stronger with noticeable reduction in the linewidths compared to the peaks of residues in well structured regions.

The topology of the global fold of LDTI-C and its secondary structure elements are shown in Fig. 4a. The core of LDTI-C consists of a short one and a half-turn 3_{10}-helix-loop and an extended strand between Asn 30 and Val 32 followed by a loop between Ser 33 and Cys 40. A fragment from Val 13 to Ala 22 may be described as a interstrand connectivity similar to a mini β-sheet with a reversal at residue 15 to 19 (Fig. 4a). The strong Cys 14(CαH)-Thr 20(CαH) NOE and the medium Val 13(NH)-Tyr 21(NH) NOE is in agreement with such a description.

In general, the conformation of the side-chains are also well defined. Fig. 3b shows that only the side-chain of residues 8, 19 and 28 have greater r.m.s. differences. This may be caused by the inability to make stereospecific assignments for these side-chain protons. A second reason is the lack of side-chain/side-chain contacts and long-range NOEs for these residues which makes it impossible to fix the side-chain of these residues.

The N-terminus, which is connected to the helix via two disulfide-bridges, consists of the binding loop between residues 6-12. Residues 4 to 10 show greater variability

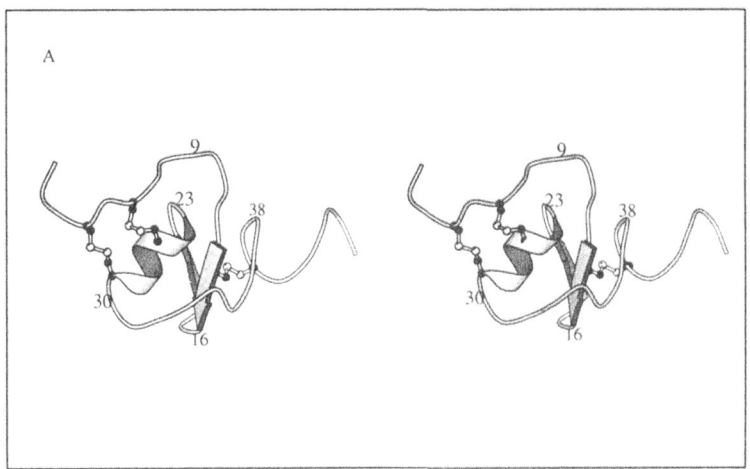

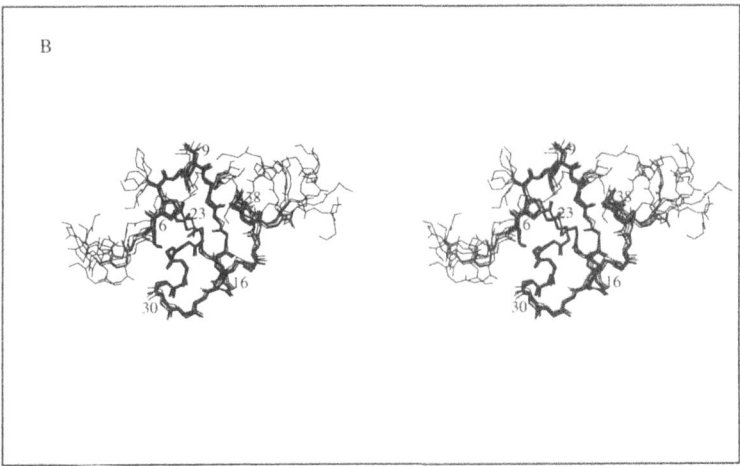

Figure 4. (A) Ribbon drawing of LDTI-C showing secondary structure and disulfide bridges. (B) set of the structures that exhibit the canonical conformation of the binding loop (residues 6-12).

compared to the core of LDTI-C in the 20 final structures, which is evident in Fig. 3a. This variability is due to the lack of non-sequential NOEs in this region. Only residue Ile 9 has long range contacts to the core of the protein. Thus this region of the strand is connected to the core exclusively via the disulfide-bridges and residue Ile 9. There was, however, a set of structures with a conformation of the binding loop shown in Fig. 4b. Although in no structures were there violations of the NOEs larger than 0.4 Å for loop residues, structures in Fig. 4b best fulfilled all the NOE and (especially) dihedral constraints. This is reminiscent of the situation found in conformations of the binding loop of CMTI-I (Holak et al., 1989a-b). In the case of CMTI-I, however, it was possible to choose a unique conformation of the trypsin binding loop from a family of different conformations because only this conformation matched the relative pattern of NOEs. For LDTI-C, other conformations of the binding loop fit the pattern of NOEs in the loop, although most can be eliminated due to unfavorable φ-ψ angles. The best fit conformation of the binding loop corresponds to the 'canonical conformation', a conformation of the residues surrounding the scissile bond (residues P3-P3') which is highly conserved among serine proteinase inhibitors (Bode and Huber, 1992; Holak et al., 1989b) and which promotes tight binding to the target protease (see Section 3 below). In LDTI-C, the following residues comprise this canonical loop: Cys 6 (P3), Pro 7 (P2), Lys 8 (P1), Ile 9 (P1'), Leu 10 (P2'), and Lys 11 (P3'). Lys 8, at position P1, is the primary determinant of specificity of the inhibitor, consistent with trypsin inhibition, and though disordered, is exposed and poised for insertion into the specificity pocket (Bode and Huber, 1992). A proline residue is found at position P2 also in CMTI (Holak et al., 1989b) and ovomucoid inhibitors (Bode and Huber, 1992) and in all cases adopts the canonical conformation.

An amino acid sequence comparison of LDTI with other protein inhibitors shows ca. 40% homology of LDTI to Kazal-type inhibitors (ovomucoids), such as turkey ovomucoid 3rd domain (OMTKY3), porcine pancreatic trypsin inhibitor (PSTI), and to the both domains of the Kazal-type-like rhodniin (Friedrich et al., 1993), with highest similarity to the non-classical Kazal-type inhibitor bdellin B3 (Fink et al., 1986). Rhodniin and bdellin B3 differ from the classical Kazal type inhibitors by their shortened NH-terminal sequences and by one or two internal deletions, respectively. The high identity of these primary structures implied similar 3D structures. Three dimensional structures of several Kazal type inhibitors are known (Bode and Huber, 1992), providing a basis for models, for example, of bdellin B3 and rhodniin (Friedrich et al., 1993; Fink et al., 1986). Common motifs of the experimentally determined and modeled structures are the exposed 'canonical' binding loop (Bode and Huber, 1992), the two to three turn α-helix, and a short three-stranded anti-parallel β-sheet. Distinct differences among the structures are found at the N-termini and at segments connecting the helix with the third β-strand.

2.2. Three-Dimensional Structure of the Bifunctional α-Amylase/Trypsin Inhibitor from Ragi

2.2.1. Introduction. The α-amylase/trypsin inhibitor family (also termed cereal inhibitor family) is a relatively new class of plant inhibitors found in cereal seeds (Laskowski and Kato, 1980; Laskowski, 1986; Laskowski et al, 1987a; Richardson, 1991). It currently comprises 24 members with known sequences and an average size of about 120 amino acids. This class of proteins is characterized by a high content of cysteines that form four or five intramolecular disulfide bridges (Maeda et al., 1983 a-b; Poerio et al., 1991). All functional members of the cereal inhibitor family are either inhibitors of mammalian α-amylases or trypsin. The bifunctional α-amylase/trypsin inhibitor from ragi (*Eleusine coracana* Gaertneri, Indian finger millet) (RBI) is the only member possessing both functions simultaneously. The amino acid sequence of RBI has been determined and consists of 122 residues with 10 cysteines forming 5 intramolecular disulfide bonds (Campos and Richardson, 1983). RBI shares between 22 and 66 percent sequence identity with the other members of the family. Biochemical data suggested that its binding to trypsin and α-amylase are located in mutually exclusive binding sites since the inhibitor is capable to form a ternary complex with trypsin and α-amylase (Shivaraj and Pattabiraman, 1981). The crystallization and preliminary X-ray investigation of RBI was published (Srinivasan et al. 1991), but the three-dimensional structure of the protein has not been reported so far. We describe here the determination of the secondary and three-dimensional structures of recombinant RBI, its disulfide bond pattern, and the sequential assignment of its NMR resonances. The secondary structure of RBI consists of both α-helical and β-structure elements. The secondary structure topology and the three-dimensional fold of the inhibitor are entirely different from known structures of the serine proteinase inhibitors. The structural features of RBI are discussed in

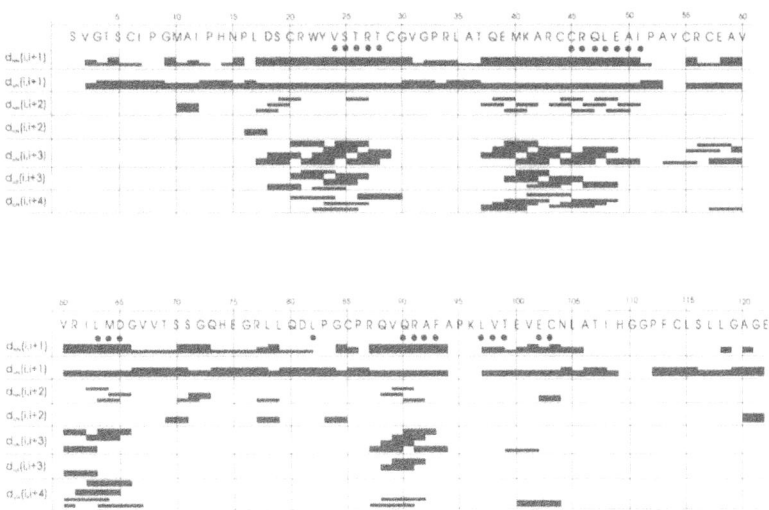

Figure 5. Summary of the short-range NOEs in RBI. The NOEs (i-j<5), classified as weak, medium, and strong, are represented by the heights of the bars and were extracted from the NOESY spectra with mixing times of 100 ms. The C$^\alpha$H(NH)(i)-C$^\delta$H(i+1) (Pro) NOE is shown along the same line as the C$^\alpha$H(NH)(i)-NH(i+1) connectivities. Filled circles indicate NHs that did not exchange against D$_2$O after 6 hours.

terms of the conserved and substrate-like inhibition mode among serine proteinase inhibitors and in terms of possible sites of interaction with α-amylase. We also show that the overall structure of the inhibitor is not affected by cleavage at the reactive peptide bond responsible for trypsin binding.

2.2.2. Results and Discussion. Secondary structure. Fig. 5 summarizes the sequential and short-range NOEs observed in the spectra of RBI. Many sequential H^N(i)-H^N(i+1), H^N(i)-H^N(i+2), and H$^\alpha$(i)-H^N(i+3) cross-peaks, typical for α-helices, were found in the spectra. β-Strands were identified by the presence of strong sequential H$^\alpha$(i)-H^N(i+1) cross-peaks, weak intraresidual H$^\alpha$(i)-H^N(i) cross-peaks, and H^N(i)-H^N(j) interstrand NOEs. As shown in Figures 6-7, four well-defined α-helices could be identified between residues 18-29, 37-51, 58-65, and 87-94. The two short β-strands, which form an antiparallel β-sheet, were found between residues 67-69 and 73-75. One interstrand H$^\alpha$(i)-H$^\alpha$(j) cross-peak and two interstrand H^N(i)-H^N(j) cross-peaks were observed, allowing the identification of three hydrogen bonds between the two different β-strands (Wüthrich, 1986). Another short β-strand may be present from residues 78 to 81. This region is characterized by strong sequential H$^\alpha$(i)-^{N}H(i+1) and weak intraresidual H$^\alpha$(i)-^{N}H(i) cross-peaks. However, since these residues do not form any hydrogen bridges to another β-strand, and as some of the α-protons show an unusual high-field shift (L78 4.27 ppm, L79 4.33 ppm, D81 4.29 ppm), this stretch of secondary structure could not be firmly described as β-strand. Based on its NOE cross-peak pattern, the turn between residues 82 and 85 could be ascribed to a type II turn.

Up to now, secondary structure analysis of the cereal inhibitor family was based on circular dichroism spectra from a few inhibitors. The secondary structure content predicted from the CD spectra was for RBI approximately 10% α-helix and 30% β-strand (Alagiri and Singh, 1993), for the rye trypsin inhibitor and the rye α-amylase inhibitor, 36-37% α-helix and 18-20% β-strand, and 39-40% α-helix and 11-13% β-strand, respectively (Lyons et al.,

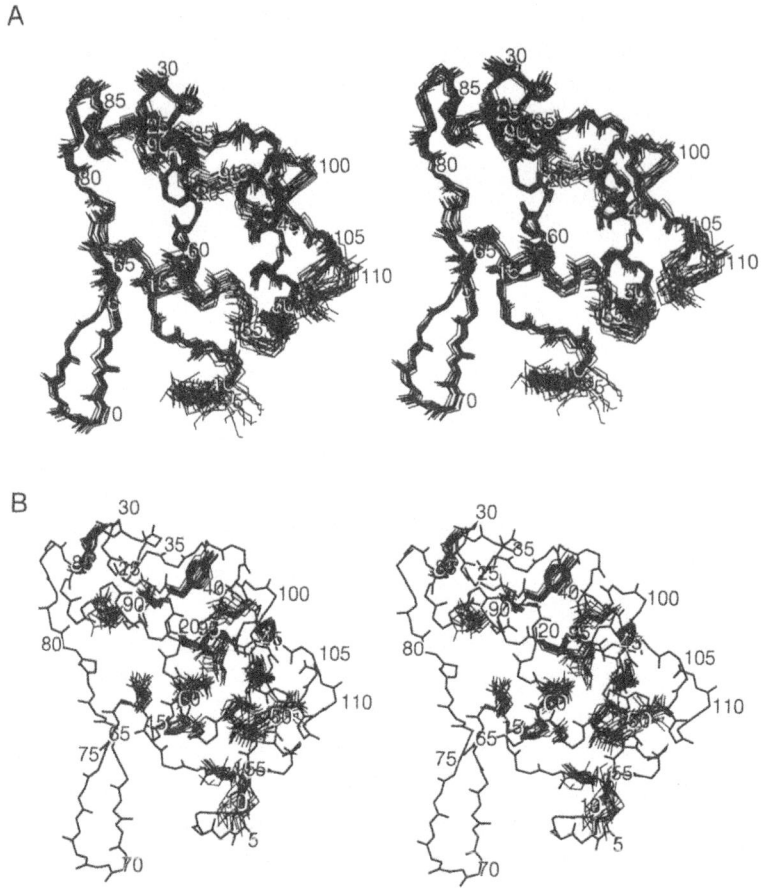

Figure 6. (A) Stereo view of the backbone atoms (N, C$^{\alpha}$, C, and O) for residues 6-111 of RBI structures best fitted to N, C$^{\alpha}$, and C atoms of the residues 13-114. (B) Stereo picture of C$^{\alpha}$ tracing of one of the RBI structures with side chains of residues important for the protein core formation.

1987), for the maize trypsin inhibitor, approximately 40% α-helix and 20% β-strand (Mahoney et al., 1984), and for the wheat α-amylase inhibitor 0.19, about 33% α-helix and 16% β-strand (Petrucci et al., 1976). Our data show unequivocally that the secondary structure of RBI contains 33% α-helices and about 7% β-structure elements. Four well defined α-helices and at least two short β-strands, which form an antiparallel β-sheet, could be determined. This result is in disagreement with the prediction of Alagiri and Singh (1993) but is in good agreement with the secondary structure predictions deduced from CD measurements for the other members of the inhibitor family. Within the pH range examined, the secondary structure elements are basically unaltered in the inhibitor cleaved at the reactive peptide bond Arg 34-Leu 35, indicating that the overall polypeptide fold is not significantly affected by the cleavage.

The overall secondary structure content of 40% is rather low in RBI. However, this is not unexpected since a number of other small protein inhibitors of serine proteinases with high disulfide bond content such as Bowman-Birk-type inhibitors (Werner and Wemmer, 1991) and hirudins (Rydel et al., 1990; Folkers et al., 1989; Haruyama and Wüthrich, 1989) have also few secondary structure elements. Obviously, in these extraordinarily stable

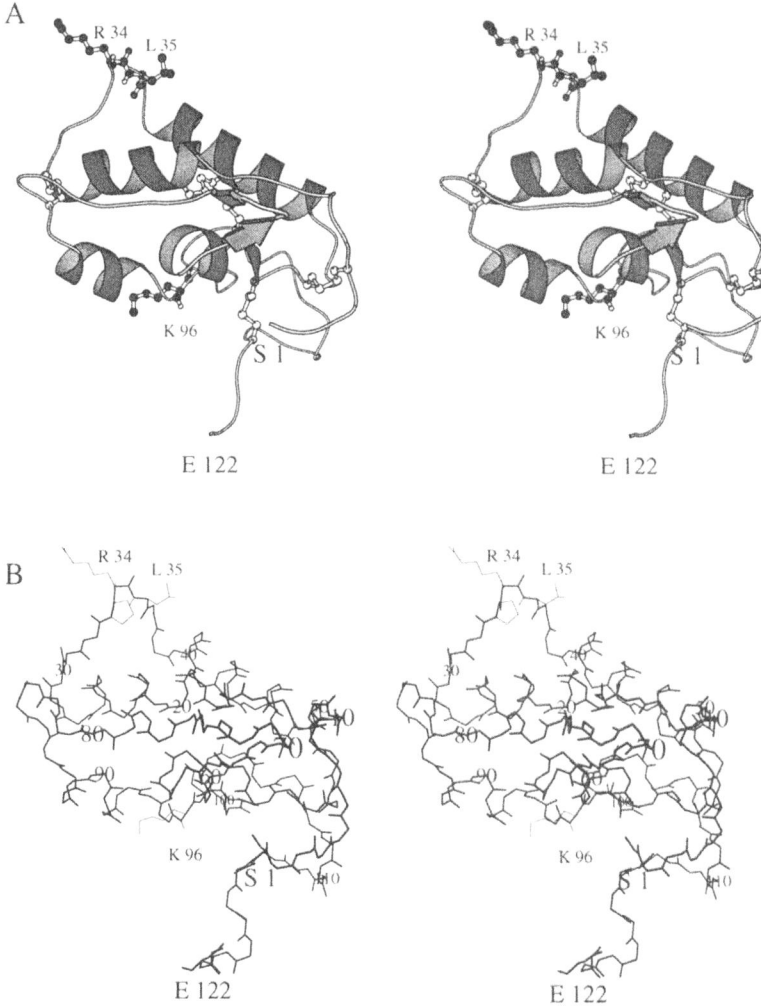

Figure 7. (A) Ribbon drawing of RBI showing the secondary structure, disulfide bridges, and side-chains of residues R34, L35 and K96. (B) The (SA)m structure with details of the canonical conformation of the binding loop (residues 32-37). All backbone atoms, with the exception of hydrogen, are shown.

secretory proteins, intramolecular disulfide bond crosslinking compensates for a low content of stabilizing secondary structure.

There are no similarities in the secondary structure topology of RBI to other members of the group of serine proteinase protein inhibitors for which three-dimensional structures have been solved up to now (Bode and Huber, 1992). The scaffold stabilizing the trypsin binding loop of RBI exhibits a new motif, quite different from the other serine proteinase inhibitors, in that its binding loop is positioned between two α-helices.

Disulfide Bridges. Ten of the 122 residues of RBI are cysteines. All of them are involved in the formation of disulfide bridges (Campos and Richardson, 1983; Wunderlich and Glockshuber, 1993). Disulfide bridges between cysteine residues 6 and 55, 20 and 44, and 29 and 85 could unambigiously be identified on the basis of the diagnostic NOE connectivities $H^\beta(i)-H^\beta(j)$, $H^\alpha(i)-H^\beta(j)$, and $H^\alpha(i)-H^\alpha(j)$, where i and j are the residues in a S-S pair. To determine the location of the remaining two disulfide bridges, NOEs between one of the cysteines with another residue located in the vicinity of the second cysteine were used, for example, Cys 103 and Leu 48 (close to Cys 45); alternatively long-range NOEs between residues which are in the vicinity of both cysteines were used, for example, Ala 53 and Pro 112, which are close to Cys 57 and Cys 114, respectively.

Cys 45 and Cys 57 are close to Cys 44 and Cys 55 in the primary sequence, respectively. In principle therefore, alternate S-S bonding could be possible (Cys 57 to Cys 6, Cys 45 to Cys 20, and Cys 114 to Cys 55, for example). However, these configurations could be excluded, because at least one of the residues in all these putative pairs already had been assigned unequivocally to another disulfide bridge. The final disulfide bonding pattern of RBI is shown in Fig. 8.

The disulfide bridge pattern of two other members of the cereal inhibitor family, WAI 0.28 and WAI 0.53, had been determined before (Maeda et al., 1983 a-b; Poerio et al., 1991). A comparison of disulfide bond patterns of RBI and WAI 0.28 reveals an identical disulfide topology (Fig. 8), which together with its 26% amino acid similarity to RBI suggests that the secondary structure elements are likely to be similar in WAI 0.28 and RBI. In contrast, the cystine pattern of WAI 0.53 differs from RBI and WAI 0.28 in that the 9 cysteine inhibitor lacks one disulfide and one of its disulfides (Cys 6/Cys 115) has a different connectivity pattern (Fig. 8). However, the secondary structure elements are likely to be also conserved in WAI 0.53. Although Cys 6 and Cys 114 in RBI, corresponding to the Cys 6/Cys 115 disulfide in WAI 0.53, do not form a disulfide, the absence of Cys 55 would allow this alternative 6-114 cystine bridge. This assumption is supported both by the three-dimensional structure of RBI and by an engineered RBI mutant lacking Cys 55 (Maskos et al., 1995). Its CD spectrum and inhibitory activity against trypsin are comparable to those of the wild type inhibitor. Moreover, like WAI 0.53, this RBI mutant forms a dimer due to oxidative dimerization via the free cysteine residue. Therefore we propose that the secondary structure content is basically conserved throughout the whole cereal inhibitor family, independent of the different disulfide bridge patterns reported for some members of the protein family.

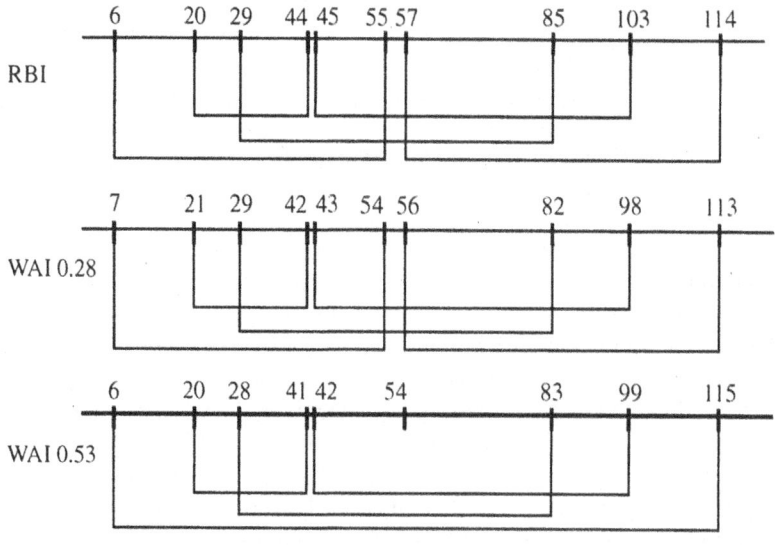

Figure 8. Schematic representation of the disulfide bond pattern of RBI and comparison with the known disulfide patterns of the related inhibitors WAI 0.28 and WAI 0.53.

Comparison Between the Native Inhibitor and the Inhibitor Cleaved by Trypsin at the Arg 34-Leu 35 Peptide Bond. The RBI samples contained a mixture of native RBI and RBI cleaved by trypsin at the reactive Arg 34-Leu 35 peptide bond. As a consequence of the disruption of the primary sequence in the cleaved form, chemical shifts of the H^N and H^α protons of residues 30 to 40 and the chemical shifts of the aromatic protons of residue 22 were different from the corresponding shifts of the native form by more than 0.01 ppm. There is no indication, however, that the secondary and tertiary structures of RBI are changed significantly beween the two forms. Despite the substantially different chemical shifts of the native and cleaved inhibitor for residues 22 and 30 to 40, and some smaller differences in chemical shifts for other residues, the NOE intensities of the cross-peaks were essentially identical for both forms, except for residues 32 to 38. Residues 32 to 38 appear to be unstructured in the cleaved form as no medium and long range NOEs could be detected in the spectra for these residues. Like native RBI, the cleaved inhibitor was fully stable during the NMR examination.

Tertiary Structure. The three-dimensional structure of RBI was calculated from 1131 approximate interresidue distance constraints.The global folding of the polypeptide chain is uniquely defined due to the large number of NOEs in the core of the protein (Figures 6,9). An average number of interresidue constraints per residue in the core of the protein was 12 (Fig. 9). All structures satisfy the experimental constraints with small deviations from idealized covalent geometry. The average atomic rms difference for heavy atoms in the core of RBI (residues 5-111) among the structures was 0.5±0.2 Å for the backbone atoms and 1.3±0.3 Å for all atoms. The average rms difference in the ϕ, ψ angles was <15° and <40°, respectively. Larger rms differences (ca. 3 Å for the backbone atoms and ca. 80° for the backbone torsion angles) were observed for the C- and N-terminus, for cysteine 55 (which makes the disulfide bridge with cysteine 5), and for Lys 96. For the two latter residues, the $J(H^N\text{-}H^\alpha)$ were ca. 6.5 Hz, preventing differentiation between positive and negative ϕ angles that were present in the calculated structures. This results in large rmsd for the ϕ angles whereas the rmsd for the atomic coordinates is small because the spatial positions of the C^α and nitrogen atoms is similar among the structures. Fig. 6 indicates that the backbone of RBI is well determined with the exception of residues 1-5 and the C-terminal residues 111-122. These two regions of the protein seem to be completely unstructured.

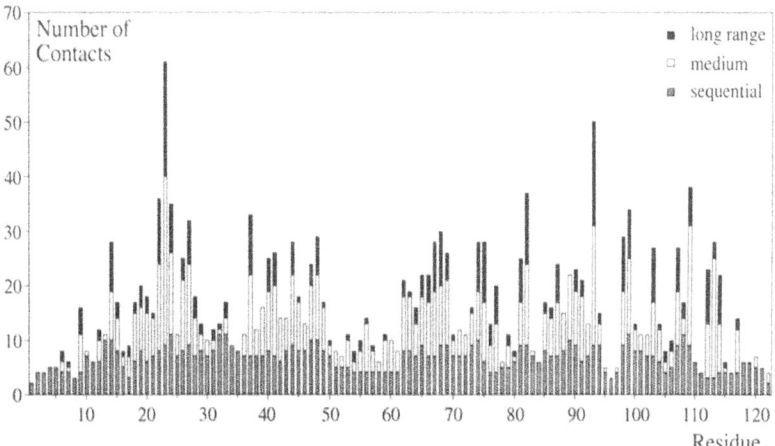

Figure 9. Plot of the number of NOE distance constraints per residue versus amino acid sequence of RBI. All constraints appear twice, once for each interacting residue. No intra residue constraints were used.

The global fold of RBI and its secondary structure elements are shown in Figure 6-7. The core of RBI consists of a globular four-helical motif with a simple 'up-and-down' topology. The helices are between residues 18-29, 37-51, 87-94, and 58-65. A fragment from Val 67 to Thr 69 and Gln 73 to Glu 75 forms a antiparallel β-sheet wi'ʰ a reversal short loop at residue 70 to 72 (Figures 6-7). The turn between residues 82 and '5 has a type II turn configuration. The RBI molecule has an approximate two fold symmᵣy with a symmetry plane perpendicular to the plane of Figure 6a and intersecting residues ₋eu 78 and Gly 110. Excepting the binding loop (see below), there are no similarities in the three-dimensional structure of RBI to other serine proteinase protein inhibitors for which structures have been solved to now (Bode and Huber, 1992).

Conserved residues present in all members of the RBI family contribute to the apolar core of the molecule (Figure 6b). None of these residues could directly form a contact with trypsin or α-amylase. Rather, residues conserved in the RBI family appear to be responsible for packing of the apolar core that defines the three-dimensional fold. The RBI fold should therefore be general for all members of the RBI family. Ten cysteines are almost completely conserved among all members of the family and appear to be primarily responsible for the stabilization of the overall fold. The only exceptions from the ten cysteine rule are the wheat inhibitor WAI 0.53[10] (lacking the fifth cysteine), barley pUp13[18] (lacking the tenth cysteine), and barley pUP13[23] (lacking the ninth cysteine). WAI 0.53 differs from RBI in that one of its disulfides (Cys 6/Cys 115) has a different connectivity pattern and one disulfide is lacking Nevertheless, the RBI fold seems to be conserved for this inhibitor too, and most probably it is also conserved for the barley pUp13[18] and pUP13[23]. Other residues from the core of the protein are also highly conserved throughout the family. These are hydrophobic residues: Pro 16, Leu 17, Val 24, Leu 48, Ile 51, Ala 59, Val 60, Leu 63, Leu 97, and Val 98, and two charged/polar residues, Arg 56 and Gln 90, which are completely buried (Figure 9b). Almost all these residues, together with Ile 12, are located at the interface between helix 1-2 and helix 3. Residues Val 24, Leu 97, and additionally Tyr 23 and Val 102, underpin the course of the polypeptide that defines packing between the end of helix 1, the middle of helix 4 and residues of the C-terminal loops.

The Conformation of the Trypsin Binding Loop. The trypsin binding loop of RBI is connected to the core of the protein via helix 18-29, the disulfide-bridge 29-85, α-helix 37-51, and disulfide bridges 44-20 and 45-103 (Figure 7). The trypsin binding loop of RBI exhibits a new motif of connection to the protein core different from all other inhibitors in that its binding loop is stabilized and positioned between two α-helices (helices 1 and 2).

3. COMPARISON OF THE LDTI AND RBI TRYPSIN BINDING LOOPS WITH X-RAY CRYSTAL STRUCTURES OF STANDARD MECHANISM INHIBITORS

The standard mechanism serine proteinase inhibitors (Laskowski and Kato, 1980) function as poorly cleaved but very tightly bound substrates. This property arises from a conserved 'canonical conformation' found in virtually all X-ray structures so far despite very different protein folds (Bode and Huber, 1992). This canonical conformation involves approximately six residues, three on either side of the scissile peptide bond (P3, P2, and P1 upstream and P1', P2', and P3' downstream). The clustering of the backbone ϕ, ψ angles is depicted in Fig. 10. This conformation is seen in isolated inhibitors and in complexes with enzymes, whereby only a relatively small increase in flexibility is apparent in the isolated inhibitors, even in spite of crystal contacts (Papamokos et al., 1982).

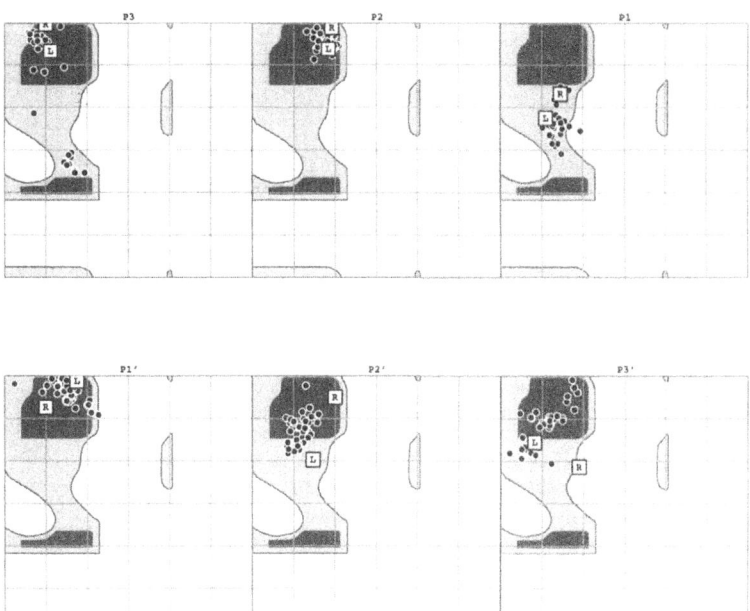

Figure 10. Ramachandran-type plots comparing LDTI and RBI reactive site loop conformations with X-ray crystal structures. Structures from the Brookhaven Data Bank include 1aap, 1acb, 1cgi, 1cgj, 1cho, 1cse, 1mct, 1mee, 1ovo (four monomers), 1pi2 (two binding loops), 1ppe, 1ppf, 1tab, 1tec, 1tgs, 1tpa, 2ci2, 2kai, 2ovo, 2ptc, 2sec, 2sic, 2sni, 2ssi, 2tec, 2tgp, 2tpi, 3sgb, 3tec, 3tpi, 4pti, 4sgb, 5pti, 6pti (1tie was eliminated due to crystallization artifacts). Squares show the conformation of the refined average structure for LDTI (L) and RBI (R). Only P3′ of RBI and P2′ of both RBI and LDTI deviate somewhat from crystal structure distributions. The energy contours plotted are for reference and only suggest folding energies; the P1 conformation is however in a less favored conformation.

The canonical conformation leaves residues P3-P3′ in an extended conformation. At position P3, the conformation usually corresponds to an antiparallel beta strand (except for proline in BPTI); a polyproline II conformation occurs at P2 and P1′; an approximate 3-10 helical conformation is typical for P1, and P2′ and P3′ each cluster near parallel beta strand conformations. The canonical conformation leaves the residues both exposed to solution in the solvated inhibitor and poised for interaction with corresponding recognition sites on the target proteinase. This is especially true for the P1 position, which is often called the primary specificity determinant (a basic residue for trypsin-like proteinases) and is most exposed. The formation of the reactive site loop into its bound conformation eliminates entropic and refolding costs of binding, and in particular enhances hydrophobic contributions to tight binding. The importance of the exposed side chains for specific inhibition is also shown by the evolutionary pressure toward diversification and hypervariability of the binding loop (Laskowski et al., 1987b)

NMR structures of serine proteinase inhibitors generally reflect the same conformations, although sparse NOE data may artifactually lead to predictions of alternate conformations or greater flexibility of the loop than actually exists (Jardetzky, 1980). The LDTI and RBI structures reported here are in close agreement with X-ray crystal structures (Tables 1-2 and Fig. 10). Deviations at P2′ are probably not significant. The unusual conformation of RBI at P3′ may be a consequence of the new protein fold represented by RBI, as the P3′ residue initiates an alpha helix. The Kazal type fold of LDTI is well represented among X-ray

Table 1. The numerical average values for φ and ψ of the NMR structures of LDTI

Amino acids in the binding loop	φ angle	φ angles in the canonical binding loop	ψ angle	ψ angles in the canonical binding loop
P4 (Ala 5)	−99±7		90±12	
P3 (Cys 6)	−113±16	−140° < φ < −120°	140±12	140° < ψ < 170°
P2 (Pro 7)	−70±1	−100° < φ < −60°	143±3	139° < ψ < 180°
P1 (Lys 8)	− 114±2	−120° < φ < −95°	45±5	9° < ψ < 50°
P1′ (Ile 9)	−75±1	−100° < φ < −60°	173±4	139° < ψ < 180°
P2′(Leu 10)	−92±1	−140° < φ < −99°	62±2	70° < ψ < 120°
P3′(Lys 11)	−131±5	−140° < φ < −99°	87±9	70° < ψ < 120°
P4′(Pro 12)	−77±1		142±18	

structures but for LDTI the N-terminus is significantly nearer the reactive site and the NMR structure indicates that only the disulphide bridge anchors the N-terminal side of the binding loop.

The nature of the protein fold and scaffolding interactions with the binding loop determine the extent of flexibility of the binding loop. Among X-ray structures, the binding loop of eglin-C is most mobile, anchored at its center to the body of the inhibitor only through polar interactions with arginine side chains. At the other extreme are the inhibitors with many disulphide bridges, such as CMTI or the Bowman-Birk inhibitors (Lin et al., 1993), which are very rigid. LDTI and RBI are intermediate between these two extremes. The shortened N-terminus of LDTI with respect to other Kazal type inhibitors seems to enhance flexibility at the upstream side of the reactive site loop, but the disulphide anchor remains restrictive. RBI's new scaffolding arrangement apparently compensates for a relative dearth of interactions with the body of the inhibitor with rigidity imposed by the downstream alpha helix.

4. MATERIALS AND METHODS

4.1. Sample Preparation

4.1.1. LDTI-C. A synthetic gene coding for form C of LDTI was designed, cloned and expressed in *Saccharomyces cerevisiae*. The secreted material was isolated after cross-flow filtration and purified by cation exchange chromatography, it is inhibitorily active and about 85% pure (Auerswald et al., 1994). The sample for NMR contained ~3 mM LDTI-C in 50 mM NaH_2PO_4/CD_3COONa buffer (pH 3.5) in 90% H_2O/10%D_2O. In addition, one sample was prepared in 100% D_2O.

Table 2. The numerical average values for φ and ψ of the NMR structures of RBI

Amino acids in the binding loop	φ angle	φ angles in the canonical binding loop	ψ angle	ψ angles in the canonical binding loop
P4 (Val 31)	−99±7		109±6	
P3 (Gly 32)	−120±3	−140° < φ < −120°	179±3	140° < ψ < 170°
P2 (Pro 33)	−65±1	−100° < φ < −60°	20±1	139° < ψ < 180°
P1 (Arg 34)	− 92±2	−120° < φ < −95°	46±5	9° < ψ < 50°
P1′ (Leu 35)	−103±6	−100° < φ < −60°	130±13	139° < ψ < 180°
P2′(Ala 36)	−79±13	−140° < φ < −99°	150±10	70° < ψ < 120°
P3′(Thr 37)	−66±15	−140° < φ < −99°	52±9	70° < ψ < 120°
P4′(Gln38)	−54±8		−103±86	

4.1.2. RBI. The RBI samples used for our NMR studies were obtained by overproduction in *E. coli* and subsequent three-step purification (Wunderlich and Glockshuber, 1993; Maskos et al., 1995). For unlabeled samples, *E.coli* JM83 cells (Yanisch-Perron et al., 1985) carrying the plasmid pRBI-PDI (Wunderlich and Glockshuber, 1993), were grown in LB medium (Sambrook et al., 1989). The uniformly ^{15}N-labeled and ^{15}N-glycine/serine-labeled samples were produced in the *E. coli* strain BL21(DE3) (Studier and Moffatt, 1986), carrying the plasmid pRBI-PDI-T7. This plasmid is a derivative of pRBI-PDI (Wunderlich and Glockshuber, 1993) and contains the T7 promoter instead of the lac promoter. The cells were grown on a M9 minimal medium (Sambrook et al., 1989) with ^{15}N-ammonium chloride (1 g/l) as the only nitrogen source (uniformly ^{15}N-labeling) and on minimal medium supplemented with ^{15}N-glycine (1 g/l) and the residual unlabled amino acids, except for serine (selective ^{15}N-glycine/serine-labeling) (Griffey and Redfield, 1985). The final yield of native RBI per liter bacterial culture was 0.41 mg for the unlabeled, 0.09 mg for the ^{15}N-labeled, and 0.10 mg for the ^{15}N-glycine/serine-labeled sample. Only the native inhibitor was used for NMR experiments. However, after extended time in the NMR tube, a significant amount (30–70%) of RBI was converted to the cleaved form, presumably due to minute amounts of trypsin released from the affinity column simultaneously with RBI.

The purified protein was dialyzed against 17 mM NaH_2PO_4, 100 mM NaCl and then concentrated to 1.7-3.0 mM in a volume of 450 μl. After addition of 50 μl D_2O (H_2O/D_2O = 9:1), the pH of the sample was 4.5. In some experiments, the pH of the sample was adjusted by addition of NaOH to 6.0. For recording spectra in D_2O, the samples were lyophilized and dissolved in D_2O. The pD was not corrected.

4.2. NMR Spectroscopy

All NMR measurements were carried out at 500 MHz and 600 MHz on the Bruker AM-500 and AMX-600 spectrometer, respectively. All spectra were measured at 27°C for LDTI-C and 37°C for RBI. The two-dimensional total correlation spectroscopy (TOCSY) was performed according to the method of Rance (Rance, 1987) with the MLEV-17 sequence (Bax and Davis, 1985) for isotropic mixing and spin-lock periods of 40, 50 and 70 ms. The TOCSY pulse sequences included presaturation of the water resonance for measurements in H_2O (Gueron et al., 1991). Nuclear Overhauser enhancement spectroscopy (NOESY) experiments (Jeener et al., 1979, Ernst et al., 1987) were recorded with a pulse sequence in which the last 90° pulse was replaced by a jump-return sequence to suppress the water resonance (Plateau and Gueron, 1982, Gueron et al., 1991). A homospoil pulse of 8 ms during the mixing times of 80 and 100 ms was also used. Essentially the same pulse sequences were used for the D_2O samples, with the exception that no water suppression was necessary. Usually 4096 complex data points were acquired in the time domain t_2 with a spectral width of 11.73 ppm in the F_2 dimension. 800 increments in the time period t_1 with an F_1 spectral width of 11.73 ppm and 96 scans per t_1 value were added. Quadrature detection in the indirectly detected dimensions was obtained with the TPPI method (Marion and Wüthrich, 1983). A CT-COSY experiment (Girvin, 1994; Bax and Freeman, 1981) with an optimized constant time delay of 36 ms and 512 t_1 increments was recorded on the unlabeled H_2O sample. The other parameters were the same as in the 2D TOCSY and NOESY spectra.

The 2D ^{1}H-^{15}N HMQC and HSQC correlation spectra were recorded for RBI as described by Summers et al. (1986) and Bodenhausen and Ruben (1980), respectively. The spectral parameters were the same as in the ^{15}N-separated 3D experiments (the number of increments in the ^{15}N time domain was larger (256)). The 3D [1]H-^{15}N-NOESY-HMQC (Zuiderweg and Fesik, 1990; Messerle et al., 1989) and [1]H-^{15}N-TOCSY-HMQC (Marion et al., 1989) spectra were recorded with a mixing time of 100 ms and 40 ms, respectively, and with 16 scans per t_1-t_2 pair. The spectral width and number of points acquired were 11.57 ppm and 90 complex points in $^1H(F_1)$, 44.11 ppm and 24 complex points in $^{15}N(F_2)$, and 5.24 ppm and 1024 real points in $^1H(F_3)$ with the $^1H(F_1)$, $^{15}N(F_2)$ and $^1H^N(F_3)$ carrier frequences placed at 4.73 ppm, 109.82 ppm and 7.83 ppm, respectively. All 3D spectra were processed with the software CC-NMR using linear prediction up to 256 data points along t_1 and 80 data points along t_2 to a final size of $256 \times 80 \times 1024$ data points (Cieslar et al., 1993). After the peak-picking routine the assignment was carried out both manually and by use of program ALFA (Bernstein et al., 1993a).

Finally, for the 3D ^{15}N separated H^N-H^α correlation spectrum HNHA, which yielded the $^3J(H^N$-$H^\alpha)$ coupling constants from the ratio of the H^N-H^α cross peak to the H^N-H^N diagonal peak intensity (Vuister and Bax, 1993), the spectral width in the $^1H(F_2)$ dimension was 9.00 ppm and 128 complex points were acquired. The $^1H(F_2)$ carrier frequency was placed at 4.73 ppm and 32 scans per t_1-t_2 pair were recorded. The spectrum was processed using linear prediction up to 64 data points along t_1 and 256 data points along t_2 to a final size of $64 \times 256 \times 1024$ data points (Cieslar et al., 1993).

An amide exchange experiments, consisting of a series of 2D NOESY experiments, were carried out with RBI and LDTI-C lyophilized from buffered H_2O and then freshly dissolved in D_2O. Within 24 h, three 2D data sets were recorded.

4.3. Assignments of NMR Resonances

4.3.1. Assignments of NMR Spectra of LDTI-C. The NMR spectra of LDTI were assigned with the conventional 2D NMR methods based on the following spectra: NOESY, TOCSY, and COSY (Ernst et al., 1987; Wüthrich, 1986). Proton resonances of all backbone and side chain atoms were assigned with the exception of the α proton of residue 43. In general the spectra had few overlaps. The side-chain NH protons of arginines and lysines were not observed in the spectra. There are three proline residues in the sequence, Pro 7, Pro 12 and Pro 41; all of them have *trans*-peptide bond conformations (as inferred from the presence of the $C^\alpha H(i)$-$C^\delta H(i+1)$ cross-peaks in the NOESY D_2O spectra).

4.3.2. Sequence Specific Assignment of the ^1H and ^{15}N NMR Spectra of RBI. The NMR studies of RBI were carried out both at pH 4.5 and 6.0. The data at pH 4.5 are presented in this review. The spectra measured at this pH were used for sequence specific assignments. The data collected from the samples at higher pH served for confirmation of assignments. The 2D NOESY spectra at both pHs studied are very similar, indicating an identical structure of the protein within the pH range studied.

Trypsin can cleave RBI at its reactive peptide bond between residues Arg 34 and Leu 35 (Campos and Richardson, 1983). The RBI samples always contained mixtures of the cleaved and native inhibitor at various ratios, presumably due to minute amounts of trypsin released from the affinity column and copurified with RBI, and to a facilitated cleavage of RBI at acidic pH. The expression yield of the ^{15}N-labeled inhibitor was low and the protein also was obtained as a 1:2 mixture of native and cleaved form. The use of the ^{15}N-labeled sample for assignments was therefore in part limited by these complications. Nevertheless, the ^{15}N edited spectra provided helpful information for the sequential assignments. The NMR spectra of RBI were assigned with 2D and 3D NMR methods based on the following spectra: NOESY, TOCSY, and COSY, ^1H-^{15}N-HSQC, ^1H-^{15}N-TOCSY-HMQC, ^1H-^{15}N-NOESY-HMQC. Proton resonances of all backbone atoms were assigned with the exception of the amide and α-protons of residue 1 and the α-proton of proline 95 (see below). The complete spin systems of Pro 13, Pro 33, Pro 52, Pro 83, Pro 86, and Pro 112 could be identified. All of them are in the *trans*-conformation because strong $H^\delta(i)^{Pro}$-$H^\alpha(i-1)$ NOE contacts were observed in the NOESY spectra. These contacts are diagnostic for the *trans*-conformation of the proline amide bond (Wüthrich, 1986). Although it was not possible to assign the entire spin systems of the other three prolines (Pro 8, Pro 16, and Pro 95), there were enough data to conclude that they are also in a *trans*-conformation. Both $H^N(i+1)$-$H^\alpha(i)^{Pro}$ and $H^\delta(i)^{Pro}$-$H^\alpha(i-1)$ connectivities could be observed for Pro 8 and Pro 16. In the case of Pro 95, neither the connectivity $H^\delta(i)^{Pro}$-$H^\alpha(i-1)$ nor a $H^\alpha(i)^{Pro}$-$H^\alpha(i-1)$ NOE (characteristic of the *cis*-conformation) were observed. There were, however, two strong cross-peaks at the Ala 94 amide proton frequency at 4.14 ppm and 3.35 ppm in the NOESY spectrum. These resonances were tentatively assigned to the H^δ frequencies of Pro 95. We assumed that the $H^\delta(i)^{Pro}$-$H^\alpha(i-1)$ cross-peaks were too weak to be observed and concluded from the presence of the $H^\delta(i)^{Pro}$-$H^N(i-1)$ cross-peaks that Pro 95 is also in the *trans*-conformation.

4.4. Stereospecific Assignments

For stereospecific assignment, $^3J_{\alpha\beta}$ were extracted from the DQF-COSY in D_2O. Stereospecific assignments of $C^\beta H$ protons and methyl groups of valines and leucines were obtained using procedures described by Wagner et al. (1987) and Hyberts et al. (1987). $^3J_{H\alpha NH}$ coupling constants were measured mostly from the 3D ^{15}N separated H^N-H^α correlation spectrum HNHA (see above) and a few from DQF-COSY spectra in H_2O. The apparent $^3J_{H\alpha NH}$ coupling constants measured from the splittings of the NH-$C^\alpha H$ cross-peaks in DQF-COSY were corrected for line width using a method of Kim and Prestegard (1989).

4.5. Interproton Distance and Torsion Angle Constraints

NOEs were mainly derived from the 2D NOESY spectra in H_2O and D_2O at mixing times of 80 and 100 ms, respectively, and from the ^{15}N-filtered 2D NOESY spectra of ^{15}N-glycine/serine labeled RBI. Some additional NOEs were obtained from the 3D ^{15}N-edited NOESY spectra. The intensities of the 2D NOE cross peaks were determined from volume integrals and converted into distance constraints according to the procedures described earlier (Holak et al., 1989a). The distance bounds of the distance constraints were set to d ±0.4 Å for the distance constraints between 2.2-3.5 Å and to d -0.6 Å/+0.8 Å for distance constraints 3.6-4.5 Å. All protons were explicitly defined in the dynamical simulated annealing calculations, in some cases however additional terms were added to the upper bounds that correspond to the pseudoatom correction introduced by Wüthrich et al. (1984).

The distance constraints were supplemented with backbone ϕ torsion angle constraints derived from the $^3J(H^\alpha H^N)$ coupling constant data and 9 ψ angles (for the *trans*-peptide bond of the X-Pro residues, Wüthrich, 1986). The ϕ constraints were introduced in two stages. At the initial stage of the structure calculation, only those ϕ constraints were used which corresponded to the $^3J(H^\alpha H^N)$ coupling constants larger than 8 Hz (ϕ = -120°±40°) (Pardi et al., 1984). The rest of the ϕ constraints were introduced at a late stage of the refinement procedure (see below). 51 χ_1 torsion angle constraints used in the calculations were derived from the stereospecifically assigned β-prochiral centers. The minimum range employed for the ϕ, ψ, and χ_1 torsion angles were ±20°, ±50°, and ±20°, respectively. 223 non-NOE distance constraints were also present in the calculations (Holak et al., 1989a).

4.6. Structure Calculations

Structure calculations were carried out essentially according to the basic protocol described previously (Holak et al., 1989a). A total of 20 structures were calculated by a hybrid distance geometry-simulated annealing method (Holak et al., 1988; Holak et al., 1989a) with the program XPLOR 3.1 (Brünger, 1993). Minor modifications included simulation of 2D NOESY spectra in the final refinement stages to check whether NOE peaks calculated from NMR structures match experimental NOESY spectra (Bernstein et al., 1993b).

For the final refinement, the NOE tables were supplemented by constraints for the hydrogen bonds identified on the basis of the examination of the structures during the secondary structure analysis. Also, the NOE tables were supplemented at this stage with the ϕ constraints derived for the $J(H^\alpha H^N)$ coupling constants equal to 8 Hz and $\leq$6 Hz (Holak et al., 1989a). These ϕ constraints were introduced for residues for which ϕ's were close to the mean value (within $\pm$20°) and fulfilled the Karplus relationship (Wüthrich, 1986) in structures calculated without the dihedral constraints.

5. ACKNOWLEDGMENTS

We thank Robert Huber, Wolfram Bode, Hans Fritz, and Milton Stubbs for stimulating discussions. This work was supported by research grants (H2, H4, H6, G11) from the DFG Sonderforschungbereich 207 of the Ludwig-Maximilians Universität.

6. REFERENCES

Alagiri, S. and Singh, T. P. (1993) Biochim. Biophys. Acta *203*, 77-84.
Auerswald, E. A., Morenwieser, R., Sommerhoff, C. R., Piechottka, G., Eckerskorn, C., Gürtler, C., and Fritz, M. (1994) Biol. Chem. Hoppe-Seyler *375*, 605-703.
Bax, A. and Davis, D. G. (1985) J. Magn. Reson. *65*, 355-366.
Bax, A. and Freeman, R., (1981) J. Magn. Reson. *44*, 542-561.
Bernstein, R., Cieslar, C., Ross, A., Oschkinat, H., Freund, J. and Holak, T. A. (1993a) J. Biomol. NMR *3*, 245-251.
Bernstein, R., Ross, A., Cieslar, C. and Holak, T. A. (1993b) J. Magn. Reson. B *101*, 185-188.
Bode, W. and Huber, R. (1992) Eur. J. Biochem. *204*, 433-451.
Bodenhausen, G. and Ruben, D. J. (1980) Chem. Phys. Lett. *69*, 185-189.
Brünger, A. T. (1993) XPLOR Version 3.1 Manual, Yale University, New Haven, CT.
Brünger, A. T. and Nilges, M. (1993) Q. Rev. Biophys. *26*, 49-125.
Campos, F. A. P. and Richardson, M. (1983) FEBS Lett. *152*, 300-304.
Caughey, G. H., Leidg, F., Viro, N. F. and Nadel, J. A. (1988) J. Pharmacol. Exp. Ther. *244*, 133-137.
Cieslar, C., Ross, A., Zink, T. and Holak, T. A. (1993) J. Magn. Reson. B *101*, 97-101.
Claman, H.N. (1989) J. Am. Med. Assoc. *262*, 1206-1209.
Eklund, K. K., and Stevens R. L. (1993) in Innovations in proteases and their inhibitors, (ed. F.X. Aviles) Walter de Gruyter, Berlin, 241-257.
Ernst, R. R., Bodenhausen, G. and Wokaun, A. (1987) Principles of NMR in One and Two Dimension, Clarendon Press, Oxford.
Fink, E., Rehm, H., Gippner, C., Bode, W., Eulitz, M., Machleidt, W. and Fritz, H. (1986) Biol. Chem. Hoppe-Seyler *367*, 1235 - 1242.
Folkers, P. J. M., Clore, G. M., Driscoll, P. C., Dodt, J., Köhler, S. and Gronenborn, A. M. (1989) Biochemistry *28*, 2601-2617.
Friedrich, T., Kröger, B., Bialojan, S., Lemaire, H. G., Höffken, H. W., Reuschenbach, P., Otte, M. and Dodt, J. (1993) J. Biol. Chem. *268*, 16216 - 16222.
Girvin, M. E. (1994) J. Magn. Reson. A 108, 99-102.
Griffey, R. H. and Redfield, A. G. (1985) Biochemistry 24, 817-822.
Gueron, M., Plateau, P. and Decorps, M. (1991) Prog.in NMR Spectr. *23*, 135-209.
Haruyama, H. and Wüthrich, K. (1989) Biochemistry *28*, 4301-4312.
Hawkins, R. A., Claman, H. N., Clark, R. A. F. and Steigerwald, J. C. (1985) Ann. Intern. Med. *102*, 182-186.
Holak, T. A., Kearsley, S. K., Kim, Y. and Prestegard, J. H. (1988) Biochemistry *27*, 6135-6142.
Holak, T. A., Gondol, D., Otlewski, J. and Wilusz, T. (1989a) J. Mol. Biol. *210*, 635-648.
Holak, T.A., Bode, W., Huber, R., Otlewski, J. and Wilusz, T.J. (1989b) J. Mol. Biol. *210*, 649-654.
Hyberts, S. G., Mäki, W. and Wagner, G. (1987) Eur. J. Biochem. *164*, 625-635.
Jardetzky, O. (1980) Biochim. Biophys. Acta *621*, 227-232.

Jeener, J., Meier, B. H., Bachman, P. and Ernst, R. R. (1979) J. Chem. Phys. *71*, 4546-4553.

Kim, Y. and Prestegard, J. H. (1989) J. Magn. Reson. *84*, 9-13.

Kokkonen, J. O. and Kovanen, P. T. (1989) J. Biol. Chem. *264*, 10749-10755

Laskowski, M., Jr. (1986) In: Toxicological Significance of Enzyme Inhibitors in Foods. Freedman, M. (ed.), Plenum Publishing Coorporation, 1-17.

Laskowski, M., Jr. and Kato, I. (1980) Ann. Rev. of Biochem. *49*, 593-626.

Laskowski, M., Jr., Kato, I., Kohr, W. J., Park, S. J., Tashiro, M. and Whatley, H. E. (1987a) Cold Spring Harbor Symp. Quant. Biol. *52*, 545-553.

Laskowski, M., Kato, I., Ardelt, W., Cook, J., Denton, A., Empie, M. W., Kohr, W. J., Park, S. J., Parks, K., Schatzley, B. L., Schoenberger, O. L., Tashiro, M., Vichot, G., Whatley, H. E., Wieczorek, A. and Wieczorek, M. (1987b) Biochemistry *26*, 202-221.

Lin. G., Bode, W., Huber, R. Chi, C. and Engh, R. A., (1993) Eur. J. Biochem. *212*, 549-555.

Lyons, A., Richardson, M., Tatham, A. S. and Shewry, P. R. (1987) Biochim. Biophys. Acta *915*, 305-313.

Maeda, K., Hase, T. and Matsubara, H. (1983a) Biochim. Biophys. Acta *743*, 52-57.

Maeda, K., Wakabayashi, S. and Matsubara, H. (1983b) J. Biochem. *94*, 865-870.

Maeda, K., Wakabayashi, S. and Matsubara, H. (1985) Biochim. Biophys. Acta. *828*, 213-221.

Mahoney, W. C., Hermodson, M. A., Jones, B., Powers, D. D., Corfman, R. S. and Reek, J. R. (1984) J. Biol. Chem. *259*, 8412-8416.

Maier, M., Spragg, J. and Schwartz, L. B. (1983) J. Immunol. *130*, 2532-2536

Marion, D., Driscoll, P. C., Kay, L. E., Wingfield, P. T., Bax, Gronenborn, A. and Clore, G. M. (1989) Biochemistry *28*, 6150- 6156.

Marion, D. and Wüthrich, K. (1983) Biochem. Biophys. Res. Commun. *113*, 967-974.

Maskos, K., Wunderlich, M. and Glockshuber, R. (1995) unpublished results.

Messerle, B. A., Wider, G., Otting, G., Weber, C. and Wüthrich, K. (1989) J. Magn. Reson. *85*, 608-613.

Papamokos, E., Weber, E., Bode, W., Huber, R., Empie, M. W., Kato, I. and Laskowski, M. (1982) J. Mol. Biol. *158*, 515-537.

Pardi, A., Billeter, M. and Wüthrich, K. (1984) J. Mol. Biol. *180*, 741-751.

Petrucci, T., Rab, A., Tomasi, M. and Silano, V. (1976) Biochim. Biophys. Acta *420*, 288-297.

Plateau, P. and Gueron, M. (1982) J. Am. Chem. Soc. *104*, 7310- 7311.

Poerio, E., Caporale, C., Carrano, L., Pucci P. and Buonocore, V. (1991) Eur. J. Biochem. *199*, 595-600.

Rance, M. (1987) J. Magn. Reson. *74*, 557-564.

Richardson, M. (1991) Methods Plant Biochem. *5*, 259-305.

Rydel, T. J., Ravichandran K. G., Tulinsky, A., Bode, W., Huber, R., Roitsch, C., Fenton II, J. W. (1990) Science *249*, 277-280.

Sambrook, J., Fritsch, E. F. and Maniatis, T. (1989) Molecular Cloning: A Laboratory Manual, 2nd ed., Cold Spring Harbor Laboratory Press, Cold Spring Harbor, N. Y.

Schwartz, L. B., Metcalfe, D. D., Miller, J. S., Earl, H. and Sullivan, T. (1987) N. Engl. J. Med. *316*, 1622-1626.

Schwartz, L. B., Bradford, T. R., Littmann, B. L. and Wintroub, B. U. (1985) J. Immunol. *135*, 2762-2767.

Shivaraj, B. and Pattabiraman, T. N. (1981) Biochem. J. *193*, 29-36.

Sommerhoff, C. P., Söllner, C., Mentele, P., Auerswald, E. A., and Fritz, M. (1994) Biol. Chem. Hoppe-Seyler, submitted.

Srinivasan, A., Raman, A. and Singh, T. P. (1991) J. Mol. Biol. *222*, 1-2.

Stevens, R. L., Somerville, L. L., Sewll, D., Swafford, J. R., Levi-Schaffer, F., Caulfield, J. P., Hubbard, F. and Dayton, E. T. (1992) Arthritis Rheum. *35*, 325-335.

Studier, F. W. and Moffatt, B. A. (1986) J. Mol. Biol. *189*, 113-130.

Summers, M. F., Marzilli, L. G. and Bax, A. (1986) J. Am. Chem. Soc. *108*, 4285-4294.

Vuister, G. W. and Bax, A. (1993) J. Am. Chem. Soc. *115*, 7772-7777.

Wagner, G., Braun, W., Havel, T. F., Schaumann, T., Go, N. and Wüthrich, K. (1987) J. Mol. Biol. *196*, 611-639.

Wenzel, S. E., Fowler, A. A. and Schwartz, L. B. (1988) Am. Rev. Respir. Dis. *137*, 1002-1008

Werner, M. H. and Wemmer, D. E. (1991) Biochemistry *30*, 3356-3364.

Wunderlich, M. and Glockshuber, R. (1993) J. Biol. Chem. *268*, 24547-24550.

Wüthrich, K. (1986) NMR of Proteins and Nucleic Acids, Wiley, New York.

Wüthrich, K., Billeter, M. and Braun, W. (1984) J. Mol. Biol. *180*, 715-740.

Yanisch-Perron, C., Vieira, J. and Messing, J. (1985) Gene *33*, 103-119.

Zuiderweg, E. R. P. and Fesik, S. W. (1990) Biochemistry *28*, 2387-2391.

3

SOLUTION STRUCTURE OF THE LONG NEUROTOXIN LSIII WITH POSSIBLE IMPLICATIONS FOR BINDING TO THE ACETYLCHOLINE RECEPTOR

Peter J. Connolly, Alan S. Stern, and Jeffrey C. Hoch

Rowland Institute for Science
100 Edwin H. Land Blvd.
Cambridge, Massachusetts 02142

LSIII is a 66-residue protein from the venom of the sea-snake *Lauticauda semifasciata*. It belongs to the class of proteins called long neurotoxins, found in the venoms of fixed-front-fanged poisonous snakes. These toxins bind competitively to the post-synaptic acetylcholine receptor (AChR), blocking nerve transmission and effectively inducing paralysis. This ability to block nerve transmission makes them pharmcologically interesting, and they have also proven to be valuable in studies of AChR. Structural studies (both NMR and X-ray) have been performed on two long neurotoxins: α-bungarotoxin and α-cobratoxin. Common structural features include a globular cysteine-rich head and three protruding loops. The central loop contains residues that have been shown by mutagenesis and chemical modification to be important for binding.

We have determined the solution structure of LSIII by ^{1}H NMR. The structure is consistent with the structures of the other long neurotoxins, but our results suggest a novel, heretofore unrecognized feature. Previous studies have shown the central loop to be disordered. Careful examination of our results reveals evidence that the loop exhibits local order, but appears disordered through rigid-body motion of the loop relative to the main body of the protein. This motion, which may be a general feature of long neurotoxins, has several implications for receptor binding.

OVERVIEW OF THE LSIII STRUCTURE

The solution structure of LSIII was determined using two-dimensional ^{1}H homonuclear NMR at 400 MHz (details are given in Connolly et al., 1995). Experimental restraints consisted of 497 NOE-derived distance restraints, 20 H-bond restraints, 31 ϕ and 14 χ_1 torsion angle restraints. The structure was refined by dynamical simulated annealing (Nilges et al., 1988), with starting structures obtained using a novel reduced representation protocol

Dynamics and the Problem of Recognition in Biological Macromolecules, edited by Jardetzky and Lefèvre
Plenum Press, New York, 1996

(Hoch and Stern, 1992). The precision of the structure, as reflected by the root-mean-square difference (RMSD) from the mean for a family of 23 structures is 1.35 Å for all heavy atoms and 0.82 Å for backbone heavy atoms.

The superimposed backbones of the 23 structures are shown in Fig. 1. The globular head, shown at the top, is held together by 4 disulfide bonds (3-20, 13-41, 45-56, and 57-62). The main element of secondary structure consists of a three-stranded β-sheet (residues 19-25, 37-42, and 52-58). The RMSD for backbone heavy atoms in the sheet is 0.26 Å. Disordered regions include the C-terminus (residues 63-66), the solvent-exposed region of Loop III (shown at right) and the prominent central Loop II (26-35), which contains several residues which have been shown to be important for binding. Loop II contains a disulfide bond (26-30) and a conserved aromatic residue (tryptophan 29).

When the structures are superimposed by least-squares matching of all backbone atoms (as in Fig. 1), Loop II appears highly disordered. When residues comprising the loop are matched, the appearance becomes much more ordered (Fig. 2): the RMSD for backbone atoms changes from 1.74 Å to 0.44 Å. (By comparison, matching based on the disordered residues 1-10 yields a backbone RMSD of 0.85 Å, only slightly reduced from the 1.18Å RMSD when all backbone atoms are matched.) This suggests that the loop undergoes rigid-body fluctuations with respect to the rest of the protein, maintaining local order. Additional evidence supporting the hypothesis of rigid-body motion of Loop II comes from

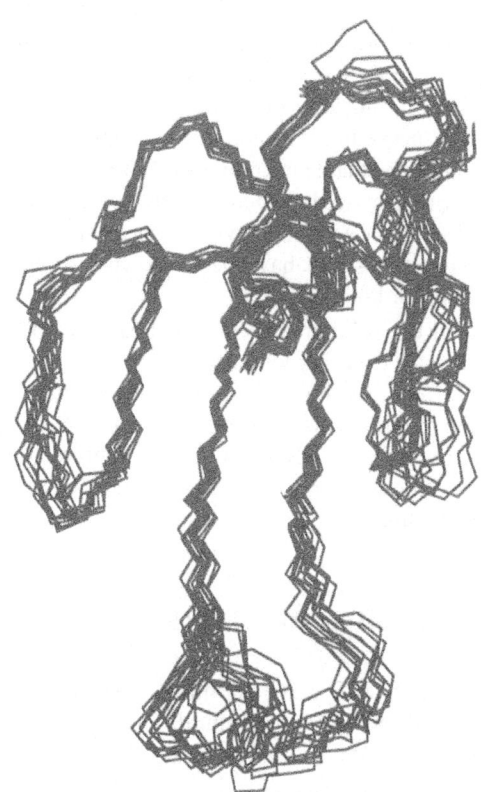

Figure 1. Superimposed family of 23 structures of LSIII. The globular head appears at the top, the C-terminal end on the right and the N-terminal end on the left.

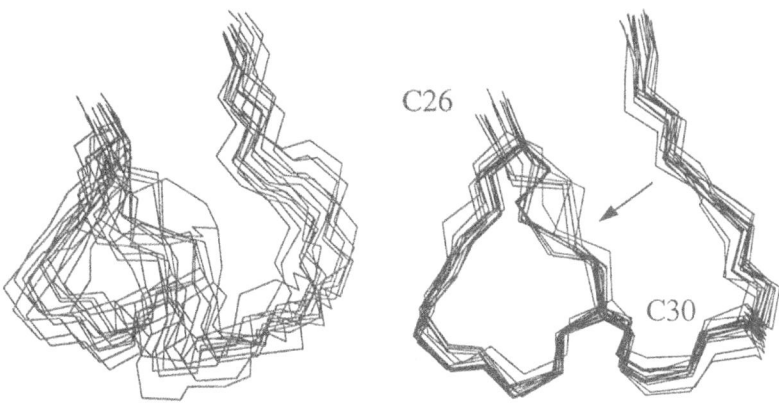

Figure 2. Detail of Loop II (the binding loop) from Fig 1. (left). On the right, the superimposed loop structures are shown when the least-squares match is restricted to residues in the loop. The arrow indicates the disulfide bond linking residues 26 and 30.

the vicinal coupling constants for the loop, which fall outside the range (6-8 Hz) that is indicative of extensive averaging.

Examination of the angular order parameters (Hyberts et al., 1992) for the main chain torsion angles reveals well ordered torsion angles for most of Loop II (Fig. 3). The "hinge" for the rigid-body motion appears to be localized at residues 27 and 28 on one side of the loop, but appears to be delocalized over several residues on the other side.

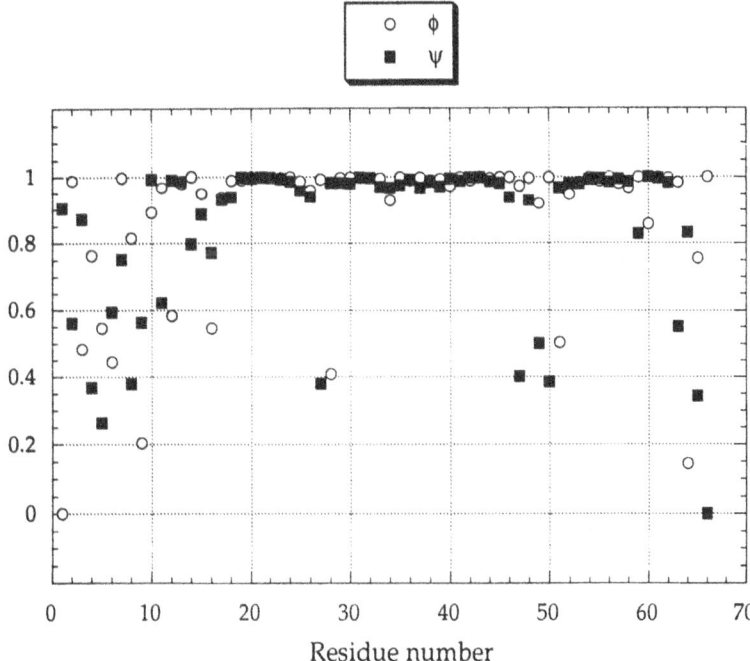

Figure 3. Angular order parameters for the main chain torsion angles of LSIII, computed from the family of 23 structures. A value of 1 indicates a well-defined torsion angle; smaller values are indicative of a distribution of torsion angles.

DISCUSSION OF THE STRUCTURE

The range of angles spanned by Loop II with respect to the body of the protein, approximately 20° wide, should not be viewed as indicative of the range of angular fluctuations in solution. The final stage of the structure determination protocol involves energy minimization, so the range of angles merely represents the distribution of local energy minima that are consistent with the experimental restraints. Quantitative assessment of the range of angular fluctuations will require additional investigation, for example relaxation measurements and molecular dynamics simulations.

Rigid-body motion of the binding loop, as opposed to a completely flexible or completely rigid loop, holds both kinetic and thermodynamic consequences for receptor binding. A completely flexible loop that becomes ordered on binding to the receptor would impose an entropic cost for each lost degree of freedom. Searle and Williams (1992) estimate the entropy loss due to completely restricting a freely rotating bond in a hydrocarbon chain to be between -1.6 and -3.6 kJ/mole (-0.4 and -0.9 kcal/mole) (TΔS at 300K). While this probably represents an overestimate of the entropic cost of restricting a bond rotation in a flexible binding loop, since the bond rotations are not completely unhindered in the free state nor completely hindered in the bound state, it nevertheless suggests that the entropic cost can be significant.

Conversely, a completely rigid binding loop would present a kinetic barrier to binding, if the orientation of the loop with respect to the body of the protein restricts access to the binding site of the receptor. We will describe below how flexibility of the loop could provide an evolutionary advantage.

There are two long neurotoxins that lack the disulfide bridging the binding loop, but they do not exhibit reduced toxicity. It has been observed that the loop disulfide can be selectively reduced by kinetic control of the reduction (Martin et al., 1983). Attempts to quantify the effects of reducing the disulfide on binding to AChR were not conclusive; the bindng affinity could be either increased or decreased, depending on the nature of the blocking group used to prevent re-oxidation. It may be possible to reduce the loop disulfide, and keep it reduced without the use of a blocking group under conditions of very low ambient oxygen concentration. This would permit investigation of the structural and dynamical role of the disulfide bond, and could allow more readily interpreted experiments to probe its influence on the kinetics and thermodynamics of receptor binding.

CHARACTERISTICS OF AChR

AChR serves as an ion channel. Normally closed, it allows ions to pass through the post-synaptic membrane when two acetylcholine molecules are bound. The atomic structure of the receptor is not known, but it has been imaged by cryoelectron microscopy at a resolution of 9Å (Unwin, 1995). The electron density reconstructed from micrographs at the position of the extracellular domain, which is believed to contain the binding sites, is shown in Fig. 4. The receptor is composed of five homologous subunits, situated pseudo-symmetrically about a cylindrical axis. Two are designated α subunits; the others are designated β, γ, and δ. (There remains disagreement concerning the order of the subunits about the cylindrical axis in the extracellular domain.) Photoaffinity labeling studies have shown that the acetylcholine binding sites are on the α subunits at the $\alpha\gamma$ and $\alpha\delta$ interfaces (Pedersen and Cohen, 1992). The two binding sites have different affinities, the $\alpha\delta$ site binding acetylcholine more tightly. Evidence for allosteric behavior, in which binding one acetylcholine to the high affinity site increases the affinity of the other site, has been reported

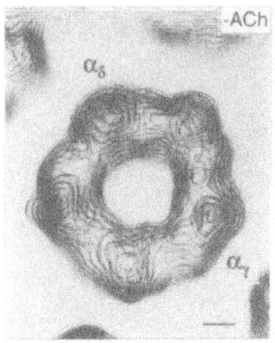 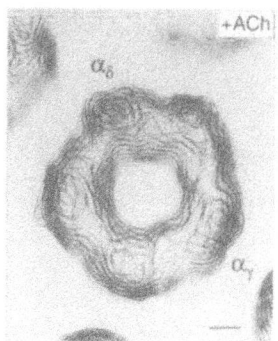

Figure 4. Electron density of the extracellular domain of AChR, reconstructed from electron micrographs in the absence (left) and presence (right) of ACh. Reprinted from Unwin 1995, Fig. 3.

(Jackson, 1994). Unwin's results suggest that there is a change in the relative orientation of some of the subunits upon binding acetylcholine.

A MODEL FOR AChR

The observation that the binding affinities of the two sites are different leads naturally to the conclusion that the differences are due to the the adjacent subunits, which are different for the two α subunits. Specifically, it has been proposed that the adjacent subunits affect the conformation of the binding site (Pedersen and Cohen, 1992). The hypothesis that a mobile binding loop allows LSIII to attack a binding site that would otherwise be inaccessible, together with the gross architecture of the extracellular portion of AChR, leads us to propose an alternative hypothesis for the difference in binding affinities of the two sites.

Instead of causing a change in the structure of the binding site, the adjacent subunit could restrict access to the binding site. The difference in binding affinities for the two sites then arises from the different accessibilities, or probabilities that the substrate — a neurotoxin or acetylcholine — could approach the binding site in an orientation suitable for binding. Figure 5 shows a schematic depiction of this scenario for one binding site and LSIII. Shown are representations of cross-sections through the extracellular subunits of AChR and LSIII. The binding loop of LSIII is depicted as a line with the binding residues represented by a dark shaded triangle, approaching the binding site on the surface of the α subunit. The AChR subunit, LSIII, and the binding loop are drawn roughly to scale. The protrusion from the surface of the subunit adjacent to the α subunit (shown with vertical hatching) is hypothetical.

The protrusion from the adjacent subunit (γ or δ) restricts direct attack of the binding residues of LSIII unless the angle θ between the binding loop and the body of the protein is less than some critical value that permits close approach of the entire protein. For angles greater than the critical value, the body of the protein bumps into the protrusion, preventing the binding residues from reaching the binding site.

This mechanism for modulating the binding affinity need only apply to one of the sites in order to account for the difference in binding affinities, but it could apply to both sites if the extent of occlusion differs. The $\alpha\gamma$ site, which has the lower binding affinity, is thus likely to be the site with greater occlusion. While the exact arrangement of the subunits

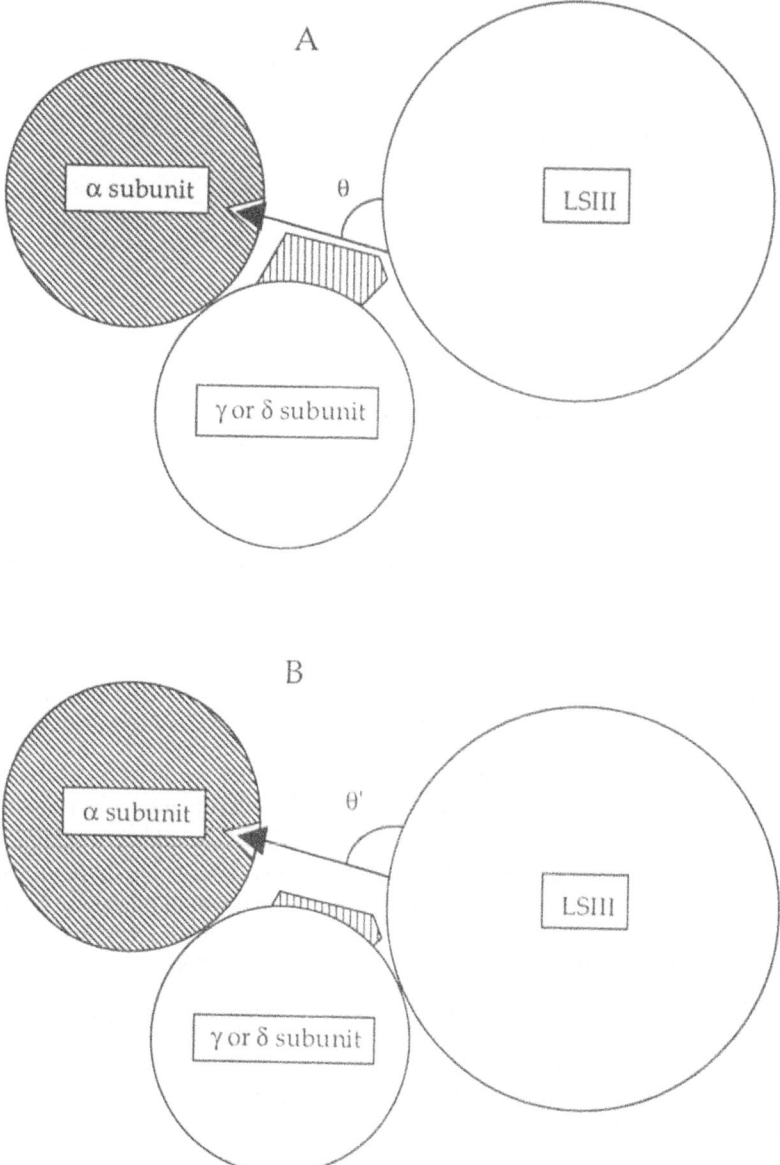

Figure 5. A schematic diagram of LSIII interacting with two subunits of the extracellular domain of AChR, depicting a hypothetical protrusion (shown in red) restricting access to the binding site located on the α subunit. In panel B, a rotation of the γ (or δ) subunit relative to the a subunit results in a retraction of the protruding surface.

is not known, this mechanism would take a particularly simple form if the low-affinity binding site is occluded by part of the subunit that lies between the two α subunits.

The mechanism has testable implications. One is that the difference in binding affinity, since it arises from a difference in the probability of productive binding collisions, rather than a difference in the structure of the binding site *per se*, should appear mainly in ΔS of binding, rather than ΔH. Conversely, if the difference were due to conformational differences of the two binding sites, then forces which contribute to ΔH — van der Waals, hydrogen bond, and electrostatic — would differ.

The model also provides a simple explanation for allosteric behavior. Unwin's conjecture of a change in the relative orientation of the extracellular subunits of AChR on binding acetylcholine would lead to a change in the extent of occlusion, by moving the protrusion away from the binding site. Panel B of Fig. 5 depicts one possible mechanism, in which the γ (or δ) subunit rotates about a horizontal axis, effectively retracting the protrusion from the binding site. Alternatively, a rotation of the subunit about the axis perpendicular to the page could accomplish a similar retraction.

CONCLUDING REMARKS

The solution structure of LSIII is consistent with previous structures of long neurotoxins, but it yields novel insight into the dynamics of the central binding loop. This dynamical behavior, a rigid-body motion of the loop with respect to the rest of the protein, may be a heretofore unrecognized general feature of the long neurotoxins. The possible role of this motion in allowing residues important for binding to reach an otherwise inaccessible binding site on the surface of AChR leads to a new model for the role of the subunits of the extracellular domain in the function of AChR, and an explanation for the difference in binding affinities for sites located on identical subunits. Although speculative, the model is both heuristic and testable, and thus can serve to focus future investigations into the function of AChR.

ACKNOWLEDGMENTS

This work is supported by the Rowland Institute for Science. Computations were performed using resources provided by the National Institutes of Health (GM-47467, Gerhard Wagner, P.I.) We are grateful to Jonathan Cohen and Abraham Szöke for helpful discussions.

REFERENCES

Connolly, P.C., Stern, A.S., and Hoch, J.C. (1995) Submitted for publication.
Hoch, J.C, and Stern, A.S. (1992) J. Biol. NMR 2, 535-543.
Hyberts, S., Goldberg, M., Havel, T., and Wagner, G. (1992) Prot. Sci. *1*, 636-751.
Jackson, M.B. (1994) Trends Biochem. Sci. October, 396-399.
Martin, B.M., Chibber, B.A., and Maelicke, A. (1983) J. Biol. Chem. *258*, 8714- 8722.
Nilges, M., Gronenborn, A.M. and Clore, G.M. (1988) FEBS Lett. *229*, 313-324.
Pedersen, S.P. and Cohen, J..B. (1990) Proc. Natl. Acad. Sci. USA *87*, 2785-2789.
Searle, M.S. and Williams, D.H. (1992) J. Amer. Chem. Soc. *114*, 10690-10697.
Unwin, N. (1995) Nature *373*, 37-43.

4

NMR STUDIES OF ENZYME-SUBSTRATE AND PROTEIN-PROTEIN INTERACTIONS

Gordon C. K. Roberts

Biological NMR Centre and Department of Biochemistry
University of Leicester
Leicester, United Kingdom

1. INTRODUCTION

NMR spectroscopy can provide information on many different aspects of protein-ligand interactions-not only structural but also dynamic, kinetic and thermodynamic (Jardetzky and Roberts, 1981; Feeney and Birdsall, 1993). For protein-ligand complexes of relatively modest size (<30kD), the use of nuclear Overhauser effect measurements and isotope-labelling can lead to the determination of a complete three-dimensional structure of the complex, in just the same way as for the structure of the protein alone (for reviews see Feeney and Birdsall, 1993; Lian et al., 1994; Wand and Short, 1994; Petros and Fesik, 1994; Wemmer and Williams, 1994). For larger complexes, the complete resonance assignment necessary for a full structure determination is not currently feasible, but very valuable structural information can nonetheless be obtained. In this article, I shall describe the use of nmr to study relatively large complexes, illustrating this by two examples of recent work from our laboratory involving the determination of the conformation of a substrate bound to an enzyme, and the location of the interaction sites in a protein-protein complex. In order to obtain this kind of information on a complex, essentially two strategies are available. Isotope labelling, either general or selective, can be employed as it is in protein structure determination, but with the added benefit that one can label either one or both of the partners in the complex. A second powerful approach is to make use of the association and dissociation of the complex.

This continuing exchange of the constituent molecules of the complex between the free and the complexed state can have profound effects on the NMR spectrum and can determine the kind of structural and kinetic information which can be obtained about the complex (Lian and Roberts, 1993). The first step in any study of a protein-ligand complex by NMR must thus be to establish the "exchange region" in which one is operating. If exchange is *slow* relative to the difference in NMR parameters between the free and complexed states, *separate signals* from the two states will be seen, whereas if exchange is *fast*, *averaged signals* will be observed. Each of these situations has its advantages and disadvantages; there are clear benefits to observing separate signals from the complex of

Dynamics and the Problem of Recognition in Biological Macromolecules, edited by Jardetzky and Lefèvre
Plenum Press, New York, 1996

interest, but averaging of signals by exchange can allow one to obtain information about complexes which are too large for detailed direct study by NMR (see below).

A significant limitation to the interpretation of nmr spectra in the fast exchange regime should always be borne in mind: the spectra represent an average across all the states which the nuclei experience, but we do not *a priori* know how many states there are, nor what their nmr parameters may be. Interpretation of fast-exchange spectra must always, therefore, be carried out in terms of a model for the exchange process, and the conclusions will only be meaningful if the model is correct. Even if exchange between the complex and the free molecules is slow, the spectrum of the bound state may be an average of more than one state. This point is discussed further below; see also Jardetzky (1980).

2. THE CONFORMATION OF BOUND SUBSTRATE: CHLORAMPHENICOL ACETYLTRANSFERASE

Chloramphenicol acetyl transferase (CAT), a trimer of total M_r 75,000, is the primary effector of bacterial resistance to the antibiotic chloramphenicol. It catalyses acetyl transfer from acetyl-coenzyme A to chloramphenicol to produce 3-acetyl-chloramphenicol. This can then undergo a non-enzymatic rearrangement to form 1-acetylchloramphenicol, which is the substrate for a second cycle of acetylation by CAT, finally yielding 1,3-diacetylchloramphenicol. The structures of the two binary complexes of CAT with chloramphenicol and coenzyme A (CoA) have been determined by crystallography, but attempts to obtain satisfactory crystals of the CAT-CoA-chloramphenicol ternary complex have so far proved unsuccessful.

This protein is very large for study by nmr, and we have used it as a system for exploring nmr methods suitable for large proteins, using the two general approaches mentioned above - selective isotope labelling in conjunction with heteronuclear "editing" experiments (Derrick et al., 1991,1992), and exchange-based magnetization transfer methods such as the transferred NOE, which will be the focus of the present discussion. Specifically, we have recently used transferred NOEs to determine the conformation of CoA bound to the enzyme in both binary and ternary complexes.

2.1. The Transferred NOE Experiment

In these experiments (Balaram et al., 1972; Albrand et al., 1979; Lian et al., 1994a), the exchange rate must be faster than the longitudinal relaxation rate of the bound ligand; if this condition is satisfied, changes in magnetization of a bound ligand proton resulting from intra- or intermolecular NOEs are transferred to the free ligand by the exchange between bound and free ligand. For small ligand molecules, cross-relaxation in the bound state is large and negative while that in the free state is small (and usually positive). The observed NOEs of the ligand protons are hence governed by the cross-relaxation in the bound state scaled down by the fraction of bound ligand. Thus, these experiments rely upon indirect characterization of ligand-protein interactions (by *inter*molecular NOEs) and of the conformation of the bound ligand (by *intra*molecular NOEs) *via* the averaged NMR spectrum of the free and bound ligand. Transferred NOEs between protons of the ligand, here CoA, can readily be measured from this averaged spectrum since an excess of ligand over protein can be used, the initial rate of build-up of the transferred NOE being independent of the ratio of ligand to protein concentrations (Lian et al., 1994a). (We have found a ratio of 15:1 to be a useful compromise between increased cross-peak intensity and the problems of t_1-noise and possible nonspecific interactions at high ligand concentrations.)

As with all NOE measurements, interpretation of transferred NOEs in terms of molecular structure requires that we differentiate direct cross-relaxation effects from spin diffusion. Many transferred NOE experiments have been interpreted either qualitatively or by using the "two spin approximation", which can at best give semi-quantitative information (Lian et al., 1994a). In a protein as large as CAT, spin diffusion is a problem which cannot be ignored, and there are a number of different ways of minimising the errors it can introduce. We have employed a combination of three approaches. When measuring NOEs between protons of a ligand bound to a protein, spin diffusion can occur *via* either protons of the ligand itself or protons of the protein. The latter route involves the *inter*molecular relaxation processes which give rise to intermolecular NOEs which are often readily observed in protein-ligand complexes; its significance can be established by the use of per-deuterated protein. In the case of the CoA-CAT system, the effect of deuteration of the enzyme on the intensities of transferred NOEs was less than a factor of two (Barsukov et al., 1995), although accurate measurement of these intensities was much easier with the deuterated enzyme, and deuteration was essential in the experiments on the ternary complex (see below). In order to deal with the problems introduced by *intra*molecular spin diffusion within the bound ligand we have used both NOESY and ROESY data, and analysis of the data using a complete combined relaxation and kinetic matrix (Lian et al., 1994a). In the ROESY (rotating frame NOE) experiment, cross-peaks arising from a direct cross-relaxation effect have the opposite sign to those arising from a two-step spin-diffusion. Thus cross-peaks in the ROESY spectrum which have opposite phase to the diagonal peaks correspond predominantly to direct magnetization transfer, and cross-peaks which are in-phase with diagonal peaks arise from indirect transfer; zero cross-peak intensity indicates either the absence of magnetization transfer due to a large interproton distance, or that direct and indirect transfer are equally effective. Only in NOESY spectra obtained with a mixing time as short as 30ms can cross-peaks arising predominantly from direct interactions be reliably identified for most of the proton pairs of CoA, and even at this mixing time, the problem of spin-diffusion remains for the geminal protons of a methylene group. In a ROESY spectrum obtained with a mixing time of 100ms all the cross-peaks have phase opposite to that of the diagonal, and thus correspond to predominantly direct interactions, and there is a clear differentiation between the intensities of the cross-peaks arising from the two protons of a methylene group. Some cross-peaks clearly seen in the NOESY spectrum are absent in the ROESY spectrum, indicating equal contributions from direct and indirect transfer (Barsukov et al., 1995). By using both NOESY and ROESY data, direct and indirect cross-relaxation can be separately measured, and by using the relaxation matrix approach to analysis, both kinds of cross-relaxation data can be used to derive structural information.

2.2. The Conformation of Coenzyme A Bound to Chloramphenicol Acetyltransferase

A quantitative analysis of a total of 71 NOESY cross-peaks and 33 ROESY cross-peaks was used to determine the conformation of CoA bound to CAT, in an iterative procedure using distance geometry and simulated annealing calculations. The inclusion of the ROESY data was found to be important in obtaining internally consistent results, with different starting structures leading to essentially the same final structure. In the final "family" of structures, none of the upper limit distance restraints were violated by more the 0.3Å and none of the experimental cross-peak intensities differ from the calculated ones by more than a factor of two. The calculated structures (Barsukov et al., 1995) are compared with the X-ray structure (Leslie et al., 1986) in Figure 1. The adenosine part of the molecule is very well defined and its conformation is close to that in the crystal. The conformation of

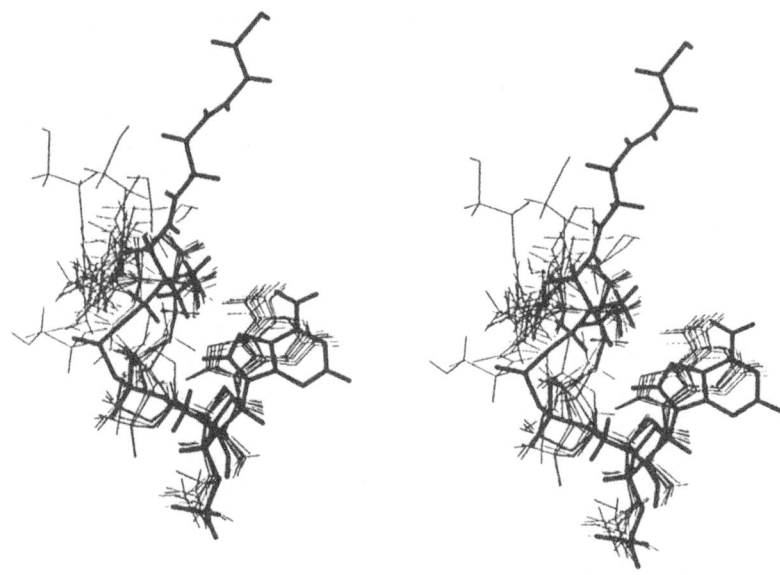

Figure 1. Stereo pair comparing the conformation of coenzyme A in its binary complex with chloramphenicol acetyltransferase calculated from distance constraints derived from transferred NOEs ((Barsukov et al., 1995); *thin lines*) with that observed in the crystal structure (*thick line*).

the pyrophosphate link is not defined by NMR as there are no NOEs in this part of the molecule. The pantetheine "arm" of CoA, however, shows a clear difference between the NMR and X-ray structures. In the crystal structure, this arm has an extended conformation which makes some of the distances between the β-mercaptoethylamine methylene protons of the pantetheine and the adenosine protons as large as 10Å. However, relatively intense NOE cross-peaks can be observed for these proton pairs, and the constraints derived from these NOEs lead to a more "folded" conformation of the pantetheine arm in the NMR structure.

Several factors may account for this difference between the two structures. First, there might be a genuine difference between the solution and crystal structures, although there is no indication in the crystal structure that crystal contacts might influence the conformation of CoA. Second, the discrepancy might arise from secondary binding sites for CoA; however, fluorescence experiments give no indication of more than one binding site per subunit, and the changes in chemical shift and linewidth of a number of CoA proton resonances as a function of ligand concentration all agree with a simple one-site model.

The most likely explanation for the discrepancy seems to be that referred to above as a general problem of fast exchange experiments - that the binding process is more complex than a simple two-site exchange. The equilibrium and kinetic data on the binary complex can be explained by postulating a two-step binding of CoA to the enzyme:

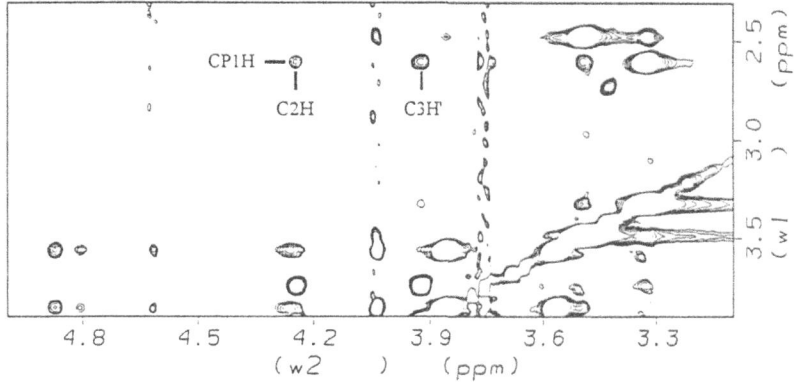

Figure 2. Part of the NOESY spectrum from a sample containing coenzyme A, chloramphenicol and per-deuterated chloramphenicol acetyltransferase (ratio 15:7:1), showing transferred NOEs between the CP1 proton of coenzyme A and the C2H and C3H' protons of chloramphenicol bound to the enzyme (Barsukov et al., 1995).

$$E + L \leftrightarrow EL \leftrightarrow EL^*$$

where EL represents an intermediate complex and EL* is the final complex after a conformational readjustment of some sort. In this situation, what one observes will depend crucially on the technique used, and on the equilibrium constant of the second step. The fact that a single clear-cut structure was observed by X-ray diffraction indicates that the equilibrium constant of the second step is well over to the right. However, if the rate of interconversion between EL and EL* is fast, the measured NMR parameters will, as noted above, be an *average* of those of the two states. Since the NOE is a non-linear function of distance, this average will be a non-linear one (Jardetzky, 1980; Jardetzky and Roberts, 1981). Suppose that in the intermediate EL complex the CoA molecule has a conformation with the pantetheine arm in the folded state, while in the final EL* complex it is, as shown by the crystal structure, in the extended state. Since in the extended conformation most of the distances to the end of the pantetheine arm from the adenine part of the ligand are too large to give rise to NOEs, the averaged NMR information will be dominated by NOEs from the folded intermediate state. This will be true even if this intermediate state is only populated to a minor extent at equilibrium. Support for this two-step binding model comes from a comparison of the estimates of the dissociation rate constant of CoA from the binary complex made by fluorescence stopped-flow methods and by NMR methods (Barsukov et al., 1995).

Similar transferred NOE experiments have been carried out on the ternary CAT-CoA-chloramphenicol complex, for which no crystallographic information is available, again using deuterated enzyme to eliminate spin diffusion through the protein. Strikingly, *interligand* NOEs were observed in the complex, between the pantothenate protons of CoA and protons of chloramphenicol; NOEs involving the C2 and C3 protons of chloramphenicol are shown in Figure 2. This provides the first experimental evidence for the relative orientation of the two ligands when both are bound simultaneously to CAT. Calculation of the structure of the ternary complex, using the NOE and ROE data together with the steric constraints imposed by the structure of the binding site, revealed that in the ternary complex the pantetheine arm of CoA adopts a more extended conformation in the ternary than in the binary complex (Barsukov et al., 1995). This suggests that in the ternary complex the equilibrium between the two states of the complex is far to the right, favouring a more extended conformation of CoA, which is necessary for the end of the pantetheine arm to be sufficiently close to chloramphenicol for acyl transfer to occur.

These experiments, while they emphasise the care that must be taken in the interpretation of transferred NOE experiments, have thus thrown light both on the structure of the ternary complex and on the kinetic mechanism of its formation.

3. LOCATION OF THE SITE OF PROTEIN-PROTEIN INTERACTIONS BY CHEMICAL SHIFT CHANGES: BACTERIAL ANTIBODY-BINDING PROTEINS

A number of species of pathogenic bacteria, notably *Streptococci* and *Staphylococci*, have proteins on their surface which bind immunoglobulins (reviewed in Boyle, 1990). Protein A from *S. aureus* and protein G from species of *Streptococci* are widely used as immunological tools and are the most extensively studied of these antibody-binding proteins. Protein A contains five highly homologous Fc-binding domains, each of about 60 amino acid residues, which bind to the Fc portion of immunoglobulin G (IgG) with an affinity which varies with the species and subclass of IgG. The C-terminal half of protein G contains three IgG-binding domains, referred to as domains I, II, and III, each consisting of 55 residues and separated from the others by short linker sequences. These domains are closely homologous to one another, but they show no sequence similarity to those of protein A. Protein G has a broader specificity than protein A for IgGs from different sources, and its IgG-binding domains are able to bind to both the Fab and the Fc portions of the antibody molecule.

The IgG-binding domains from protein A and protein G have different three-dimensional structures. The B domain of protein A is a three-helix bundle (Torigoe et al., 1990; Gouda et al., 1992), whereas the several protein G domains which have been studied all consist of an α-helix packed against a four-stranded antiparallel–parallel–antiparallel β-sheet (Lian et al., 1991, 1992; Derrick and Wigley, 1994; Gronenborn et al., 1991; Orban et al., 1992; Achari et al., 1992; Gallagher et al., 1994) (Figure 3). The binding of protein G to Fc fragments is competitive with respect to protein A (Stone et al., 1989; Frick et al., 1992), suggesting that the binding sites for protein A and protein G on Fc overlap, notwithstanding the lack of sequence or structural similarity. At the same time, the small antibody-binding domains of protein G are able to bind to both Fab and Fc fragments of IgG. There are thus interesting questions about the structural origins of the specificity of protein-protein interactions in this system.

The molecular mass of the complexes between a domain of protein G or protein A and an Fab or an Fc fragment of IgG is approximately 58 kDa, so that a complete structure determination by nmr is not feasible. However, since the structure of each component of the complex is known, it is possible to obtain a model of the complex if the residues in each component involved in the interaction can be identified. The best model would be obtained if assigned intermolecular NOEs could be identified, for example by selective isotope labelling; however, in the absence of a complete resonance assignment of both partners, this requires a preliminary model to allow the appropriate residues for selective labeling to be identified. The simplest approach to obtaining a medium-resolution model is to identify residues whose chemical shifts are affected by complex formation, by using generally or specifically isotope-labelled proteins. The changes in chemical shift may reflect direct protein-protein contacts and/or changes in conformation on complex formation. The *unaffected* resonances, on the other hand, can be clearly identified and must arise from residues that are *not* involved in intermolecular contacts. Identification of the regions of the protein surface *not* involved in complex formation obviously leads in turn to the identification of

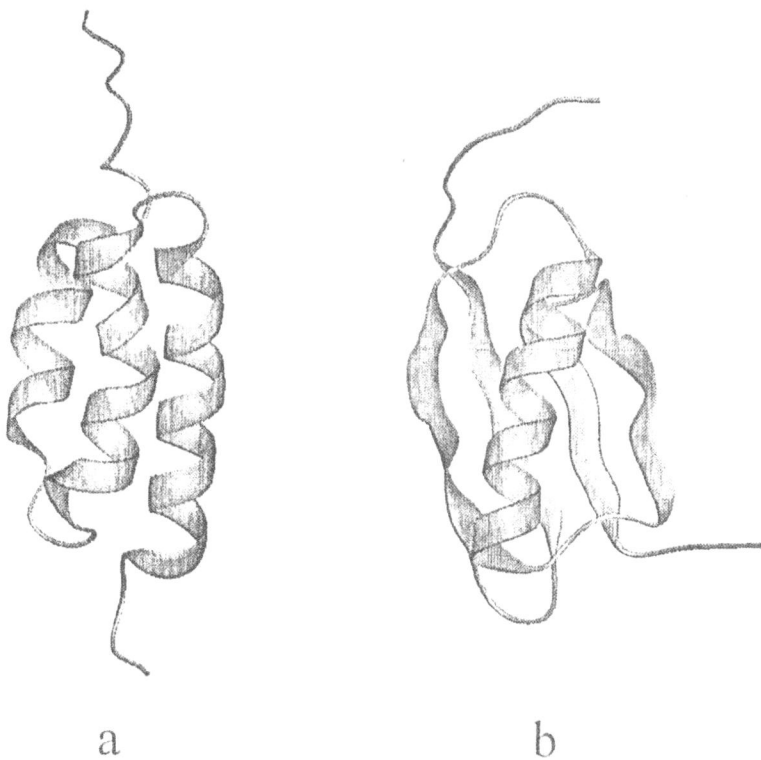

a   b

Figure 3. Comparison of the solution structures of IgG-binding domains of **(a)** Protein A (Gouda et al., 1992) and **(b)** Protein G (Lian et al., 1992).

the region(s) which are likely to be so involved, and hence to an approximate model of the complex

3.1. Residues of Domain II of Protein G Involved in Binding Fab and Fc

Heteronuclear correlation spectra provide a convenient means of identifying resonances affected by the binding of protein G to IgG fragments. ^{1}H-^{15}N and ^{1}H-^{13}C HSQC spectra of protein G domain II, alone and in its complex with an Fab fragment of mouse IgG show that resonances of rather more than a third of the residues either undergo significant chemical shift changes (greater than the linewidth of the crosspeak) or are so broadened as to be undetectable. This marked line-broadening is specific to these residues, and distinct from the smaller increase in linewidth seen for all resonances due to the increase in correlation time on complex formation. It can most easily be explained if the chemical shift changes of these resonances on complex formation are such that the exchange rate between the free and bound states is in the intermediate range on the nmr timescale, leading to substantial exchange broadening.

The regions of the protein G molecule that are involved in Fab binding can thus be identified as the turn between the strands 1 and 2 of the β-sheet and the first two-thirds of β-strand 2, and the loop following the helix, but not the helix itself (Lian et al., 1994b) (Figure 4A). As discussed previously (Lian et al., 1994b), the changes observed in the nmr spectrum on complex formation in solution are entirely consistent with the crystal structure (Derrick

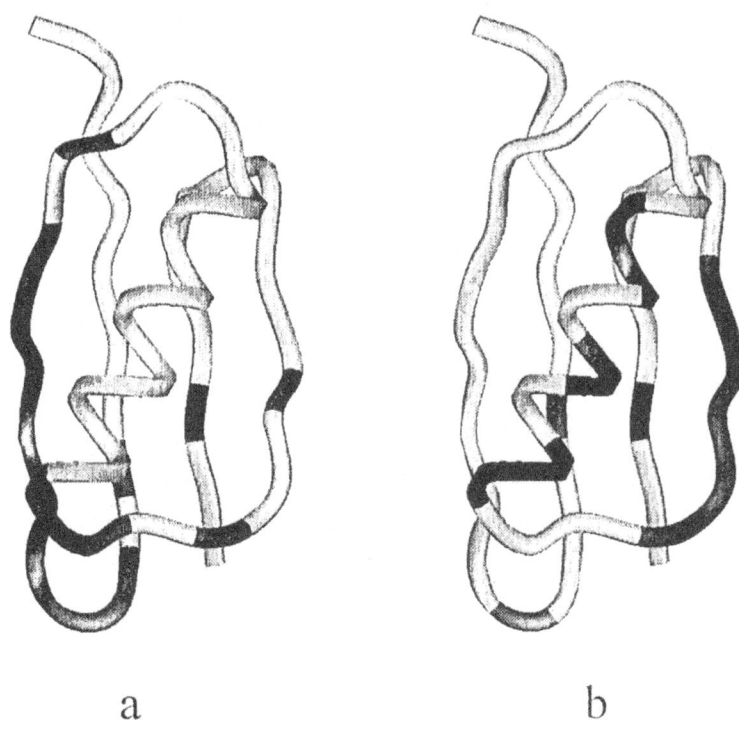

a   b

Figure 4. Comparison of the residues of the IgG-binding domains of protein G affected by binding to (*a*) Fab and (*b*) Fc fragments of IgG. Those residues whose amide $^1H/^{15}N$ or methyl $^1H/^{13}C$ chemical shifts and/or linewidths are altered on formation of the respective complexes are shaded (Lian et al., 1994; Kato et al., 1995). The protein G domain is oriented with its N-terminus at the top.

and Wigley, 1992, 1994); it is clear that the crystal structure does correspond to the structure of the complex in solution. Figure 4B similarly shows the residues affected when protein G domain II binds to an Fc fragment of mouse IgG (Kato et al., 1995); these are primarily in the α-helix and in the third strand of the β-sheet. This pattern of affected amide resonances is almost identical to that observed for the binding of a slightly different protein G domain (from *Streptococcus* strain GX7809) to a human Fc fragment (Gronenborn and Clore, 1993). It is clear that the same part of the surface of protein G is involved in binding to the two Fc fragments, although it binds them with markedly different affinity; equally, it is clear that a different region of domain II is involved in binding to the Fab and the Fc fragment of IgG. In the complex with Fab, the *second* strand of the β-sheet plays a major role, together with the loop at the C-terminal end of the helix, while in the complex with Fc, the regions most affected are the helix and the *third* strand of the β-sheet, but not the intervening loop. The IgG-binding domains of protein G thus interact with the Fab and Fc regions of the antibody in quite different ways (Lian et al., 1994b; Kato et al., 1995).

3.2. Residues of Fc Involved in Binding Protein G

To locate the binding site for protein G on Fc, we have used the ^{13}C resonances of the carbonyl carbons of histidine, leucine, methionine, tryptophan, and tyrosine residues as "probes". These experiments gave a total of 35 residues whose behaviour could be monitored. Of these, only Met-252, His-433, His-435 and His-436 were affected by the binding of domain II of protein G (Kato et al., 1995). These four residues lie in the 'groove' between the C_H2 and C_H3 domains of Fc, indicating that this region is primarily responsible for the binding of protein G. A comparison with the results of similar experiments with domain B of protein A reveals that all these residues, 252, 433, 435 & 436, are also perturbed by the binding of protein A (Kato et al., 1993). However, the binding of protein A also affects the chemical shifts of residues 310, 314 and 429, which are not affected by protein G. Thus, protein A and protein G bind to overlapping but not identical sites on Fc. The differences between the effects of protein G and protein A suggest that the former interacts more with residues from the C_H3 domain than with those from the C_H2 domain.

These experiments thus identify the regions of the surfaces of the two proteins which are involved in the interaction. Taking the known structures of the IgG-binding domains of protein G and of the Fc fragment, it is possible to go a step further and to construct an approximate model for the structure of the protein G domain - Fc complex. This was done by Monte Carlo minimisation (details are given in Kato et al., 1995). The nmr information was introduced into the calculations by defining the interacting region on each protein as including all surface residues within 8Å of one or more residues whose chemical shift was affected by complex formation and using a pseudo-potential which constrained the affected residues in either partner to lie close to one or more residues of the other partner which were within this interacting region. In Figure 5, the model calculated in this way is compared with that of the complex between Fc and domain B of protein A (Deisenhofer, 1981). In this model the protein G domain is located in the 'groove' between the C_H2 and C_H3 domains, with the helix lying more or less in the groove, and the third strand of the sheet making contact with the C_H3 domain. The helix of domain II of protein G is found to lie in an essentially identical position to that occupied by helix 1 of domain B of protein A in its complex with human Fc (Deisenhofer, 1981). However, the orientation of these two helices differs by 180º, so that the third strand of the β-sheet of protein G interacts only with the C_H3 domain, while protein A in its complex has more extensive interactions with the C_H2 domain. This accounts for the observed differences in the residues of Fc affected by the binding of protein G and protein A. The procedure used to arrive at this model depends on the assumption that there is no significant change in the conformation of either partner on formation of the complex, and hence that changes in chemical shift reflect only intermolecular contacts. It therefore leads only to an approximate, medium-resolution model. Since this work was completed, a 3.5Å resolution crystal structure of a protein G - Fc complex has been reported (Sauer-Eriksson et al., 1995). In this structure the location of the helix of protein G is very similar to that in the model shown in Figure 5, but its orientation is different, its axis being tilted at angle of about 60º with respect to that in our model. Since there are extensive intermolecular contacts between the protein G domains bound to neighbouring Fc molecules in the crystal, further work will be required to establish the exact structure of the complex with certainty.

The interactions of these antibody-binding proteins with their 'target' immunoglobulins show two interesting features. First, although the IgG-binding domains of protein A and protein G have quite different three-dimensional structures, they bind to closely similar sites on the Fc fragment of IgG. In both cases an α-helix plays an important role in the recognition, but in a very different orientation - two different structural solutions to the recognition of the same region of a protein surface. Secondly, although the constant domains in Fab and Fc are of course structurally related, protein G binds quite differently to them. This small domain is able to recognise specifically two quite different protein surfaces, by employing an almost completely different set of residues on its surface.

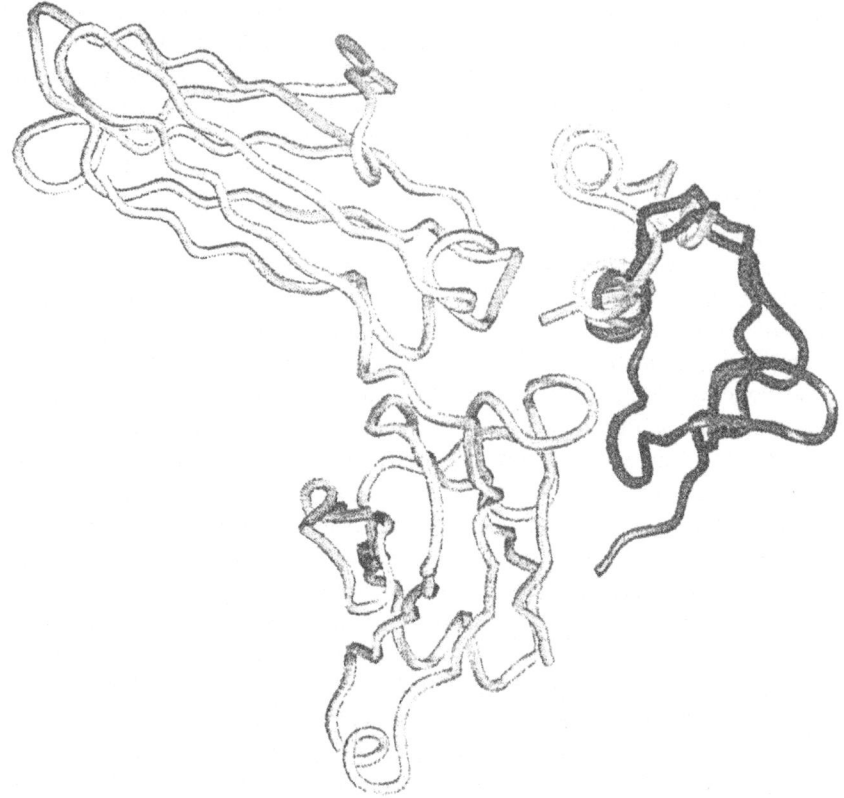

Figure 5. A model of the complex between Fc and domain II of protein G derived from the chemical shift changes observed in solution (Kato et al., 1995), superimposed on the crystal structure of the complex between Fc and domain B of protein A (Deisenhofer, 1981). The two structures were superimposed by least-squares superposition of the backbone atoms of the Fc fragments only; for simplicity, only half the molecule is shown. The protein G domain is shown with dark shading, and the protein A domain with light shading.

4. CONCLUSIONS

These examples, together with a substantial number of others in the literature (see Jardetzky and Roberts, 1981; Feeney and Birdsall, 1993; Lian et al., 1994a), show that it is possible to use NMR to obtain valuable structural and kinetic information about protein complexes which are too large for complete structure determination by this technique. The resonances of interest can be resolved and identified either by isotope labelling (general or selective) or by making use of the association and dissociation of the complex. The best structural information comes from NOE measurements, but chemical shift changes can be used to identify the regions of a protein surface involved in an interaction. Whatever parameter is measured, on must always be alert to the potential difficulties in interpretation arising from the exchange process(es) involved.

ACKNOWLEDGMENTS

I am most grateful to the members of my laboratory, especially Drs. L.-Y. Lian and I.L. Barsukov, who have carried out the work summarised here, and to our collaborators, for their essential contributions: for protein G, Drs. L.-Y. Lian, I.L. Barsukov, M.J. Sutcliffe, J.P. Derrick, K. Kato, I. Shimada and Prof. Y. Arata; and for CAT Drs. L.-Y. Lian, I.L. Barsukov, J. Ellis and Prof. W.V. Shaw. My thanks also to A. Prescott for excellent technical assistance. The work in Leicester was supported by the Biotechnology and Biological Sciences Research Council and the Medical Research Council.

REFERENCES

Achari, A., Hale, S. P., Howard, A. J., Clore, G. M., Gronenborn, A. M., Hardman, K. D., & Whitlow, M. (1992) Biochemistry *31*, 10449-10457.

Albrand, J.P., Birdsall, B., Feeney, J., Roberts, G.C.K., and Burgen, A.S.V. (1979) Int. J. Biol. Macromol. *1*, 37.

Balaram, P., Bothner-By, and Breslow, E. (1972) J. Am. Chem. Soc. *94*, 4015-4017.

Barsukov, I.L., Lian, L.-Y., Ellis, J., Sze, K.-H., Roberts, G.C.K., and Shaw,W.V. (1995) submitted for publication.

Boyle, M. D. P. (1990) Ed: Bacterial Immunoglobulin Binding Proteins, Academic Press, San Diego.

Deisenhofer, J. (1981) Biochemistry *20*, 2361-2370.

Derrick, J. P., & Wigley, D. B. (1992) Nature *359*, 752-754.

Derrick, J. P., & Wigley, D. B. (1994) J. Mol. Biol. *243*, 906-918.

Derrick, J.P., Lian, L.-Y., Roberts, G.C.K., and Shaw, W.V. (1991) FEBS Lett. *280*, 125-128.

Derrick, J.P., Lian, L.-Y., Roberts, G.C.K., and Shaw, W.V. (1992) Biochemistry *31*, 8191.

Feeney, J., and Birdsall, B. (1993) In: NMR of Biological Macromolecules (Roberts, G.C.K. ed.). IRL Press at Oxford University Press. pp.183-216.

Frick, I.-M., Wikström, M., Forsén, S., Drakenberg, T., Gomi, H., Sjöbring, U., & Björck, L. (1992) Proc. Natl. Acad. Sci. USA *89*, 8532-8536.

Gallagher, T., Alexander, P., Bryan, P., & Gilliland, G. L. (1994) Biochemistry *33*, 4721-4729.

Gouda, H., Torigoe, H., Saito, A., Sato, M., Arata, Y., & Shimada, I. (1992) Biochemistry *31*, 9665-9672.

Gronenborn, A. M., Filpula, D.R., Essig, N.Z., Achari, A., Whitlow, M., Wingfield, P.T., & Clore, G. M. (1991) Science *253*, 657-661.

Gronenborn, A. M., & Clore, G. M. (1993) J. Mol. Biol. *233*, 331-335.

Jardetzky, O. (1980) Biochim. Biophys. Acta *612*, 227.

Jardetzky, O., and Roberts, G.C.K. (1981) NMR in Molecular Biology. New York, Academic Press.

Kato, K., Gouda, H., Takaha, W., Yoshino, A., Matsunaga, C., & Arata, Y. (1993) FEBS Lett. *328*, 49-54.

Kato, K., Lian, L.-Y., Barsukov, I. L., Derrick, J.P., Kim, H., Tanaka, R., Yoshino, A., Shiraishi, M., Shimada, I., Arata Y., and Roberts, G.C.K. (1995) Structure *3*, 79-85.

Leslie, A.G.W., Liddell, J.M., and Shaw, W.V. (1986) J. Mol. Biol. *188*, 283.

Lian, L.Y., Yang, J.-C., Derrick, J. P., Sutcliffe, M. J., Roberts, G. C. K., Murphy, J. P., Goward, C. R., & Atkinson, T. (1991) Biochemistry *30*, 5335-5340

Lian, L.-Y., Derrick, J. P., Sutcliffe, M. J., Yang, J. C., & Roberts, G. C. K. (1992) J. Mol. Biol. *228*, 1219-1234.

Lian, L.-Y., and Roberts, G.C.K. (1993) In: NMR of Biological Macromolecules (Roberts, G.C.K. ed.). IRL Press at Oxford University Press. pp.153-182.

Lian, L.-Y., Barsukov, I.L., Sutcliffe, M.J., Sze, K.H., and Roberts, G.C.K. (1994a) Methods in Enzymology *239*, 657-700.

Lian, L.-Y., Barsukov, I. L., Derrick, J. P., & Roberts, G. C. K. (1994b) Nature Structural Biology *1*, 355-357.

Orban, J., Alexander, P., & Bryan, P. (1992) Biochemistry *31*, 3604 - 3611.

Petros, A.M., and Fesik, S.W. (1994) Methods in Enzymology *239*, 717-739.

Sauer-Eriksson, A.E., Kleywegt, G.J., Uhl, M., and Jones, T.A. (1995) Structure *3*, 265-278.

Stone, G. C., Sjöbring, U., Björck, L., Sjöquist, J., Barber, C. V., & Nardella, F. A. (1989) J. Immunol. *143*, 565-570.

Torigoe, H., Shimada, I., Saito, A., Sato, M., & Arata, Y. (1990) Biochemistry *29*, 8787-8793.

Wand, A.J., and Short, J.H. (1994) Methods in Enzymology *239*, 700-717.

Wemmer, D.E., and Williams, P.E. (1994) Methods in Enzymology *239, 739-767*.

5

THE DIRECT DETERMINATION OF PROTEIN STRUCTURE FROM MULTIDIMENSIONAL NMR SPECTRA WITHOUT ASSIGNMENT

An Evaluation of the Concept

R. Andrew Atkinson and Vladimir Saudek

Marion Merrell Dow Research Institute
16 rue d'Ankara, 67080 Strasbourg Cedex, France
Email: VladimirSaudek@mmd.com, AndrewAtkinson@mmd.com

1. INTRODUCTION

Since the advent of multidimensional NMR (Jeener, 1971; Aue et al., 1976; Ernst et al., 1986) and its application to the study of biological macromolecules (Wüthrich, 1986), detailed analysis of the spectra of such molecules has allowed the investigation of the structure, dynamics and interactions of proteins, nucleic acids, etc. at the atomic level (Oppenheimer and James, 1989). It is widely assumed that assignment of NMR spectra is a prerequisite for the interpretation of spectra in structural and dynamic terms. Here, we show that structural information may be extracted *directly* from NMR spectra without assignment and that the molecular structure is independent of the assignment. The assignment of the resonances is then a useful, but not essential, by-product of the procedure providing additional information for refinement and subsequent investigation of dynamics and interactions. We discuss the implications and the problems that need to be overcome for the application of the approach and we suggest that some of the approaches to the interpretation of data used in X-ray crystallography may find parallels in the determination of structures by NMR.

Detected nuclear Overhauser effects (NOE) indicate the close spatial proximity of pairs of nuclei in the molecule of interest (Neuhaus and Williamson, 1989). A large number of such pieces of information are required to define the three-dimensional structure of a biological macromolecule. A systematic approach to determining protein structure from NMR spectra has been developed (Wüthrich, 1986) in which it is necessary to first determine to which nuclei in the known chemical structure particular resonances belong (assignment), before interpreting detected NOEs as distance constraints between the corresponding atoms. A wide range of methods (distance geometry, molecular dynamics simulation, etc.) are then

Dynamics and the Problem of Recognition in Biological Macromolecules, edited by Jardetzky and Lefèvre
Plenum Press, New York, 1996

available to calculate structures of the protein consistent with the experimental data (Havel and Wüthrich, 1985; Braun and Go, 1985; Brünger et al., 1986). Such methods generally use a full structure of the molecule and seek to satisfy simultaneously the chemical constraints on the system (bond lengths, bond angles, chirality, etc.) and the experimental measurements (NOE-derived distances, J-derived torsion angles, inferred hydrogen bonds). It should be stressed, however, that the calculated structure is defined principally by the NOE information. The question then arises as to whether the assignment is at all necessary for determining the structure?

The direct use of structural information from unassigned spectra has been proposed previously to accurately position backbone amide protons and thus trace the protein backbone (Malliavin et al., 1992); to identify elements of secondary structure when spin systems have been identified (Oshiro & Kuntz, 1993); and to assign ^{13}C- and ^{15}N-separated NOE data using calculations similar to those presented here, coupled to chemical shift databases (Kraulis, 1994). Here we wish to emphasise the use of structural information from unassigned spectra to directly deteremine the structure, rather than as an alternative means of assignment.

Since NOEs detected between pairs of unassigned resonances indicate close spatial proximity between pairs of nuclei (or sets of nuclei), then by letting one atom represent each resonance in the spectrum, each NOE defines a distance between two such atoms. Application of standard methods of structure determination (from NMR data) to a set of unconnected and unidentified atoms, between which only distance constraints are specified, yields a structure containing only those atoms for which structural information could be derived from the experiments performed. The problem then becomes akin to that faced by crystallographers: placing the known chemical structure of the molecule into an experimentally determined map of detected atoms (Blundell and Johnson, 1976; Drenth, 1994). This could be achieved by similar means, that is, graphically building a model to match the positions of the detected atoms, with subsequent refinement to incorporate chemical structure information. In practice, the task may be rendered easier, since the type of nucleus is known and the atoms are labelled by their chemical shift value. This gives substantially greater chemical information than is available to the crystallographer when interpreting an electron density map. Further simplification of the task may come from the inclusion of data from experiments that detect through-bond interactions (e.g. TOCSY, COSY (Kessler et al., 1988)), since this information identifies those atoms that belong to the same amino acid residue.

Here we illustrate the method using data sets derived from the crystal structure of BPTI and published chemical shifts to simulate the data expected from two-dimensional ^{1}H NMR experiments (Havel and Wüthrich, 1985; Braun and Go, 1985). Readily available software has been applied to test the feasibility of the approach. The hydrogen atom structure is recreated relatively well and may be used to construct a model. Graphical display of TOCSY data renders assignment straightforward and refinement may proceed through inclusion of chemical structure information.

2. CALCULATIONS

2.1. Reference Structure

Coordinates of BPTI were taken from the Brookhaven Protein Data Bank (entry 6PTI, Wlodawer et al., 1987) and water atoms, the phosphate ion and the C-terminal nitrogen atom removed. Disulphide bridges were defined, hydrogens added and the structure energy minimised using X-PLOR (Brünger, 1992). All heavy atoms were removed to generate a structure file containing 430 hydrogen atoms. Hydrogen atoms not normally observed by NMR and those for which no assignment was reported (Wagner et al., 1987; Berndt et al., 1992) were removed and magnetically equivalent hydrogen nuclei (e.g., methyl groups) were collapsed to the average position. The resulting reference structure contained 296 hydrogen atoms labelled by chemical shift. Overlap in the reported assignments was removed by small adjustments to the chemical shift values.

2.2. Distance Constraints

Distance constraints were generated between pairs of chemical shift labelled atoms if the distance, d_o, in the all-hydrogen structure (430 atoms) between any of the atoms with the particular chemical shift values was equal to or less than 0.40 nm (1255 entries). In separate calculations, constraints were specified either as (i) 0.18 nm $\leq$ d $\leq$ 0.40 nm, or (ii) (d_o - 0.02 nm) $\leq$ d $\leq$ (d_o + 0.02 nm). Additional constraints (effectively non-NOEs) were defined as 0.40 nm $\leq$ d $\leq$ 5.00 nm between all other pairs of atoms and were removed in the final stage of calculation. Overlap was simulated by allowing only one occurrence of each reported chemical shift value in the starting structure file and specifying distinct sets of distance constraints to the same atom in the starting structure (255 atoms, 1239 distance constraints).

Distance constraints were defined as square well potentials. Standard X-PLOR protocols for the calculation of protein structure using NMR data (Brünger, 1992) were modified such that the substructure included all atoms and only the E_{NOE} term was used while creating the bounds matrix, while E_{VDW} was included during embedding. Following energy minimisation (1000 cycles, force constant, k_{NOE}, 420 kJ.mol^{-1}.nm^{-2}, E_{NOE} only; 2000 cycles, k_{NOE} 4200 kJ.mol^{-1}.nm^{-2}). The structure was submitted to a short restrained molecular dynamics simulation (200 steps, 1 fs step, 300K, $E_{NOE}+E_{VDW}$) and energy minimised (2000 cycles, E_{NOE} only) before final minimisation with no (0.40 nm $\leq$ d $\leq$ 5.00 nm) constraints (2000 cycles).

2.3. Results

Resulting hydrogen atom structures (Figure 1) were analysed by fitting to the reference structure and displayed using SYBYL (Tripos Associates, St. Louis, MO, USA). Data from COSY and TOCSY spectra were simulated by generating SYBYL command files to connect atoms expected to give crosspeaks in these experiments (Figure 2). The average RMSD values between calculated and reference structures are listed in Table 1, with details of the residual violations and energies. The set of coarser distance constraints (0.18 nm $\leq$ d $\leq$ 0.40 nm) results in structures with no residual violations. The average RMSD value with respect to the reference structure is quite low (0.191 nm) and the average pairwise RMSD value between calculated structures is 0.155 nm. When distance constraints are specified more precisely, (d_o - 0.02 nm) $\leq$ d $\leq$ (d_o + 0.02 nm), a large number of violations result, reflected in the high E_{NOE} value (1010 kJ.mol^{-1}). This strain arises from the distance constraints applied to single atoms that correspond to more than one atom (e.g. HD and HE atoms of aromatic rings). The average RMSD value with respect to the reference structure is quite reasonable (0.169 nm) and again the structure could be used for model building. Removal of the (0.40 nm $\leq$ d $\leq$ 5.00 nm) constraints allows the structure to collapse somewhat to give fewer violations but a higher RMSD value. It would seem preferable to begin with relatively imprecise definition of distance constraints, increasing the precision (through measurement of crosspeak intensity or NOE buildup rates) as the structure is refined.

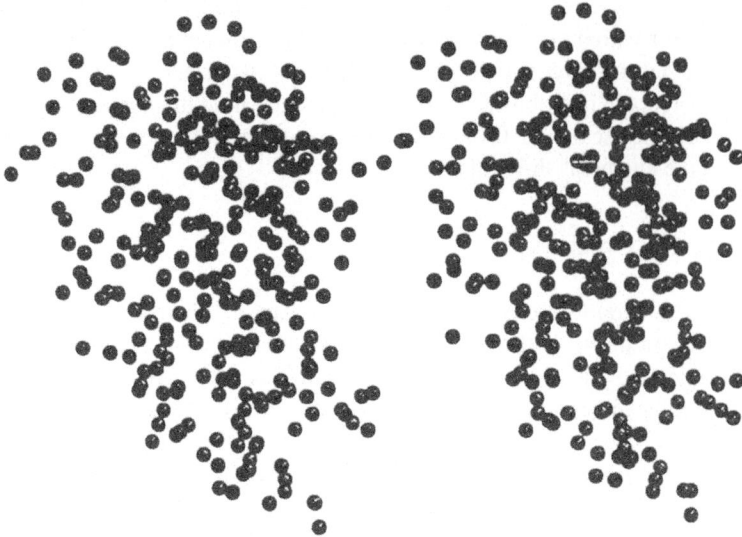

Figure 1. Hydrogen atom structure of BPTI, calculated using distance constraints derived from the crystal structure. Each atom may be labelled with the chemical shift value of the corresponding resonance in the spectrum (for clarity, these labels have been omitted).

2.4. Overlap

Simulation of a single case of resonance overlap results in violations of some distance constraints, reflected in the E_{NOE} value (Table 1). Inclusion of all (36) cases of overlap in the reported resonance assignments causes more severe problems and many distance constraint violations. The resulting structures fit only poorly to the reference structure.

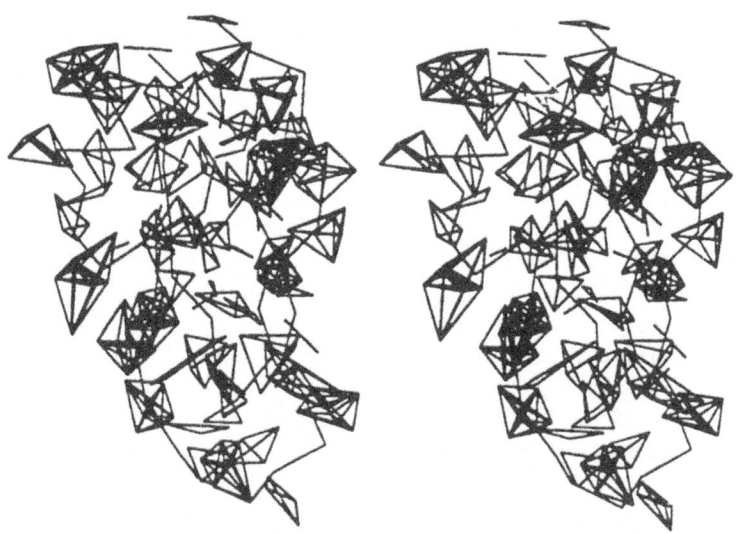

Figure 2. Hydrogen atom structure of BPTI with atoms related by scalar couplings, detectable in TOCSY experiments, connected. The continuous line represents the Cα trace of BPTI.

Table 1. Results of calculations using only distance constraint terms as input for distance geometry and subsequent energy minimisation

	I		II		III		IV	
	(a)	*(b)*	*(a)*	*(b)*	*(a)*	*(b)*	*(a)*	*(b)*
Total E_{NOE} (kJ.mol^{-1})	114.99	0.00	1171.40	194.30	1612.37	0.00	45833.95	0.00
Sum of constraint violations (nm)	0.125	0.000	0.527	0.216	0.620	0.000	3.311	0.000
Maximum E_{NOE} (kJ.mol^{-1})	18.74	0.00	32.35	12.42	139.48	0.00	381.69	0.00
Maximum constraint violation (nm)	0.050	0.000	0.087	0.055	0.182	0.000	0.301	0.000
RMSD (nm)	0.192	0.191	0.169	0.361	0.298	0.303	0.854	0.846

All data are mean values of 10 separate calculations. For each set of calculations, data are given (a) before, and (b) after the final minimisation without the (0.40nm $\leq$ d $\leq$ 5.00nm) constraints. I. Distance constraints specified as 0.18nm $\leq$ d $\leq$ 0.40nm. II Distance constraints specified as (d_o-0.02nm) $\leq$ d $\leq$ (d_o+0.02nm). III. Distance constraints specified as 0.18nm $\leq$ d $\leq$ 0.40nm, with one case of chemical shift overlap (1.03 ppm) included in the input data. IV. Distance constraints specified as 0.18nm $\leq$ d $\leq$ 0.40nm, with all cases of chemical shift overlap in the reported assignments included in the input data.

3. CONCLUSIONS AND PERSPECTIVES

It is evident that there is a direct correspondence between the NOE spectrum of a molecule and its structure that is independent of the assignment of the resonances. It follows that it should be possible to determine protein structure directly without recourse to chemical shift and scalar coupling data. The above calculations show that it is indeed possible to reconstruct the spatial positions of the hydrogen atoms of a protein using NOE information alone *and without resonance assignments*, to yield a starting model for further refinement. For the practical implementation of this approach, the problem of spectral overlap (resolution) must be overcome. A small number of overlapping resonances can be easily detected in analysis of the calculated structures and the constraint violations. Conservative omission of ambiguous data would give a poorer structure that could be refined subsequently. Approaches such as that developed to treat ambiguous NOEs (Bernstein et al., 1993; Nilges, 1995) will prove very useful in such cases. A larger number of cases of overlap, however, severely impair the calculations and may require other solutions that reliably separate NOEs from overlapping resonances. The necessary resolution could be obtained through homonuclear three-dimensional experiments, provided efficient routines for handling the input data can be developed, and three- and four-dimensional heteronuclear experiments, where enrichment in ^{13}C and/or ^{15}N is possible (Ernst, 1992; Clore and Gronenborn, 1993). Furthermore, heteronuclear experiments provide additional distance constraints and chemical information while increasing the number of observed nuclei.

Calculation of the structure directly from spectra opens up many far-reaching possibilities. The approach to the experimental data becomes analogous to that used in X-ray crystallography. Existing methods for building a model into the experimentally determined positions of the atoms and for evaluating the quality of the model could be adopted (Drenth, 1994). Discussion of missing or insufficient data or ill-defined sidechains becomes clearer than at present since the interpretation relies solely on experimental data, that is, the determined molecular structure contains only what is observed by experiment. The chemical information content of the NMR spectra is clearly separated from the conformational information and is not initially used in defining the *observed* molecular structure since including this information (bond lengths, bond angles, chirality, etc.) imposes *inferred* structural constraints, not obtained from the spectra.

As in X-ray crystallography, the observed structure should probably be refined as far as possible before inferring the positions of unobserved atoms. Several approaches may be envisaged. The structure could be re-solved by established means (Havel and Wüthrich, 1985; Braun and Go, 1985; Brünger et al., 1986) now using the resulting assignment to relate the distance information to atoms in the full chemical structure. It might prove advantageous, however, to refine the *observed* atom structure including additional information gained during the previous stage. Crosspeaks detected at longer mixing times may be added to the constraint list and intensities used in defining distances. Corrections to inter-atomic distances (Wüthrich et al., 1983) may be applied once an atom has been identified as corresponding to a set of degenerate resonances (e.g. methyl group, aromatic ring). Certain distances may be set to known values and the observed structure refined before resorting to full molecule calculations. Full relaxation matrix methods could also be readily applied (Mertz et al., 1991; Nilges et al., 1991). In the absence of all unobserved atoms and the constraints that define the chemical structure, calculations are rapid. Inevitably, however, the mirror image of the structure is an equally valid solution to the problem (Pastore et al., 1991).

Large proteins often suffer from poor TOCSY transfer, such that little or no scalar coupling information may be acquired. This may be compensated by more efficient NOE transfer and a richer NOE data set since a larger proportion of the molecule constitutes the core of the protein (Wagner, 1993). Experiments such as 3D NOESY-NOESY should give access to sufficient resolved data to determine the structure without scalar coupling information, using crystallographic tools to build the structure into the hydrogen atom cloud. Furthermore, it should prove possible to determine rapidly the structure of a protein for which an approximate model is available (e.g. NMR structure determined under different conditions, X-ray structure, homologous protein, mutant, model structure) through molecular replacement, again adopting and adapting existing crystallographic methods.

Much effort is currently being concentrated in the development of automated or semi-automated methods to speed up structure determination (Hare and Prestegard, 1994; Olson and Markley, 1994; Xu et al, 1994). Clearly, circumventing the need for assignment should greatly accelerate NMR analysis since only a list of NOE-related pairs of atoms is required and the experimental data set is exploited as a whole. Intra-residual and sequential NOEs are commonly considered as having little or no structural information content, but here are crucial since they define the spatial proximities of atoms within the amino acid residues.

In conclusion, we wish to draw attention to an alternative approach to the interpretation of NMR spectra for structure determination. The direct relationship between spectrum and structure is evident but has so far been obscured by the assumption that assignment was essential for interpretation. Analysis of the spectra will still be required until the practical difficulties of the approach have been overcome but we can aim to reach the stage where it will be no more necessary to inspect NMR spectra to determine structure than it is to study X-ray diffraction patterns.

4. ACKNOWLEDGMENTS

We would like to thank Drs. C. Brockel, A. Pastore, J.T. Pelton, J-M. Rondeau, H.A. Schreuder and V. Sklenar for useful discussions.

5. REFERENCES

Aue, W.P., Bartholdi, E. and Ernst, R.R. (1976) J. Chem. Phys. *64*, 2229-2246.

Berndt, K.D., Güntert, P., Orbons, L.P.M. and Wüthrich, K. (1992) J. Mol. Biol. *227*, 757-775.

Bernstein, R., Schnuchel, A., Cieslar, C. and Holak, T.A. (1993) J. Magn. Reson. Series B *102*, 116-119.

Blundell, T.L. and Johnson, L.N. (1976) Protein Crystallography, Academic Press, New York.

Braun, W. and Go, N. (1985) J. Mol. Biol. *169*, 611-626.

Brünger, A.T., Clore, G.M., Gronenborn, A.M. and Karplus, M. (1986) Proc. Natl. Acad. Sci. USA *83*, 3801-3805.

Brünger, A.T. (1992) X-PLOR Software Manual, version 3.1, Yale Univ. Press, New Haven.

Clore, G.M. and Gronenborn, A.M. (1993) Science *252*, 1390-1399.

Drenth, J. (1994) Principles of Protein X-Ray Crystallography, Springer-Verlag, New York.

Ernst, R.R., Bodenhausen, G. and Wokaun, A. (1986) Principles of Nuclear Magnetic Resonance in One and Two Dimensions, Oxford University Press, Oxford.

Ernst, R.R. (1992) Angew. Chem. *104*, 817-836.

Hare, B.J. and Prestegard, J.H. (1994) J. Biomol. NMR *4*, 35-46.

Havel, T.F. and Wüthrich, K. (1985) J. Mol. Biol. *182*, 281-294.

Jeener, J. (1971) Ampère International Summer School, Basko Polje, Yugoslavia.

Kessler, H., Gehrke, M. and Griesinger, C. (1988) Angew. Chem. Int. Ed. Engl. *27*, 490-536.

Kraulis, P. (1994) J. Mol. Biol. *243*, 696-718.

Malliavin, T.E., Rouh, A., Delsuc, M.A. and Lallemand, J.-Y. (1992) C. R. Acad. Sci., Ser II *315*, 653-659.

Mertz, J.E., Güntert, P., Wüthrich, K. and Braun, W. (1991) J. Biomol. NMR *1*, 257-269.

Neuhaus, D. and Williamson, M.P. (1989) The Nuclear Overhauser Effect in Structural and Conformational Analysis, VCH Publishers Inc., New York.

Nilges, M., Habazettl, J., Brünger, A.T. and Holak, T.A. (1991) J. Mol. Biol. *219*, 499-510.

Nilges, M. (1995) J. Mol. Biol. *245*, 645-660.

Olson Jr., J.B. and Markley, J.L. (1994) J. Biomol. NMR *4*, 385-410.

Oppenheimer, N.J. and James, T.L. (eds.) (1989) Meth. Enzymol. *176-177*.

Oshiro, C.M. and Kuntz, I.D. (1993) Biopolymers *33*, 107-115.

Pastore, A., Atkinson, R.A., Saudek, V. and Williams, R.J.P. (1991) Proteins, *10*, 22-32.

Wagner, G., Braun, W., Havel, T.F., Schaumann, T., Go, N. and Wüthrich, K. (1987) J. Mol. Biol. *196*, 611-639.

Wagner, G. (1993) J. Biomol. NMR *3*, 375-385.

Wlodawer, A., Nachman, J., Gilliland, G.L., Gallagher, W. and Woodward, C. (1987) J. Mol. Biol. *198*, 469-480.

Wüthrich, K., Billeter, M. and Braun, W. (1983) J. Mol. Biol. *169*, 949-961.

Wüthrich, K. (1986) NMR of Proteins and Nucleic Acids, J. Wiley, New York.

Xu, J., Straus, S.K., Sanctuary, B.C. and Trimble, L. (1994) J. Magn. Reson. Series B *103*, 53-58.

6

ACCURACY OF NUCLEAR MAGNETIC RESONANCE DERIVED MOLECULAR STRUCTURES

Quantitative Uncertainty Analysis of Distance Constraints

Ricardo González Méndez

Physics and Radiobiology Division
Department of Radiological Sciences
University of Puerto Rico School of Medicine
San Juan, Puerto Rico

1. INTRODUCTION

In this paper we will look at some of the issues regarding the intrinsic accuracy of the Nuclear Magnetic Resonance (NMR) derived biomolecular structure models in solution by asking the question: How close is the NMR derived structure model to the true but unknown value (or true but unknown ensemble average value) of the macromolecule being studied.

There is a need to explore the fundamental limits of the accuracy of NMR derived structures with regards to the true but unknown value (or ensemble average value) from the perspective of the accuracy and precision of the determination of the distance constraints and dihedral angle constraints that can be obtained from multidimensional NMR spectroscopic data (Sutcliffe, 1993).

NMR data is used to derive distance constraints, and dihedral angle constraints which define a low-resolution structure model. This low-resolution model is then refined by one of several methods based on molecular mechanics or restrained molecular dynamics to derive a distribution of high-resolution structure models (James, 1994). To date, the main thrust of the research in the field has been first, to ascertain the accuracy and precision of the calculational models both at the theoretical level and the computational models used (Jardetzky, 1991; Thomas et al., 1991; Liu et al., 1992; Post, 1992; Clore et al., 1993, Zhao and Jardetzky, 1994), and second to do model validation by comparison of the NMR derived solution structures to crystal structures (Wagner et al., 1992).

Our approach in this paper will be to use statistical methods, error theory, probability theory, and Monte Carlo simulations combined in a specific methodology — quantitative uncertainty analysis — (Hoffman and Hofer, 1989; Morgan and Henrion, 1990; Shlyakhter, 1994) to compare experimentally derived values for distance constraints (including the errors

Dynamics and the Problem of Recognition in Biological Macromolecules, edited by Jardetzky and Lefèvre
Plenum Press, New York, 1996

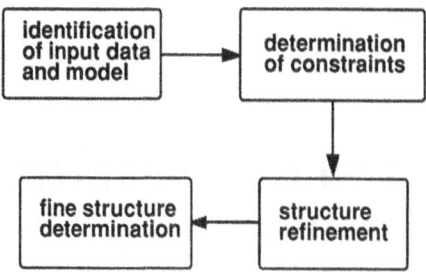

Figure 1. Major steps used in structural determination by NMR.

and uncertainties inherent to their determination) to the true but unknown values of the molecular structure(s) that we are trying to model.

The approach and methodology presented here will be particularly useful when there are no "gold-standard" structures, such as a crystal structure, for comparison with the NMR derived structure. This approach could also be useful in contributing to the understanding of the intrinsic accuracy of the determination of crystal structures beyond the standard calculations of F-values and B-factors that are typically found in this type of structural determination.

1.1. Deriving Molecular Structures by NMR

The derivation of a solution structure by NMR is comprised of several distinct steps: a) data acquisition and assignments; b) determination of distance constraints; c) refinement; and d) structure determination. Most of the discussion has centered around steps c) and d) (Jardetzky, 1991; Thomas et al., 1991; Liu et al., 1992; Post, 1992; Clore et al., 1993, Zhao and Jardetzky, 1994), and has concentrated on model validation using crystal structures as an accurate representation of the true but unknown structure(s). The discussion assumes that step a) has been properly carried out, i.e. no errors ("blunders"(Taylor, 1982)) have occurred at this stage. The determination of distance constraints has been addressed from the perspective of its effect on the accuracy and precision of the structures derived (Thomas et al., 1991; Liu et al., 1992; Clore et al., 1993, Zhao and Jardetzky, 1994). The distance constraints are usually derived by one of four methods: isolated spin pair approximation (ISPA) (Jardetzky and Lane, 1988), iterative solution to the Bloch equations and fitting to the experimental data (Lefèbre et al., 1987), relaxation matrix analysis (CORMA or MARDI-GRAS) (Borgias and James, 1989) or an empirical classification setting upper and lower bounds (see (Kuntz et al., 1989) for a discussion). Some discussion of the uncertainties implicit in the ISPA has been presented (Jardetzky and Lane, 1988; Barsukov and Lian, 1993). The analysis of the iterative solution to the Bloch equations, and the relaxation matrix methods, has been carried out in terms of structure refinement and inherent systematic errors, without directly addressing the inherent uncertainties in the input data or the uncertainty on the output structural distances (Madrid and Jardetzky, 1988; Borgias and James, 1989). The empirical method is the most commonly used procedure because "it works well" with refinement strategies (Kuntz et al, 1989; Barsukov and Lian, 1993). We will use quantitative uncertainty analysis (Hoffman and Hofer, 1989; Morgan and Henrion, 1990), including the more sophisticated techniques involving Monte Carlo simulations, to address the uncertainties in the analysis of the NMR data and the generation of distance constraints. In addressing the uncertainty in the distances constraints we will be referring to the accuracy of the distances derived. We define accuracy as how close we have come to the true but unknown value. This will be given by deriving a 95% confidence interval from the quantitative uncertainty analysis. We define precision as how reproducible a determination is. This is usually given by a standard deviation, or a standard error of the mean. In our analysis precision enters only as the error of the cross relaxation intensities or the error of the correlation times.

1.2. Quantitative Uncertainty Analysis

Quantitative uncertainty analysis is a method used to assess the reliability of models in the physical and environmental sciences when data is sparse and/or there is lack of knowledge about parameters (Hoffman and Hofer, 1989; Morgan and Henrion, 1990). Here, reliability is defined as a measure of confidence in model predictions, which may be qualitative or quantitative. The assessment of models will require some amount of subjective judgment based on the scientific knowledge (or lack thereof) about the problem. Essentially one can attempt to answer two questions: a. how large is the uncertainty associated with the model prediction, e.g. what is the 95% confidence interval?; and b. is this uncertainty estimate acceptable for "decision making", that is, assigning distance constraints. The procedure is performed as follows: 1. A conceptual model is identified; 2. A computational model is described (i.e., a set of equations); 3. Estimation of parameter values is carried out; 4. Calculation of results. There will be two types of uncertainty that need to be taken into account: that due to stochastic variability (type A) versus that due to lack of knowledge (type B). Type A uncertainties will require probabilistic answers to our questions, but in the presence of type B we will only obtain ranges of alternative or possible true answers to our questions (Hoffman and Hofer, 1989; Morgan and Henrion, 1990; Shlyakhter, 1994). In general this type of analysis will yield answers in terms of probabilities. But for type A uncertainties probability will mean relative frequency of values from a specified interval, while for type B uncertainties it will mean a degree of belief that a vaguely or imprecisely known value is within a specified interval (i.e., subjective probability - Bayesian interpretation). The selection of appropriate probability distributions is critical to the uncertainty analysis, and in general, in the case of minimum knowledge the distribution should be a uniform one over the maximum conceivable range. The results will typically be presented as confidence intervals and/or 'best estimates". These problems can be solved exactly by generating the analytical solution to the problem - a simple thing for the two spin case (ISPA) but virtually impossible for a protein with a several thousand spins combined into multiple spin systems; or they can be solved approximately by statistical distribution parametrization and Monte Carlo simulations. In general the results from NMR structure model determination yield Type B uncertainties because of the scientific judgments involved in the selection of distance constraints and refinement models.

1.3. The Problem

Distance constraints derived from NMR experiments (i.e., from the measurement of the multidimensional Nuclear Overhauser Effect (NOE)) are usually generated by an empirical classification, for example:

As can be seen from Table 1, the ranges of cross relaxation intensity versus distance are derived by the investigator applying a series of scientific value judgments based on his experience or expertise. We have identified at least twenty three such classifications from the NMR literature.

The distance can also be calculated by the ISPA:

Table 1. Empirical classification of NOE values

NOE Type	NOE Intensity,[†] %	Distance[†] (Conservative), Å	Distance* (Restrictive), Å
Strong	> 2	1.8–2.8	1.8–2.8
Medium	0.9–2	1.8–3.5	2.8–4.1
Weak	0.3–0.9	1.8–5.0	4.1–6.0

[†]From Table 1 in Thomas et al., 1991.
*From Liu et al., 1992.

$$\frac{a_{ij}}{a_{ref}} = \left(\frac{r_{ref}}{r_{ij}}\right)^6 \times \frac{\tau_{ij}}{\tau_{ref}} \tag{1}$$

where a_x indicates cross peak intensity, r_x indicates distance in Å, and τ_x indicates correlation time, the index $_{ref}$ indicates the calibration values, and the index $_{ij}$ indicates experimental values.

Furthermore, if one assumes a rigid molecule the correlation time terms cancel out from equation 1:

$$\frac{a_{ij}}{a_{ref}} = \left(\frac{r_{ref}}{r_{ij}}\right)^6 \tag{2}$$

Therefore we ask the following questions:

a. Assuming the NOE intensity ranges in Table 1, how accurate are the distance ranges used if one carries out an uncertainty analysis using the ISPA as an analytical model?

b. What kind of distance ranges can we obtain from the ISPA through a quantitative uncertainty analysis?

c. For a set of experimentally determined NOE's what distances can one calculate from a quantitative uncertainty analysis using the ISPA?

2. METHODS

A quantitative uncertainty analysis was carried out using a Macintosh computer and Crystal Ball v.2.0 to carry out a Monte Carlo simulation (5,000 trials per run) and obtain subjective probability estimates. All distributions used were assumed as uniform or normal given the minimal knowledge implicit in the empirical classification and ISPA, and lack of knowledge about correlation times. Published error ranges were assumed to represent upper and lower bounds of allowable values for the distribution limits in the uniform distribution or a standard deviation in the normal distribution. Calculations were carried out as follows:

a. To test the empirical approach we used the NOE ranges listed in column 2 of Table 1.

b. Both the full ISPA and the rigid molecule approximation were used. For correlation times both empirical values or the full range of possible values (no knowledge assumption) were used.

c. Data and parameters published on Basic Pancreatic Trypsin Inhibitor (BPTI) as described in the literature (Jardetzky and Lane, 1988; Ribeiro et al., 1980 , Madrid and Jardetzky, 1988) were used:

     a. Tyr 35 Hδ2-Hϵ2 was used as reference:

       i. a_{ref} = 2.5±0.5

       ii. r_{ref} = 2.49Å with a 10% error range (±0.25Å) (Wlodawer et al., 1984)

       iii. τ_{ref} = 0.3±0.15 ns, or 0.01 - 3.0 ns (no knowledge assumption)

    d. For the full ISPA τ_{ij} was allowed to vary from 0.01 - 3.0 ns.

    e. A preliminary assessment of ISPA predictions was carried out using the data for three BPTI NOE's and the distances generated were compared to the crystal structure distances as presented by Madrid and Jardetzky (1988). For this set of calculations the residue τ_{ij} was assumed to be 0.8±0.4 (typical for protein backbone - Lane and Jardetzky, 1988), or a rigid molecule was assumed.

Results are presented as 95% confidence intervals. For calculations in e. above the central (most probable) estimate taken as the median of the resulting distribution (50% probability of the derived distance to be longer or shorter than the chosen value) is presented also. Root mean square deviation between the predicted distances and the crystal structure were also calculated for e. above.

3. RESULTS

The subjective confidence intervals are shown in the Table 2. Results for the rigid molecule assumption, the full ISPA, and the no-knowledge assumption are shown. The results of the medium intensity NOE's and the weak intensity NOE's are the same from a statistical standpoint.

A sensitivity analysis carried out on each of the individual variables used for the empirical model analysis showed that the dominating sources of uncertainty are the correlation time τ_{ij} , accounting for ≈50-60% variability around the calculated central estimate (data not shown). All other variables account for ≈3-10% variability around the central estimate, with the exception of the "strong NOE" range which accounts for 15% (due to the width of that range when compared to others) (data not shown). Similar results are seen for calculations using a normal distribution (data not shown).

The subjective confidence intervals are shown in Tables 3 and 4 below. It can be seen that the estimates obtained from the rigid molecule ISPA have an RMSD of 2 0.7 Å when compared to the crystal structure. The estimates obtained from the full ISPA show an RMSD of 2 1.38 Å. In general the subjective confidence intervals include the distance from the

Table 2. Evaluation of the empirical classification and the ISPA

	NOE range (%)	Lower bound (Å)	Upper bound (Å)	Confidence interval (%)
Weak/ rigid	0.3–0.9	2.68	3.66	95
Medium/rigid	0.9–2.0	2.36	3.09	95
Strong/rigid	2.0–15	1.76	2.70	94
Weak/full	0.3–0.9	2.75	5.45	95
Medium/full	0.9–2.0	2.39	4.68	95
Strong/full	2.0–15	1.78	3.76	95
Weak/no knowledge	0.3–0.9	1.77	6.00	99
Medium/no knowledge	0.9–2.0	1.77	5.50	99
Strong/no knowledge	2.0–15	1.78	4.00	88

All distances are given in Angstroms.

Table 3. RMSD calculations for selected NOE's of BPTI using the full ISPA

NOE identity	Crystal distance	NOE-best distance	RMSD	Low bound	RMSD	High bound	RMSD
HBPhe45-HNSer47-I	2.55	3.55	0.71	2.69	0.10	4.37	1.29
HBPhe45-HNSer47-II	2.42	3.55	0.80	2.69	0.19	4.37	1.38
HAArg17-HGVal34-I	2.77	3.06	0.21	2.38	0.28	3.67	0.64
HAArg17-HGVal34-II	2.38	3.06	0.48	2.38	0.00	3.67	0.91
HAPro9-HNTyr10-I	2.27	2.77	0.35	2.24	0.02	3.28	0.71
HAPro9-HNTyr10-II	1.80	2.77	0.69	2.24	0.31	3.28	1.05

All distances are given in Angstroms.

crystal structure, but in the full ISPA the values obtained seem to be biased towards longer distances, and the crystal distances are at the lower bound of the range of distances in the calculated distribution of distance constraints.

4. DISCUSSION

The results presented in Table 2 above are compatible with the conservative empirical interpretation of the NOE's. Indeed, one can conclude that a medium range interval is not supported by the uncertainty analysis shown here. This statement is true whether one uses the rigid molecule approximation or takes into account the existence of internal motion with or without any knowledge of the effective correlation times. Further exploration of the type of distributions used is needed before firmly establishing this conclusion

The calculations for the three test NOE's show that the rigid molecule approximation is a better predictor than the full ISPA. This may not be the case if better estimates of the correlation times were available. This is an example of how inadequate assumptions may yield less accurate estimates. Furthermore it shows that the scientific value judgments used for setting distance constraints may actually lead to incorrect distances, and therefore to an incorrect structural determination.

It is worthwhile to note that the predictions of the uncertainty analysis for the rigid molecule ISPA yielded upper and lower bounds that included the distances from the crystal structure within the confidence interval in all of three examples, although in one it was at the lower bound (99% confidence)(HBPhe45-HNSer47).

Table 4. RMSD calculations for selected NOE's of BPTI using the rigid molecule approximation ISPA

NOE identity	Crystal distance	NOE-best distance	RMSD	Low bound	RMSD	High bound	RMSD
HBPhe45-HNSer47-I	2.55	3.01	0.33	2.55	0.00	3.43	0.62
HBPhe45-HNSer47-II	2.42	3.01	0.42	2.55	0.09	3.43	0.71
HAArg17-HGVal34-I	2.77	2.60	0.12	2.20	0.40	2.98	0.15
HAArg17-HGVal34-II	2.38	2.60	0.16	2.20	0.13	2.98	0.42
HAPro9-HNTyr10-I	2.27	2.36	0.06	2.07	0.14	2.63	0.25
HAPro9-HNTyr10-II	1.80	2.36	0.40	2.07	0.19	2.63	0.59

All distances are given in Angstroms.

Further work is needed to explore the use of quantitative uncertainty analysis in setting NOE distance constraints for structure refinement. It will be important to compare an ISPA uncertainty analysis with one for the relaxation matrix approach, or the solution to the Bloch equations approach and determine the accuracy of all three methods in setting NOE constraints. This approach may be a useful way to answer the questions of accuracy of NMR derived structures when there is no crystal structure against which to validate our models.

Finally, we can also conclude from this work that in the absence of accurate NOE's, *and* accurate knowledge of the individual effective correlation times, one is better off in allowing the structural refinement algorithm to search over all of configurational space by setting the distance constraints to 1.8 - 6.0 Å, which basic physics tell us is the possible distance range. This approach has been shown to yield more *accurate* estimates of model structures than the usual empirical approach used in the literature (Zhao and Jardetzky, 1994).

Future research must look at whether the complete relaxation methods or the iterative solution to the Bloch equations are more accurate predictors of distance constraints by carrying this type of uncertainty analysis. Only then can we approach the problem of which are the more accurate of the refinement and structural determination schemes, if any is to be shown to be superior to the others. When this questions are answered we can then proceed to determine a method for how to choose one of the many structural models determined by NMR as the "best model estimate" of the true but unknown structure.

5. REFERENCES

Barsukov I.L. and Lian L-Y. (1993) in: NMR of Macromolecules - A Practical Approach. (Roberts G.C.K. ed.), pp. 315-357, IRL Press, Oxford University Press,NY.
Borgias B.A. and James T.L. (1989) Meth. Enzymol. 176:169-183.
Clore G.M., Robien M.A. and Gronenborn A.M. (1993) J Mol. Biol. 231:82-102.
Hoffman F.O. and Hofer E. (1989) Safety Series No. 100, IAEA, Vienna.
James T.L. (1994) Meth. Enzymol. 239:416-439.
Jardetzky O. (1991 in: Computational Aspects of the Study of Macromolecules by Nuclear Magnetic Resonance Spectroscopy (Hoch J.C. et al. eds.) Plenum Press, NY, p. 375-390.
Jardetzky O. and Lane A N. (1988) in: Physics of NMR Spectroscopy in Biology and Medicine (Maraviglia B. ed.) North Holland Publishing, Amsterdam, pp. 267-300.
Kuntz I.D., Thomason J.D. and Oshiro C.M. (1989) Meth. Enzymol. 177:159-204.
Lefèvre J-F., Lane A.N. and Jardetzky O. (1987) Biochemistry 26:5076-5090.
Liu Y., Zhao D., Altman R and Jardetzky O. J. (1992) Biomol. NMR 2:373-388.
Madrid M. and Jardetzky O. (1988) Biochim. Biophys. Acta 953:61-69.
Morgan M.G. and Henrion, M. (1990) Uncertainty, Cambridge University Press, NY.
Post C.B. (1992) J. Mol. Biol. 224:1087-1101.
Ribeiro A.A., King R., Restivo C. and Jardetzky O. (1980) J. Am. Chem. Soc. 102:4040-4051.
Shlyakhter A.I. (1994) in: Uncertainty Modeling and Analysis: Theory and Applications (Ayuub B.M. and Gupta M.M. eds.), North Holland, NY.
Sutcliffe M.J. (1993) in: NMR of Macromolecules - A Practical Approach (Roberts G.C.K. ed.), pp. 358-390, IRL Press, Oxford University Press,NY.
Taylor J.R. (1982) An Introduction to Error Analysis, Oxford University Press, pp. 3.
Thomas P.D., Basus V.J. and James T.L. (1991) Proc. Natl. Acad. Sci. USA 88:1237-1241.
Wagner G., Hyberts S.G. and Havel T.F. (1992) Annu. Rev. Biophys. Biomol. Struct. 21:167-198.
Wlodawer A., Walter J., Huber R. and Sjölin L. (1984) J. Mol. Biol. 180:301-329.
Zhao D. and Jardetzky O. (1994) J. Mol. Biol. 239:601-607.

7

WHAT LIMITS PROTEIN FOLDING

Tobin R. Sosnick and S. Walter Englander

The Johnson Research Foundation
Department of Biochemistry and Biophysics
University of Pennsylvania School of Medicine
Philadelphia, Pennsylvania 19104-6059

1. ABSTRACT

Recent work shows that the barriers that produce slow protein folding are not intrinsic to the folding process. It appears that slow folding is caused by off-pathway misfolding steps, and that the rate-limiting step is the reorganization necessary to correct the misfolding. When folding is fast in the absence of a misfolding-reorganization barrier, the rate-limiting step appears to be the initial molecular collapse. Available evidence suggests that these same conclusions hold for many other proteins.

2. CYTOCHROME C AS A MODEL FOR FOLDING STUDIES

The obligate steps in protein folding include molecular collapse, secondary and tertiary structure formation, water extrusion, the tight packing and interdigitation of side chains, and in larger proteins the establishment of intersubunit and quaternary interactions. Models that have been proposed for the folding process picture that one or the other of these steps is intrinsically difficult and limits folding rates to the usually observed time scale of seconds. For example in many folding experiments, well-structured collapsed intermediates, often termed molten globules, are rapidly formed but fold slowly to the native state. Consequently, it has been generally assumed that the later steps in folding are intrinsically slow.

To test these and other assumptions and models for protein folding, we have studied the folding of cytochrome c, taking advantage of the special opportunity it provides for manipulating the limiting barriers directly. Cytochrome c (cyt c) is a 104 residue protein built up of 3 major helices and several omega loops (Takano and Dickerson, 1981; Bushnell et al., 1990) (Fig. 1). The central heme group is held by covalent thioether linkages to Cys14 and Cys17, and by two axial ligands provided by His18 and Met80. The heme provides some most useful spectroscopic probes for folding studies. The single tryptophan in cyt c, Trp59, fluoresces strongly in the unfolded protein, but when the protein folds Trp59 is brought close to the heme and fluorescence is totally quenched by Forster energy transfer (Forster, 1965).

Dynamics and the Problem of Recognition in Biological Macromolecules, edited by Jardetzky and Lefèvre
Plenum Press, New York, 1996

65

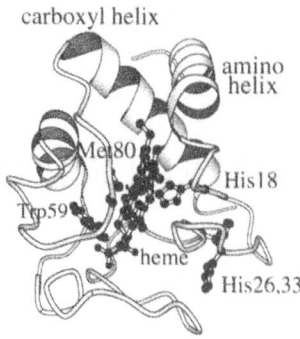

Figure 1. Cytochrome c molecule (Takano and Dickerson, 1981; Bushnell et al., 1990). Molecular diagrams were drawn using Molscript (Kraulis, 1991).

The fluorescence is a strong function of the Trp59 to heme distance and thus provides a good probe for molecular collapse (Tsong, 1976). The weak Met80 ligation, which is present only in the fully native protein, produces a characteristic and unusual absorbance band at 695 nm (Schechter & Saludjian, 1967) that provides a specific probe for acquisition of the fully native state (Ridge et al., 1981)(Myer, 1984).

3. FAST FOLDING

Fig. 2 shows the results of some experiments in which cyt c was unfolded by guanidinium chloride (GdmCl) and then was allowed to refold by diluting the denaturant in a stopped-flow experiment (Sosnick et al, 1994). When the protein was initially unfolded at pH values above 6, dilution into low GdmCl (pH 5, 10°C) led to the same kind of slow folding (see also (Elove et al, 1994)) seen for many other proteins (Fig. 2B). Surprisingly, when the protein was initially unfolded at pH 5 or lower and then diluted into the same refolding condition as before, the protein reached the native state (A_{695}) 100 times more quickly, on the same time scale as molecular collapse (fluorescence quenching) (Fig. 2A).

Experiments were done to determine whether cyt c actually reaches the authentic native state on this fast time scale (Sosnick et al, 1994). We measured the formation of secondary and tertiary structure by circular dichroism measurements in the far and the near ultraviolet and the formation of secondary and tertiary structural hydrogen bonding by hydrogen exchange pulse labeling detected by 2D NMR (Fig. 2A). Also we tested for the acquisition of native-like stability by hydrogen exchange pulse labeling at pH values up to pH 12 and in double jump experiments that measured the unfolding rate of the refolded protein. All these experiments show that cyt c is able to fold to the authentic native state with a 15 msec time constant.

The fast folding capability is by no means unique to cyt c. About 7 other proteins have now been recognized to fold on a similar time scale (~15 msec) (Radford et al., 1992; Jackson and Fersht, 1991; Khorasanizadeh et al., 1993; Alexander et al., 1992; Houry et al., 1994; Kuszewski et al., 1994; Viguera et al., 1994; Huang and Oas, 1995). This establishes that all the obligate folding steps are inherently fast processes. Therefore the much slower folding usually observed cannot be due to kinetic barriers that are intrinsic and fundamental to the folding process.

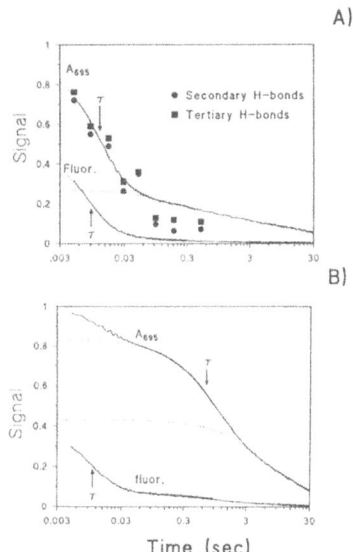

Figure 2. Cyt c in fast folding (A) and slow folding (B) modes. Folding behavior was measured in terms of molecular condensation (fluorescence quenching of Trp59) and final Met80 to heme ligation (absorbance at 695 nm). The ordinate gives the ratio of the time-dependent signal to the unfolded protein signal. Dotted lines indicate single exponential components. Folding was at 0.7 M GdmCl pH 4.9, 10°C in sodium acetate buffer for all cases, but the folding behavior depends on the initial, pre-mix condition, namely pH 2.4 in 3 M GdmCl for panel A (2-state behavior), and pH 6.2 in 4.2 M GdmCl for panel B (3-state behavior). Panel A also shows the formation of secondary (solid circles) and tertiary (squares) structural H-bonds obtained using hydrogen exchange/NMR labelling methods (Sosnick et al., 1994).

4. WHAT CAUSES SLOW FOLDING

Why then do proteins fold so slowly? Experiments with cyt c suggest an answer. It is widely believed that the slow step in protein folding is represented by the late molten globule to native state transition (Kuwajima, 1989; Ptitsyn et al., 1990; Baldwin, 1993; Matthews, 1993). Cyt c at low pH and high salt concentration adopts the molten globule state, a partially destabilized form intermediate between the native and the unfolded states that has substantial secondary structure and compactness but disordered side chain packing (Ohgushi & Wada, 1983; Ptitsyn, 1987; Kuwajima, 1989; Jeng et al., 1990). The existence of this equilibrium form permits the explicit study of the transition from the molten globule to the native state.

When the equilibrium molten globule was diluted into native conditions, the native-like packing of aromatic side chains was fully recovered within the 6 msec deadtime, as indicated by near ultraviolet circular dichroism (CD) (Fig. 3A). The high quality of the 695 nm absorbance data (Fig. 3B) makes it clear that the native absorbance was completely recovered within the deadtime of 3 msec. This sets an upper limit of 1 msec on the time constant for folding from the equilibrium molten globule. This extremely fast folding directly establishes that the later steps in folding, including desolvation and tight side chain packing (Ptitsyn, 1987; Potekhin & Pfeil, 1989; Jeng et al., 1990), which have been generally assumed to be slow, are in fact intrinsically very fast processes. These processes cannot even account for the 15 millisecond refolding rate observed under fast conditions.

In cyt c, the barrier that makes folding slow must occur earlier than the molten globule form. In the unfolded cyt c molecule, the weak Met80 ligand dissociates from the heme iron. The empty ligation position can then be occupied by other amino acid side chains (Jones et al., 1993; Elove et al, 1994). Cyt c has two non-ligand histidines, His26 and His33, shown in Fig. 1 on the right side of the protein. Above pH 6 the histidines titrate to the neutral form and either one can ligand the heme iron at the empty Met80 position. This necessarily pulls

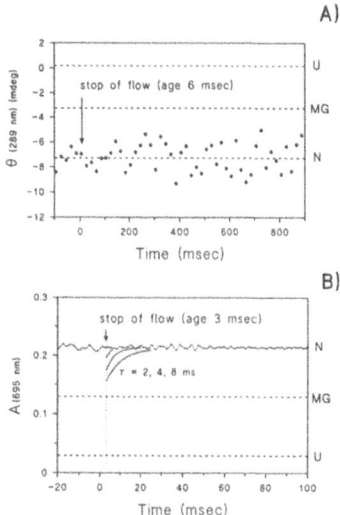

Figure 3. Folding of the cyt c molten globule to the native state monitored by CD at 289 nm (A) and absorbance at 695 nm (B) in stop-flow experiments. Spectral values for the unfolded (U), the molten globule (MG) and the native (N) states, indicated by the dashed lines, were measured separately. The disparity between the data and the solid curves in (B) illustrating hypothetical traces assuming 2, 4 and 8 msec time constants implies the reaction takes under 1 msec. Folding was initiated by a 1:1 dilution with acetate buffer at 10° C to yield pH 5.

a segment of the main chain around to the wrong side of the heme. On dilution of the GdmCl, the protein begins to refold. It collapses rapidly, independently of the initial unfolding conditions, as seen by the fluorescence quenching measurement in both Figs. 2A and 2B. Subsequent events depend on the presence or absence of the initial false ligation and accompanying misfold.

When the false ligation is present, the refolding protein encounters a barrier and folding is slow (compare A_{695} in Figs. 2A and 2B). It is important to note that the strength of the histidine-heme bond alone is not responsible for the slow folding since the histidines can dissociate from the heme on a 30 ms time scale when the molecule is unfolded (Sosnick et al, 1994). This is more than ten fold faster than the time required to refold to the native state. Thus, chain misorganization and not the false ligation is the important factor in the observed slow folding.

These results suggest a general explanation for slow protein folding (Fig. 4). If the initial chain collapse leads to some significant misorganization in the collapsed core, or allows some incipient misfolding influence to exist in peripheral segments, for example a non-native proline isomer, then the chain may be steered into a significantly misfolded state. The sizeable rearrangement necessary to continue productive folding can then constitute a kinetic trap that makes folding slow.

5. WHAT LIMITS FAST FOLDING

At the fast folding condition, cyt c folds in an apparent 2-state manner. The early folding probe, fluorescence quenching, and the late folding probe, the 695 nm absorbance band, both develop at the same fast rate with a 15 msec time constant. One knows that different folding events occur along the folding pathway, perhaps in some distinct order. However in 2-state folding all the events are kinetically indistinguishable so that one cannot discern the order of folding events. With cyt c, it is possible to separate the early and late

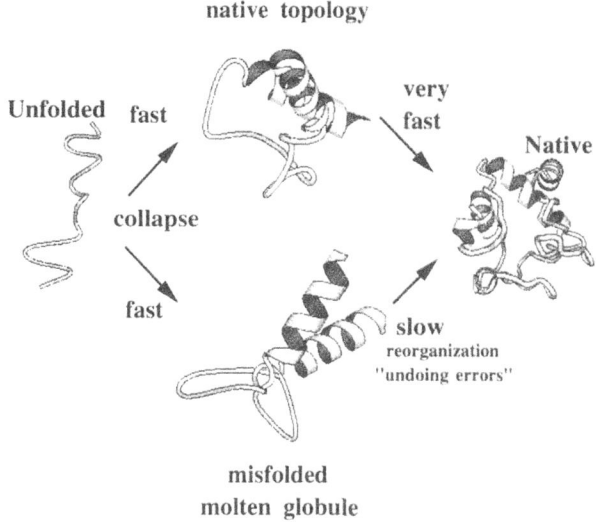

native topology

Unfolded   fast

very
fast

Native

collapse

fast

slow
reorganization
"undoing errors"

misfolded
molten globule

Figure 4. Model for fast and slow folding.

folding steps by exploiting the capability for inserting the misfolding barrier at will. We used this approach to gain information on the rate-limiting step in fast folding.

Fig. 2 compares data for 2-state and 3-state cytochrome c folding. The time scale for final acquisition of the native state (A_{695}) is very different in these two cases, but molecular collapse (fluorescence quenching) occurs at the same fast rate in both. It appears that the collapse process is independent of subsequent events and represents the same early process in both 2-state and 3-state folding. That is, the barrier that limits fast 2-state folding is the same barrier that limits the formation of the collapsed intermediate in 3-state folding. Thus, the ultimate barrier in fast folding must come before the collapsed intermediate and *reside in the collapse process itself.*

6. THE CHARACTER OF MOLECULAR COLLAPSE

Cyt c collapse occurs at a rate that is orders of magnitude slower than the diffusion-limited rate that one expects for the simple compaction of a polymer when placed in a poor solvent. Experiments that measured the dependence of cyt c collapse rate on ambient conditions show that collapse is a barrier crossing process. The dependence of collapse rate on GdmCl (Matthews, 1987) shows that the transition state buries 45% of the total surface that is buried in the U to N transition. The temperature dependence (Chen et al., 1989) indicates that 65% of the hydrophobic surface is buried in the collapse transition state. Evidently assembling the transition state for collapse requires some large scale structural events. This view is at odds with the presumption that protein collapse is a simple energetically downhill process.

These and related observations point to a nucleation model for the molecular collapse process, which appears to represent the rate-limiting step in protein folding. Formation of the transition state represents an energetically uphill search process to find a critical level of clustering of hydrophobic residues, as suggested in Fig. 5. Many trial clusters are formed and lost in the search for a successful transition state, where the forward folding reactions

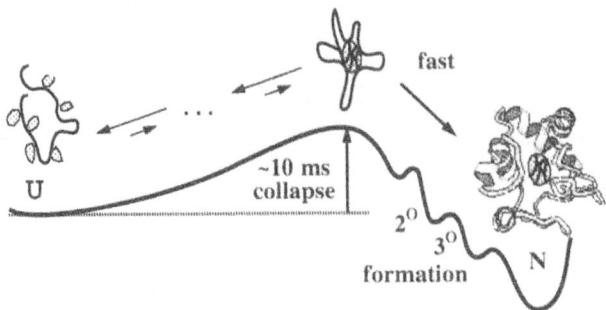

Figure 5. Search-nucleation model for collapse and folding. In the collapse process, an energetically uphill search consisting of many unfavorable association reactions ultimately finds a rare combination of interacting groups that nucleates subsequent energetically downhill condensation of the exposed loops onto the core to form secondary and tertiary structures, perhaps proceeding through predetermined intermediate species. In the collapse transition state, preferentially hydrophobic groups form a large collapsed core while some main chain loops remain exposed to solvent.

become equal to the back dissociation reactions. Core condensation appears to occur prior to the formation of significant amounts of secondary structure. Beyond the transition state, folding is energetically downhill and rapid as exposed loops condense onto the core. An attractive feature of this model is that non-local hydrophobic contacts introduce the crucial, native-like tertiary interactions directly into the fold at the beginning of the pathway.

The formation of the transition state for collapse is the time consuming portion of the conformational search problem. In this sense the transition state "solves" the Levinthal paradox. It reduces the global search process to the folding of individual elements such as helices and omega loops as discussed in the previous chapter (see also (Bai et al, 1995)). These smaller scale events occur on the native, downhill side of the rate-limiting collapse barrier. Even though these partially unfolded intermediate forms that the protein then traverses do not accumulate during a refolding experiment, they are essential to rapid folding.

7. REFERENCES

1. Alexander, P., Orban, J. and Bryan, P. (1992) *Biochemistry 31*, 7243-7248.
2. Bai, Y., Sosnick, T., Mayne, L. and Englander, S.W. (1995) *Science*, in press.
3. Baldwin, R.L. (1993) *Curr. Op. Struct. Biol. 3*, 84-91.
4. Bushnell, G.W., Louie, G.V. and Brayer, G.D. (1990) *J. Mol. Biol. 213*, 585-595.
5. Chen, B., Baase, W.A. and Schellman, J.A. (1989) *Biochemistry 29*, 691-699.
6. Elove, G.A., Bhuyan, A.K. and Roder, H. (1994) *Biochemistry 33*, 6925-6935.
7. Forster, T.H. (Ed. O. Sinanoglu) (1965) pp. 93-137 (Academic Press, N.Y.)
8. Houry, W.A., Rothwarf, D.M. and Scheraga, H.A. (1994) *Biochemistry 33*, 2516-2530.
9. Huang, G.S. and Oas, T.G. (1995) *Biochemistry 34*, 3884-3892.
10. Jackson, S.E. and Fersht, A.R. (1991) *Biochemistry 30*, 10428- 10435.
11. Jeng, M.F., Englander, S.W., Elöve, G.A., Wand, A.J. and Roder, H. (1990) *Biochemistry 29*, 10433-10437.
12. Jones, C. M., Henry, E. R., Hu, Y., Chan, C. K., Luck, S. D. et al. (1993) *Proc. Natl. Acad. Sci. USA 90*, 11860-11864.
13. Khorasanizadeh, S., Peters, I.D., Butt, T.R. and Roder, H. (1993) *Biochemistry 32*, 7054-7063.
14. Kraulis, P.J. (1991) *J. Appl. Crystallogr. 24*, 946-950.
15. Kuszewski, J., Clore, G.M. and Gronenborn, A.M. (1994) *Protein Science 3*, 1945-1952.
16. Kuwajima, K. (1989) *Proteins: Struct. Funct. Genet. 6*, 87- 103.
17. Matthews, C.R. (1987) *Meth. Enzymol. 154*, 498-511.

18. Matthews, C.R. (1993) *Annu. Rev. Biochem. 62*, 653-683.

19. Myer, Y.P. (1984) *J. Biol. Chem. 259*, 6127-6133.

20. Ohgushi, M. and Wada, A. (1983) *FEBS 164*, 21-24.

21. Potekhin, S. and Pfeil, W. (1989) *Biophys. Chem. 34*, 55-62.

22. Ptitsyn, O.B. (1987) *J. Protein Chem. 6*, 273-293.

23. Ptitsyn, O.B., Pain, R.H., Semisotnov, G.V., Zerovnik, E. and Razgulyaev, O.I. (1990) *FEBS Lett. 262*, 20-24.

24. Radford, S.E., Dobson, C.M. and Evans, P.A. (1992) *Nature 358*, 302-307.

25. Ridge, J.A., Baldwin, R.L. and Labhardt, A.M. (1981) *Biochemistry 20*, 1622-1630.

26. Schechter, E. and Saludjian, P. (1967) *Biopolymers 5*, 788-790.

27. Sosnick, T.R., Mayne, L., Hiller, R. and Englander, S.W. (1994) *Nature Struct. Biol. 1*, 149-156.

28. Takano, T. and Dickerson, R.E. (1981) *J. Mol. Biol.* 79-94.

29. Tsong, T.Y. (1976) *Biochemistry 15*, 5467-5473.

30. Viguera, A.R., Martinez, J.C., Filimonov, V.V., Mateo, P.L. and Serrano, L. (1994) *Biochemistry 33*, 2142-2150.

8

APPROACHES TO THE DETERMINATION OF MORE ACCURATE CROSS-RELAXATION RATES AND THE EFFECTS OF IMPROVED DISTANCE CONSTRAINTS ON PROTEIN SOLUTION STRUCTURES

Charles G. Hoogstraten[*] and John L. Markley[†]

Department of Biochemistry
University of Wisconsin, Madison
Madison, Wisconsin 53706

1. INTRODUCTION: MACROMOLECULAR STRUCTURE DETERMINATION FROM NMR DATA

In the past decade, nuclear magnetic resonance (NMR) has become an accepted and widely-used technique for studying the structure and dynamics of small- to moderate-sized proteins, nucleic acids, oligosaccharides, and molecular complexes (Wüthrich, 1986; Clore and Gronenborn, 1991). The fundamental phenomenon that allows the extraction of three-dimensional structural information from NMR studies is cross relaxation between protons (Neuhaus and Williamson, 1989). While other NMR parameters such as chemical shifts and J-coupling constants also contain structural information (Case et al., 1994), cross relaxation as manifested in the nuclear Overhauser effect (NOE) is unique in providing pairwise information: each cross peak in a two-dimensional NOESY (Nuclear Overhauser Effect SpectroscopY) spectrum provides evidence of a short-range through-space interaction between two identified protons, and a macromolecular NOESY dataset may contain thousands of such peaks. Typically, each assigned cross peak is used to derive a distance "constraint," or pair of bounds between which the distance is assumed to lie, and the dataset of bounds (perhaps along with additional information) is fed into one of several complex computer algorithms for conversion to a set of three-dimensional coordinates. Thus, the accuracy and precision of distances derived from NOESY spectra, and the effect of such

---

[*] The author to whom correspondence should be addressed.

[†] Present Address: Department of Chemistry and Biochemistry, Campus Box 215, University of Colorado at Boulder, Boulder, CO 80309-0215.

Dynamics and the Problem of Recognition in Biological Macromolecules, edited by Jardetzky and Lefèvre
Plenum Press, New York, 1996

accuracy and precision on the derived macromolecular structures, are key considerations in NMR structure determinations. In this Chapter, we review the development of experimental and calculational NMR techniques that allow improved accuracy in the determination of cross-relaxation rates and assess the usefulness of these techniques in structure determinations in proteins and nucleic acids. In Section 2, we develop the theoretical basis of the NOESY experiment with emphasis on those aspects necessary to understand the advantages and limitations of the new experiments. We describe the phenomenon of spin diffusion, which is an important cause of inaccuracies in NOESY-type experiments and is the problem addressed by the new experiments, and discuss some previously developed techniques that attempt to overcome these inaccuracies. In Section 3, we describe in detail the principles, usefulness, and limitations of a particular approach, Magnetization Exchange Network Editing (MENE), in which the basic NOESY pulse sequence is modified so as to cancel selected cross-relaxation contributions and thereby attenuate some spin-diffusion pathways. Finally, in Section 4, we summarize a number of studies examining the effect of such efforts to overcome multispin errors on the accuracy of macromolecular structural models.

The usual process of macromolecular structure determination by NMR is summarized in Figure 1. First, the NOESY spectral data must be analyzed to yield peak intensities: The interproton interaction giving rise to each cross peak must be identified, the surrounding baseplane reliably corrected, and the peak volume determined as accurately as possible. The conversion of peak volumes a_{ij} to cross-relaxation rates normally relies on the neglect of multispin effects, and thus is the step at which the isolated spin-pair approximation (discussed below) is applied and spin-diffusion effects cause inaccuracies. Distances may be estimated from cross-relaxation rates by means of an assumed motional model: This step usually involves the assumption that the molecule is rigid so that all interproton vectors are in motion only according to the overall molecular tumbling (rigid-molecule approximation). Distances are reported as distance constraint ranges, with lower and upper distance bounds playing the role of error bars on the determined distance. The preceding two steps (and corresponding assumptions) are often combined in a direct calibration against a cross-peak corresponding to a fixed distance (see Eq. 7 below); in a semiquantitative version of this process, cross-relaxation peaks are classified into three categories (strong, medium, and weak) and very broad constraint ranges are applied to each peak in a given category. This conservative data interpretation is intended to obviate the inaccuracies inherent in the above analysis by setting very loose error limits on each observed distance and relying on the large number of distances normally obtained to define the molecular structure (Clore and Gronenborn, 1991; Jardetzky, 1991; James, 1994). It should be emphasized, however, that these procedures still rely on the two fundamental assumptions above to ensure that the derived constraint range, however broad, contains the actual interproton distance. Finally, distance constraints may be converted to three-dimensional structures by means of a variety of algorithms, including metric matrix distance geometry (DG) and restrained molecular dynamics (rMD) (Brünger and Nilges, 1993). Since the NMR data, in general, are consistent with a range of structures, the results are generally reported as a collection, or "ensemble," of several coordinate sets that represent the conformational space compatible with the constraints. The quality of the final structures obtained obviously depends on a large number of parameters throughout the process, but, most fundamentally, the precision and accuracy of the structural model rely on the appropriateness of the two-spin approximation and the rigid-molecule approximation, or the ways in which deviations from these approximations have been taken into account. Of the two, the two-spin approximation generally is the more troublesome in protein structure determinations (Keepers and James, 1984), so long as multiple macroscopically different conformational minima are not populated (James, 1994). The term "spin diffusion" commonly is used to refer to the component of cross relaxation that does not fit the two-spin approximation. Such effects generally are treated as experi-

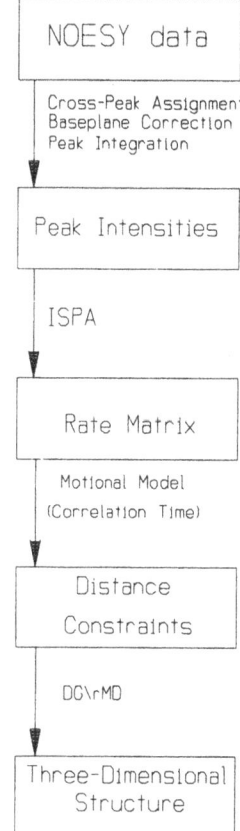

Figure 1. Structure determination by NMR. The procedures used to derive three-dimensional macromolecular structures from cross-relaxation data are shown in schematic form. Assumptions and sources of error at each step are indicated. Abbreviations: ISPA, isolated spin-pair approximation; DG, distance geometry; rMD, restrained molecular dynamics.

mental error, although they are fully justified theoretically by the multispin formalism (see Eq. 3 below). The significance of the spin diffusion problem has been appreciated since the early days of macromolecular NMR (Kalk and Berendsen, 1976) and a large amount of effort has been put into overcoming it, as discussed in detail in the following.

2. THE SPIN DIFFUSION PROBLEM IN BIOMOLECULAR NMR

2.1. NMR Cross Relaxation and the Nuclear Overhauser Effect

The phenomenon of cross relaxation is general to systems that contain two or more nuclear spins that interact by dipolar coupling (for example, neighboring protons that are close enough in space so that the magnetic field of one proton significantly affects the value of the total magnetic field seen by the other). The effect was used in studies of small organic molecules and natural products well before it was applied to macromolecules (Noggle and Schirmer, 1971). The theory of cross relaxation, and the nuclear Overhauser effect to which it gives rise, was worked out in a classic paper by Solomon (1955) and is a generalization of the concept of longitudinal relaxation. An isolated spin, when placed in a magnetic field such as that provided by an NMR instrument, develops an equilibrium magnetization with a component parallel to the static field (longitudinal magnetization along the z-axis). If the spin is perturbed (by a radio-frequency pulse, for example), the magnetization vector returns to its equilibrium value following a single-exponential time course characterized by the longitudinal relaxation time T_1. In the case of two dipolar-coupled spins I and S, neither spin relaxes in single-exponential fashion; instead, as shown by Solomon (1955), the two spins relax according to the coupled differential equations

$$\frac{dI_z}{dt} = -\rho_I(I_z - I_0) - \sigma(S_z - S_0)$$

$$\frac{dS_z}{dt} = -\rho_S(S_z - S_0) - \sigma(I_z - I_0) \tag{1}$$

Here, I_z and S_z are the longitudinal magnetizations for the two spins, I_0 and S_0 are the equilibrium magnetizations, ρ_I and ρ_S are the autorelaxation rates (the equivalent of T_1^{-1}) for the two spins, and σ is the cross-relaxation rate that expresses the dependence of the relaxation of each spin on the magnetization state of the other. All experiments that measure interproton distances are based on some variation of the idea of perturbing the magnetization of one spin and observing the effect on the relaxation or steady-state magnetization value of another, as the size of the observed effect is determined by the auto- and cross-relaxation rates in the system. The cross-relaxation rates are related to both the structure and the dynamic properties of the system, but may be interpreted directly in terms of interproton distances if a simple dynamical model is assumed (see below). In small molecules, the simplest experiment is the one-dimensional steady-state experiment, in which a saturating RF field is applied to one spin for a time long compared to all relaxation rates. Protons near in space to the irradiated proton will reach a steady-state magnetization different from their equilibrium value due to the action of the cross-relaxation rate(s) σ (Solomon, 1955), allowing a variety of useful structural correlations (Noggle and Schirmer, 1971; Neuhaus and Williamson, 1989). This fractional change in the intensity of one signal, or enhancement, upon the saturation of a second is the original precise meaning of the term "nuclear Overhauser effect"; now, any experiment that gives information on the cross-relaxation rates is said to reflect the NOE.

For large macromolecules, the steady-state NOE experiment gives little useful information. This is because, in the "spin-diffusion" or "slow-motion" regime of molecular Brownian motion (roughly, any molecule with a correlation time for overall rotation of at least a few ns), the high efficiency of cross relaxation relative to leakage to the lattice causes an NOE enhancement to be observed from any proton to essentially any other proton in the molecule. More useful in this motional regime are one-dimensional kinetic NOE experiments, which allow the observation of the development of the NOE as a function of time (Wüthrich, 1986). One-dimensional kinetic NOE experiments can take the form of the truncated driven NOE, in which the resonance of interest is irradiated for a defined length of time shorter than necessary to reach steady state, or the transient NOE, in which the spin of interest is inverted and the response of other spins observed after a period during which no further perturbation is applied (Neuhaus and Williamson, 1989). Because both are one-dimensional experiments, however, their application to macromolecules, which have crowded and badly overlapped proton spectra, is severely limited.

The use of the NOE to systematically study macromolecular structures, therefore, only became possible with the development of the two-dimensional NOESY experiment by the laboratory of Richard Ernst (Macura and Ernst, 1980). NOESY is a true transient experiment in which cross relaxation occurs during a selected experimental delay known as the mixing time. Because NOESY is a two-dimensional experiment, the information on interproton distances is contained in cross peaks whose location depends on the resonance frequencies of both of the interacting protons, and the problem of overlap is significantly reduced. For proteins of up to approximately 100 residues, the two-dimensional NOESY spectrum is often sufficiently resolved that nearly all visible cross peaks can be analyzed

(Wüthrich, 1986). For still larger systems, the NOESY spectra again become badly crowded and difficult to analyze. A large and growing variety of three- and four-dimensional heteronuclear-edited pulse sequences combined with isotope labeling, which spread the NOESY peaks into additional spectral dimensions without changing the underlying physics of the cross-relaxation measurement, can resolve this difficulty; for recent reviews, see Clore and Gronenborn (1991), Markley and Kainosho (1993), and LeMaster (1994).

Macura and Ernst (1980) described the processes that occur during the mixing time of a NOESY experiment by generalizing the coupled differential equations of Solomon. The longitudinal relaxation of every proton in the molecule is expressed as a function of the magnitudes and signs of the z-magnetization of every other proton. This is best cast as a matrix equation, in which the magnetization vector $\boldsymbol{m}$, with elements $m_1, m_2, \ldots, m_n$ for a molecule of n spins, represents the deviations from equilibrium z-magnetization. The multispin Solomon equation is then:

$$d\frac{\boldsymbol{m}}{dt} = -\boldsymbol{R}\boldsymbol{m} \tag{2}$$

in which $\boldsymbol{R}$ is the relaxation matrix comprising autorelaxation rates R_{ii} and cross-relaxation rates R_{ij} (for the two-spin case, the connection with Solomon's equations is made by $R_{ii} = \rho_i$ and $R_{ij} = \sigma_{ij} = \sigma$). The results of a NOESY experiment are described by a matrix $\boldsymbol{a}$ whose elements are the integrated volumes of the cross and diagonal peaks. This matrix is given by (Macura and Ernst, 1980):

$$\boldsymbol{a}(\tau_m) = \exp\left(-\boldsymbol{R}\tau_m\right) \cdot \boldsymbol{a}(0) \tag{3}$$

in which $\boldsymbol{a}(\tau_m)$ is the spectral matrix at mixing time τ_m.

The results of a NOESY experiment are thus interpreted most directly in terms of the elements of the matrix $\boldsymbol{R}$. Often, however, the quantities of interest are structural parameters such as interproton distances. The extraction of distances from cross-relaxation data relies on the relation between cross-relaxation rates and quantum-mechanical transition probabilities given by (Solomon, 1955):

$$R_{ij} = W_2^{ij} - W_0^{ij} \tag{4}$$

in which W_2^{ij} and W_0^{ij} are the quantum-mechanical transition probabilities of simultaneous parallel (double-quantum) and antiparallel (zero-quantum) spin flips, respectively. These transitions are stimulated primarily by fluctuating magnetic fields induced by the overall rotation of the molecule (Bloembergen et al., 1948), and, therefore, depend on parameters characterizing the molecular motion. Time-dependent perturbation theory gives expressions for these probabilities that, in the slow-motion regime appropriate for macromolecules, reduce to:

$$R_{ij} = -q\tau_c^{ij} \tag{5}$$

in which τ_c^{ij} is the correlation time for isotropic reorientation for the interproton vector and q is a combination of fundamental constants with the value $5.69*10^4*r_{ij}^{-6}$ s^{-2} (r_{ij} in nm) (Macura et al., 1994). The term r_{ij} in this expression refers to the appropriately-averaged distance between protons i and j. This expression is the source of the common statement that cross-relaxation rates are proportional to the inverse sixth power of the (average) interproton distance. This is true; however, it should be noted that the proportionality constant involves the rotational correlation time for the particular proton vector. In principle, there is no reason this parameter should have the same value for all pairs of protons in a biological macromole-cule with complex internal dynamics. While some progress has been made in taking a distribution of correlation times into account in the analysis of cross relaxation (Liu et al., 1992; Baleja and Sykes, 1991; Koning et al., 1991; Dellwo and Wand, 1993), it is an almost-universal practice to assume that all of these vectors are characterized by the correlation time for the isotropic reorientation of the macromolecule. This assumption amounts to neglecting the presence of internal motion, and is referred to here as the rigid-molecule approximation.

If the entire matrix a is known, the cross-relaxation rates, and thus the desired distances, can be obtained from straightforward matrix transformations (Olejniczak et al., 1986) based on Eq. 3. The complete matrix a, however, is generally not experimentally accessible in biological macromolecules due to spectral overlap (particularly on the diago-nal) and sensitivity problems. Alternatively, Eq. 3 may be expanded in a power series for short mixing times to eliminate the dependence of each peak volume on every element of R. Expressed in scalar notation, this expansion yields:

$$\frac{a_{ij}(\tau_m)}{a_{ii}(0)} = -R_{ij} \cdot (\tau_m) + \frac{1}{2}\sum_k R_{ik} \cdot R_{kj} \cdot (\tau_m)^2 - \frac{1}{6}\sum_k \sum_l R_{ik} \cdot R_{kl} \cdot R_{lj} \cdot (\tau_m)^3 + \dots \tag{6}$$

in which $a_{ij}(\tau_m)$ is the peak volume at mixing time τ_m and the summations are performed over all spins in the relaxation network.

If Eq. 6 is truncated after the first term, each cross-relaxation rate may be calculated from the corresponding cross-peak volume, given appropriate normalization. This is the familiar isolated-spin-pair approximation (ISPA), or two-spin approximation, which is widely used for macromolecular structure determinations by NMR (James and Basus, 1991). The direct extraction of cross-relaxation rates using the truncated version of Eq. 6 would assume knowledge of $a_{ii}(0)$, the volume of the corresponding diagonal peak in a hypothetical zero-mixing time experiment acquired under the same experimental conditions. In practice, NOESY cross peaks are interpreted by calibration within a single spectrum using cross peaks corresponding to known distances, such as covalently fixed distances between ortho protons on aromatic rings or backbone proton-backbone proton distances in regions of regular secondary structure (Wüthrich, 1986). On combining Eqs. 5 and 6 for an unknown distance r_{ij} and a known distance r_{fix}, we obtain:

$$r_{ij} = r_{fix}\left(\frac{R_{fix}}{R_{ij}}\right)^{\frac{1}{6}} = r_{fix}\left(\frac{a_{fix}}{a_{ij}}\right)^{\frac{1}{6}} \tag{7}$$

which allows r_{ij} to be calculated from cross peaks within a NOESY spectrum at a single mixing time. This is a practical and widely-used relationship, since the desired interproton distances can be determined from two easily measured quantities (cross-peak intensities) and one value that can be looked up in a table (r_{fix}). Note that the first equality in Eq. 7 depends on the direct relationship between the cross-relaxation rate and the interproton distance, and therefore on the rigid-molecule approximation. The second equality depends on a version of the two-spin approximation, in that all quadratic and higher terms in Eq. 6 are assumed to be negligible. Terms in the expansion containing diagonal elements of the matrix R represent the autorelaxation contributions that cause the downward trend in the cross-peak intensity at longer mixing times. Nonlinear terms in Eq. 6 not involving autore-laxation represent multispin magnetization exchange, referred to as spin diffusion. For example, transfer of magnetization from spin i to spin j via a third spin k contributes to $a_{ij}(\tau_m)$

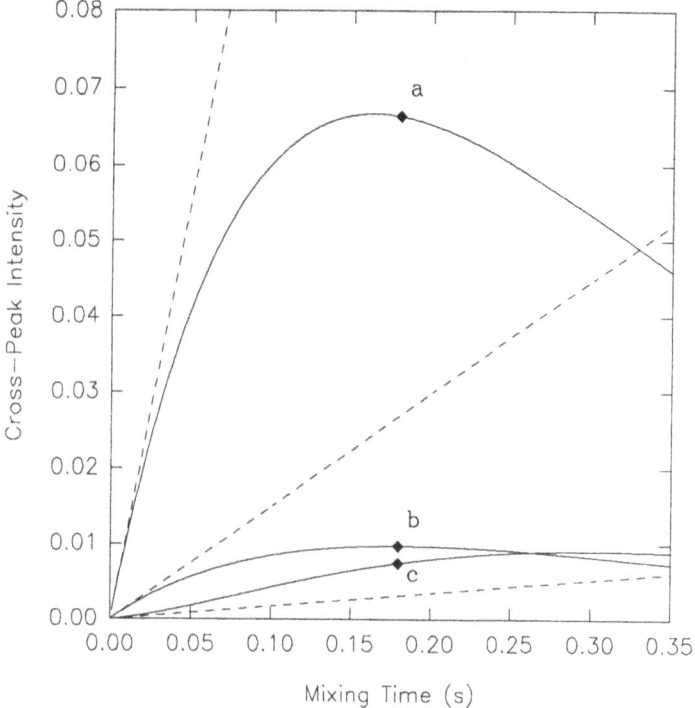

Figure 2. Simulated NOESY buildup curves. NOESY cross-peak intensities in various geometries were calculated as a function of mixing time using Eq. 3. The intensity of a single-proton diagonal peak at mixing time zero was set equal to unity. For curves *a* and *b*, an isolated three-spin geometry with interproton distances of 2.5 Å (curve *a*), 3.5 Å (curve *b*), and 5.5 Å (curve not shown) was used. For curve *c*, a linear three-spin geometry with adjacent protons separated by 2.5 Å was used, and the peak intensity connecting the two outer protons was plotted. All calculations assumed an isotropic correlation time of 4.8 ns and an external contribution to autorelaxation of 5 s^{-1}. Tangents to the buildup curves at mixing time zero, which have slope equal to the cross-relaxation rate for the proton pair, are shown as dashed lines. The cross-peak intensities obtained in a NOESY spectrum at 180 ms mixing time are indicated by diamonds.

according to $R_{ik}R_{kj}\tau_m^2$. If the two-spin approximation is applied, this intensity will be falsely interpreted as arising from the linear term, $R_{ij}\tau_m$, and the rate constant R_{ij} will be overestimated, causing the determined distance r_{ij} to be underestimated.

In Figure 2, we plot simulated curves for the mixing-time dependence of NOESY cross-peaks corresponding to a short distance, such as would be used for calibration (curve *a*), and a longer distance, such as would be desirable to measure (curve *b*). The ratio (R_{fix}/R_{ij}) is given accurately by the ratio of the initial slopes of these curves, shown as tangents. At a given mixing time τ_m, however, the experimental quantities are the cross-peak intensities, represented by diamonds at $\tau_m = 180$ ms. If the ratio of the cross-peak intensities is equal to the ratio of the initial slopes, the second equality in Eq. 7 is satisfied. In cases where the shapes of the buildup curves are different for the reference and unknown proton pairs, this assumption can lead to inaccuracies. Curve *c* shows an example of severe spin-diffusion effects; multispin effects cause upward curvature in the buildup, and the cross peak ratios for curve *c* and curve *a* at the indicated mixing time would give a very poor indication of the ratio of the cross-relaxation rates (initial slopes). Cross peaks *c* and *b* are at nearly the same intensity at the indicated mixing time, whereas the corresponding cross-relaxation rates differ by almost an order of magnitude.

In some spatial distributions of protons, spin-diffusion contributions completely dominate the cross-peak intensity of a given pair. Take, for example, three protons equidistant along a line: cross relaxation between the outer two will be dominated by spin diffusion mediated by the central proton at any practical mixing time. If cross peaks are calibrated using Eq. 7, spin diffusion effects will cause distances longer than the reference distance to be underestimated and distances shorter than the reference distance to be overestimated. Since reference distances tend to be short (2.2 - 2.5 Å is typical), the overall effect is a systematic underestimation of interproton distances, with the worst problems arising in the critical long-range distances that define the tertiary structure of the macromolecule. The systematic underestimation of distances in macromolecules caused by spin diffusion, that is, by the neglect of multispin effects in the application of the isolated-spin-pair approximation (see Figure 1) can lead to inaccuracies in derived structures, as will be discussed in detail in Section 4. When cross-relaxation rates are calculated explicitly, inappropriate use of the two-spin approximation can cause the rate constants to be inaccurate, but the calculation of these rates from NOESY data is independent of any motional model, and thus does not rely on the rigid-molecule approximation (Figure 1). In the remainder of the present section, we survey various techniques designed to deal with spin diffusion, and critically examine their advantages and limitations.

2.2. Experimental Techniques for the Analysis of Spin Diffusion

2.2.1. NOESY Buildup Analysis and the Initial Rate Approximation. Although solution structures commonly are determined on the basis of NOESY data collected at single mixing times, it is clear that this approach is flawed. Macura (1994) recently used analytical error propagation techniques to demonstrate that the mixing time at which the most accurate determination of a given interproton distance is possible varies considerably with the interproton distance itself, independently of multispin effects. Thus, analysis of a mixing time series is needed for the accurate determination of the full spectrum of cross-relaxation rates of interest. It was recognized at an early point that the dependence of the cross-peak intensity on the mixing time can be diagnostic of spin diffusion (Anil Kumar et al., 1981). The first term in Eq. 6, which describes the direct NOE contribution, is linear in mixing time; all subsequent terms, representing spin diffusion or autorelaxation contributions, are quadratic or higher in mixing time. This has two major implications: (1) contributions due to spin diffusion will be minimized at short mixing times, and (2) spin-diffusion peaks may be identified by taking a series of NOESY experiments at varying mixing times. The application of these two principles leads to the initial-rate approach, in which a series of NOESY experiments is taken at short mixing times and the cross-peak intensities are plotted as a function of mixing time. Cross-peak intensities are interpreted as due to direct cross relaxation at mixing times for which this dependence is linear, and the slope in the linear regime is directly proportional to the cross-relaxation rate. Using this approach, Anil Kumar et al. (1981) identified several spin-diffusion cross peaks in bovine pancreatic trypsin inhibitor (BPTI) by the nonlinear shape of their buildup curves. The initial-rate approach greatly increases the confidence with which NOESY data may be interpreted. The most serious problem is that the initial-rate condition is only satisfied at very short mixing times, whereas the cross-peak intensity increases to longer mixing times, after which it declines due to overall longitudinal relaxation (see Figure 2) (Macura and Ernst, 1980; Anil Kumar et al., 1981). Thus, taking initial rates of buildup curves requires taking NOESY spectra at mixing times for which the experiment's sensitivity is much less than optimal. The procedure is time-consuming and difficult to apply to cross peaks that become visible only at longer mixing times because of low signal-to-noise. In addition, since NOESY buildup curves may contain both upward curvature from spin diffusion and downward curvature from autorelaxation, it is at least in principle possible for buildup curves to appear linear when significant indirect contributions are present, due to a cancellation of these two effects. The derivation of buildup curves is a straightforward technique that can dramatically improve the reliability of NOESY distances; it currently is a standard technique for the careful analysis of cross-relaxation rates (Neuhaus and Williamson, 1989). Many workers will at least discard those peaks that give nonlinear buildup curves from their dataset and proceed with the interpretation of a single NOESY spectrum (Wüthrich, 1986).

A technique to overcome the confounding effects of autorelaxation and extend the region of applicability of the initial-rate approach involves scaling the cross-peak intensities by those of the corresponding diagonal resonances (Macura et al., 1986). Appropriately scaled buildup curves retain the linear relationship of the initial slope with the cross-relaxation rate but, to a first approximation, are independent of the autorelaxation contributions. This allows the unambiguous determination of the linearity or nonlinearity of the buildup, and thus the applicability of the initial-rate approach. In addition, since the earliest source of nonlinearity in buildup curves is very often autorelaxation rather than spin diffusion, this procedure can give buildup curves that remain linear to much longer mixing times than the standard plot of absolute cross-peak intensity, allowing increased sensitivity and the measurement of smaller cross-relaxation rates (Macura et al., 1986). An additional advantage of normalization is that comparisons of absolute intensities between spectra are unnecessary, eliminating errors due to instrumental instabilities and other complicating factors. The most serious problem with this technique is that it requires an accurate estimate of the diagonal intensity; in two-dimensional NOESY spectra of macromolecules, individual peaks on the diagonal generally are not resolved, making such an analysis impossible except for isolated instances. The recent development of three- and four-dimensional heteronuclear-edited versions of NOESY (Clore and Gronenborn, 1991), however, permits the acquisition of spectra in which the "direct" peaks (the equivalent of the diagonals in a homonuclear NOESY) are reasonably well resolved. These approaches extend the feasibility of normalized buildup analyses to larger molecules (Wagner, 1990), but few studies have taken advantage of this for the analysis of cross relaxation.

Another approach to extending the applicability of buildup analysis is to mathematically fit the data past the linear region and extract the linear term representing the cross-relaxation rate. The most straightforward manner to do this is to fit the buildup series, either at the level of the spectra or of extracted peak intensities, to a second-order polynomial (Hyberts and Wagner, 1989; Majumdar and Hosur, 1989; Fejzo et al., 1989). This fit takes into account the first two terms in Eq. 6. The quadratic equation is likely to fit the data well to longer mixing times than the linear approximation, allowing the use of moderately long mixing times. Of course, if normalized buildup data are available, these also may be fitted to a quadratic polynomial, which should improve the determination of the cross-relaxation rates still further.

2.2.2. Random Fractional Deuteration. It is possible to label proteins prepared in the laboratory from bacterial or other sources with deuterium by growing the source organism on a medium appropriately enriched in deuterium (LeMaster, 1990). Because the largest contributions to the linewidth of proton NMR resonances come from dipolar interactions with other protons, uniform deuteration at levels of 50% to 80% (random fractional deuteration) can sharpen the proton lines dramatically and improve resolution. Further improvement in spectral quality is seen because this pattern of labeling simplifies the structure of cross-peak multiplets. Reduction in the effective concentration of the observed protons leads to reduced sensitivity in spectra of random fractionally deuterated proteins, but this is partially compensated by the line-sharpening, which has the effect of increasing the spectral signal-to-noise ratio in terms of peak height. In addition, because the backbone amides of deuterated proteins dissolved in H_2O become fully protonated by exchange with solvent, these sites are observed at nearly full intensity (LeMaster and Richards, 1988).

Random fractional deuteration also affects the spin diffusion pathways present in a molecule. For example, in a 75% deuterated sample, a particular third-spin effect will be present in only 25% of the molecules, meaning that the spin-diffusion contribution is decreased by 75% (LeMaster and Richards, 1988). The major disadvantage of this approach, especially for nonexchangeable protons, is the accompanying decrease in sensitivity. Le-Master and Richards (1988), for example, observed three-fold lower sensitivity for cross peaks between two carbon-bound protons in a 75% deuterated sample of *E. coli* thioredoxin. To recover the original level of sensitivity arising from direct cross-relaxation (neglecting autorelaxation) in this case, it would be necessary to operate at a three-fold increased mixing time. This would increase the relative contribution of two-step terms by a factor of three, for a very small net improvement of 4/3 in the direct-to-spin diffusion contribution ratio. For cross peaks involving exchangeable protons, the situation is better, and larger improvements in cross-peak accuracy are realized. For at least some classes of cross peaks, then, random fractional deuteration provides a practical method for observing cross peaks with reduced (but not eliminated) contributions from spin diffusion. Clearly, this method is only applicable to proteins that can be produced in the laboratory with the requisite labeling pattern. In addition, a measure of care is required, since certain metabolic pathways can lead to nonrandom incorporation that confounds the interpretation of the results.

Another labeling pattern of interest is high-level selective deuteration (Markley et al., 1968; LeMaster, 1994). Selective deuteration often is performed with the goal of simplifying the assignment process, but substantial effects on cross-relaxation are also seen, and improvements in both sensitivity and the reduction of spin diffusion over random fractional deuteration have been claimed (Pachter et al., 1992; Tsang et al., 1990). In general, the cross-peaks remaining in the NOESY spectrum have some degree of enhanced intensity due to the loss of autorelaxation components (the magnitude of the diagonal elements of the matrix **R** depends on the presence of neighboring protons), but the calibration of cross-peaks may become complicated because of the differing environments of various proton pairs, and distance measurements in these samples must be interpreted quite conservatively (Pachter et al., 1992). This is a powerful technique, but is time- and material-intensive compared to uniform labeling procedures. For a systematic analysis of cross relaxation using selective deuteration techniques, a different labeling pattern would have to be prepared (including the possible need for an appropriate strain of auxotrophic microorganism) for each type of interaction analyzed.

Much recent attention has focused on the combination of random fractional deuteration with uniform $^{13}C, ^{15}N$-labeling (Grzesiek et al., 1993; Yamazaki et al., 1994b; Yamazaki et al., 1994a). The primary application so far has been to assignment-directed techniques, particularly in combination with high-power ^2H-decoupling to narrow the linewidth of the C^α resonance and extend the applicability of the main-chain directed strategy to larger systems (Yamazaki et al., 1994a). Such labeling patterns will also undoubtedly have useful applications to the analysis of cross-relaxation. A recent example is the procedure of dissolving perdeuterated and uniformly-^{15}N labeled proteins in protonated solvent; in conjunction with four-dimensional double-^{15}N-edited NOESY spectroscopy, this labeling pattern allows the observation of amide proton-amide proton cross peaks in the absence of spin diffusion mediated by any non-amide proton, along with a considerable degree of line sharpening. This procedure has allowed, in proteins with otherwise-prohibitive linewidths, an efficient analysis of secondary (Grzesiek et al., 1995) and tertiary (Venters et al., 1995) structure.

2.2.3. Three-Dimensional NOESY-NOESY Spectroscopy. An interesting NOESY variant that requires no isotopic labeling is the three-dimensional NOESY-NOESY, or 3D NOE-NOE, experiment (Boelens et al., 1989b). In this experiment, two mixing times, normally of equal length, are used, and an additional frequency labeling period is inserted between the two mixing times. The three frequency axes therefore represent the location of magnetization at the beginning of the experiment, after the first mixing time, and after both mixing times. This feature of the experiment is very useful for the diagnosis of spin diffusion, because a peak at three different frequencies (a_{ijk}) represents the multistep pathway $i \rightarrow j \rightarrow k$, whereas direct magnetization transfer gives rise to peaks of types a_{iik} or a_{ikk} (Boelens et al., 1989b). The contribution of a given spin-diffusion pathway to a given direct peak thus may be diagnosed directly (Habazettl et al., 1992a). The greatest difficulty with this experiment is the large number and variety of peaks obtained: For a symmetric 22-bp DNA fragment, approximately 10,000 peaks were observed (Boelens et al., 1989b). The practical difficulties of analyzing this large number of resonances in a 3D spectrum with (typically) poor digital resolution are considerable. The increase in information content about cross-relaxation rates in comparison to a standard NOESY upon complete interpretation, however, is substantial (Habazettl et al., 1992b). With highly-automated procedures for data extraction and analysis, structures of 29- and 118-residue proteins have been derived exclusively from 3D NOE-NOE spectra (Holak et al., 1991; Habazettl et al., 1992b). Since the number of peaks is greatly increased over the number present in 2D NOESY spectra, the interpretation of 3D NOE-NOE data for larger systems is more likely to be hindered by spectral overlap than that of heteronuclear 3D NOESY data (Clore and Gronenborn, 1991), which contain no additional peaks compared with 2D NOESY spectra but gain the dispersion of the third dimension. In cases where it can be applied, however, the 3D NOE-NOE technique has the potential to provide considerable power for the analysis of cross-relaxation pathways. It also lends itself in a natural way to incorporation into relaxation matrix refinement schemes (Bonvin et al., 1991a), as described below (Section 2.3.4).

2.2.4. Cross-Relaxation Spectroscopy in the Rotating Frame. The NOESY experiment measures cross relaxation between protons with magnetization vectors parallel to each other and to the external (static) magnetic field. Another regime for investigations of cross relaxation is one in which the spins are perpendicular to the external magnetic field. This condition can be achieved by forced precession in the x-y plane if an applied RF field is used to "spin lock" the protons. Cross relaxation in the rotating frame gives rise to the rotating-frame Overhauser effect, or ROE. The two-dimensional analogue of NOESY under spin-locked conditions is known as ROESY (Rotating-Frame Overhauser Enhancement Spectroscopy) (Bothner-By et al., 1984; Bax and Davis, 1985). ROESY is the primary structural experiment for mid-size (ca. 500 - 2000 molecular weight) molecules, for which NOESY is extremely insensitive. When applied to larger macromolecules, it yields data that can be used to analyze the presence of spin diffusion in corresponding NOESY spectra. Whereas direct and indirect cross-relaxation give rise to cross-peak contributions of the same sign in NOESY, direct and two-step indirect contributions have opposite signs in ROESY (Farmer et al., 1987). Thus, pure spin-diffusion cross peaks may be identified immediately in ROESY by a change of sign; composite peaks containing direct and spin-diffusion contributions will generally lose intensity or be unobservable due to a subtraction of the two contributions. ROESY data may thus be used to identify peaks in NOESY that are suspect for spin-diffusion effects (Farmer et al., 1987).

More quantitative use of ROESY to analyze spin diffusion in NOESY is also possible. ROESY and NOESY buildup curves may be acquired and analyzed simultaneously using a quadratic fit to extract a cross-relaxation rate that best accounts for all of the data (Fejzo et al., 1989). This approach has been used to derive more accurate distance constraints for the refinement of the NMR solution structure of a small protein (Krezel et al., 1994). A method has also been demonstrated in which a linear combination of NOESY and ROESY spectra at the same mixing time is calculated to eliminate contributions due to two-step magnetization transfer (Fejzo et al., 1991a). This method, known as D. NOESY (Direct NOESY), depends on normalization of the two spectra on the diagonal intensity, and thus fails when diagonal peaks with different autorelaxation rates overlap. Nevertheless, dramatic improvements in cross-peak accuracy are possible in favorable cases.

The most serious problem with all methods relying on the application of ROESY to large macromolecules is sensitivity. The diagonal elements of the cross-relaxation matrix R in ROESY are not the reciprocal T_1 values, as in NOESY, but the reciprocal transverse autorelaxation rates $T_{1\rho}$. Unlike T_1, which increases with molecular correlation time (and therefore with molecular weight), leading to enhanced sensitivity for NOESY at larger molecular sizes, proton $T_{1\rho}$ values decline rapidly at molecular weights in excess of 10 kDa. The sensitivity of ROESY spectra thus is greatly diminished as molecular size increases, and the application of ROESY-based techniques to very large macromolecules is unlikely to yield favorable results. For simple diagnostic analysis, a ROESY spectrum at half the NOESY mixing time of interest is appropriate, since cross-relaxation rates in the rotating frame are twice as rapid as in the laboratory frame (Farmer et al., 1988). A D.NOESY spectrum, however, must be acquired at the same mixing time as the corresponding NOESY spectrum.

2.2.5. Elimination of Spin-Diffusion Effects by Means of Network Editing. Finally, a variety of experimental techniques known as Magnetization Exchange Network Editing (MENE), or relaxation network editing, have been developed that allow the observation of selected cross-peaks from a NOESY spectrum with the removal of certain spin-diffusion contributions. The key distinction from the above methods is that the NOESY spectrum itself is perturbed in such a way as to strengthen the connection between the intensity of individual cross-peaks and the corresponding cross-relaxation rate; in other words, the two-spin approximation is better satisfied for network-editing spectra than for NOESY spectra. The development and application of network-editing pulse sequences are the major topics of this Chapter, and the principles of this approach are discussed at length in Section 3.

2.3. Relaxation Matrix Refinement of Cross-Relaxation Data

As mentioned above, when the entire spectral matrix a is available experimentally, the cross- and auto-relaxation rates may be calculated directly (Olejniczak et al., 1986). This calculation relies on the inversion of Eq. 3 to obtain:

$$R = \frac{-\ln(a(\tau_m) \cdot a(0)^{-1})}{\tau_m} \tag{8}$$

We refer to this transformation as full matrix analysis (FMA). In macromolecules, the complete matrix a is not adequately determined from NOESY data, and FMA cannot be applied directly to calculate the cross-relaxation rates. However, the reverse transform, or calculation of a simulated intensity matrix from the rate matrix R using Eq. 3, may in general be performed. The elements of R, in turn, may be calculated directly from a set of coordinates using Eq. 5 if a motional model is assumed. Usually, the rigid-molecule approximation is applied, and all the τ_c^{ij} are assumed to be equal to the isotropic correlation time for the molecule, τ_o, which may be determined by a number of methods. This relatively straightforward calculation makes it possible to generate a simulated set of NOESY intensities from a derived set of cross-relaxation rates, distances, or coordinates, compare these to the experimental data, and adjust in some way until agreement is achieved. This procedure can be carried out in several ways, as discussed in the following; the common idea for all is that NOESY cross-peaks that do not obey the two-spin approximation nevertheless contain information on cross relaxation, and the derived quantities should be consistent with all of this information.

2.3.1. Direct Least-Squares Refinement of the Cross-Relaxation Matrix. As mentioned above, the derived quantities most closely related to the experimental NOESY intensities are the cross- and autorelaxation rates. This direct relationship was exploited by Koehl and Lefèvre (1990), who introduced an algorithm to refine the cross-relaxation matrix **R** against the experimental data. The procedure involves the estimation of an initial rate matrix on the basis of a model structure; this "starting" structure, which incorporates a motional model, provides only the initial values for the adjustable parameters. NOESY intensities calculated from the model matrix are compared with the available experimental data, and the model rate matrix is adjusted in least-squares fashion until the best fit is obtained. Two features of this algorithm are noteworthy. First, such a refinement lends itself naturally to treating data from multiple mixing times simultaneously, as the rate matrix can predict NOESY intensities at arbitrary mixing times. This is advantageous for several reasons, as discussed above, and the authors describe improved convergence of their method with the availability of data from multiple mixing times. Second, while a motional model is necessary to provide the initial guesses of the relaxation rates, subsequent refinement can proceed independently of motional models and therefore of the rigid-molecule approximation. This is true because the calculation of NOE intensities from cross-relaxation rates, in contrast to calculation from distances or coordinates, does not depend on the value of the correlation time for the interproton vector. Application of this algorithm thus amounts to a separation of the analysis of spin diffusion and the analysis of molecular motion; the result is a cross-relaxation matrix that best accounts for the experimental data. Potential variations in the correlation times due to internal motions may be analyzed subsequently. In the initial report, the method was used to detect variations in τ_c^{ij} for intra-residue distances along the sequence of a DNA oligomer, allowing conclusions to be drawn concerning the internal dynamics of the molecule.

Despite the apparent generality of this approach, little subsequent work has been done in this area. One problem might be that the original algorithm (Koehl and Lefèvre, 1990) was introduced with optimization using a version of the Marquardt-Levenberg algorithm (Press et al., 1988), which is vulnerable to becoming trapped in local minima in complicated parameter spaces. This problem, however, could be alleviated in a straightforward way by the introduction of a technique such as simulated annealing to locate the global minimum. This algorithm is the only version of matrix refinement yet described that allows the separation of the analysis of spin-diffusion effects from internal molecular motions, and thus deserves further study.

2.3.2. Iterative Refinement of Interproton Distances. One solution to the lack of a complete NOESY intensity matrix *a* in experimental situations is to supplement the experimentally determined elements with values derived from an initial structural and motional model according to Eq. 3. The complete cross-relaxation matrix may then be calculated from this "hybrid" matrix using Eq. 8. If the new cross-relaxation matrix is altered to reflect knowledge about the covalent structure of the molecule and the relationships between off- and on-diagonal elements and the refined matrix is used in another cycle of supplementation of the original data, the cross-relaxation rates corresponding to experimentally determined cross-peaks will converge to accurate estimates of the true rates in a small number of cycles (Borgias and James, 1990). This process is the heart of the MARDIGRAS (Matrix Analysis of Rates for Discerning GeometRy in Solution) algorithm of James and coworkers (Borgias et al., 1990; Liu et al., 1994); a number of similar algorithms appear in the literature (Madrid et al., 1991; Edmondson et al., 1991), but MARDIGRAS is the best-developed and best-known.

The appeal of MARDIGRAS is based in part on its simplicity; the only input parameters necessary are the NOE intensities, a starting structure, and a motional model (normally an isotropic correlation time estimate). In a relatively short amount of computer processing time on the scale of an NMR structure determination, MARDIGRAS-type algorithms produce a set of refined distances that, given a sufficient dataset, can show substantial improvements in accuracy as compared to the input data (Borgias et al., 1990). It has also been shown experimentally that the results of MARDIGRAS calculations are relatively independent of starting structure, although a better structure (i.e., a protein model with the correct global fold, as opposed to a random coil) will produce better results. The residual model-dependence, while not a severe handicap in practice if MARDIGRAS is used for structural refinement, is likely due to the replacement of cross-relaxation rates not corresponding to experimental data with the rates from the model at every iteration (Borgias and James, 1990), which is claimed to improve convergence. In another implementation of this basic scheme (Madrid et al., 1991), this replacement is not performed, and essentially no dependence on the model structure is seen.

MARDIGRAS has been used in the refinement of a number of structures, including the neurotoxic peptide ω-conotoxin (Davis et al., 1993), a DNA sequence from the HIV-1 genome (Mujeeb et al., 1993), and a DNA octamer containing the Pribnow box (Schmitz et al., 1992). In addition, results of a study of data simulated from a given structure demonstrated that structures calculated using the MARDIGRAS output were closer to the target coordinate set than structures calculated from unimproved distance measurements; this indicates that MARDIGRAS (and, by implication, other strategies for overcoming multispin effects) can improve the accuracy of derived structures in experimental situations (Thomas et al., 1991). More recent modifications to the MARDIGRAS algorithm include extensions to take account of local motions in methyl groups and aromatic rings, the inclusion of internal motion effects as analyzed by heteronuclear relaxation, and the effects of chemical exchange including solvent exchange (James, 1994). MARDIGRAS is designed to operate on a single NOESY matrix at a given mixing time; input at multiple mixing times, if available, is handled independently and used to set error bars on a given distance (Liu et al., 1994). This sacrifices a significant amount of the advantage gained by acquiring data at multiple mixing times, since the time course of a given peak as a function of mixing time is not analyzed. Because the algorithm is straightforward to apply and the CPU time requirements are extremely modest, however, MARDIGRAS has found a solid place in the array of NMR structural refinement techniques.

2.3.3. Iterative Refinement of Three-Dimensional Structures. An alternative to MARDIGRAS-type algorithms is to actually calculate, using distance geometry (DG) or restrained molecular dynamics (rMD) techniques, a refined three-dimensional structure from each iterated cross-relaxation matrix. This new structure is then used to generate the simulated portions of the hybrid spectral matrix for the next iteration. The essential difference from MARDIGRAS-type calculations, therefore, is that structure calculations are included within the iterative loop, instead of being performed as a separate step after a converged set of distance constraints is obtained. This procedure was introduced by the Kaptein group as IRMA (Iterative Relaxation Matrix Approach), using rMD as the structure determination step (Boelens et al., 1989a; Boelens et al., 1988). IRMA and similar algorithms have been used to refine a number of macromolecular structures determined by NMR (Bonvin et al., 1994; Nikonowicz et al., 1991).

The source of the structural improvement in IRMA-type procedures is more intuitively obvious than in refinement at the level of distances, since the actual molecular coordinate set is visibly refined at each step of the procedure. Structure calculations, however, are much more calculation-intensive than any of the matrix operations involved in distance refinement; application of the IRMA algorithm, therefore, involves a heavy penalty in computation time compared to MARDIGRAS. In practice, the effect of this is to decrease the size of the structural ensemble that can be calculated. In addition, the treatment of data from multiple mixing times is unsatisfactory. In the IRMA program, a relaxation matrix is calculated from the hybrid spectral matrix at each mixing time; the set of relaxation matrices is then averaged, and the result used to generate distance constraints for the next iteration of structure calculation (Boelens et al., 1988). Because each cross-relaxation rate is best defined by experimental data at a distinct mixing time (Macura, 1994), this procedure indiscriminately averages data from mixing times at which a given matrix element is well defined with data from mixing times at which it is poorly defined. It is not clear that such a procedure will yield better results than the analysis of data from a single compromise mixing time. Adoption of the procedure used in MARDIGRAS, in which each mixing time is analyzed independently and the spread of output is used to derive error bars, is impractical because of the nature of the output (coordinates instead of distances) and for the considerations of CPU time outlined above.

Both the IRMA-type and MARDIGRAS-type procedures have been found to yield more accurate structures from cross-relaxation data containing spin diffusion than procedures involving distance determination by use of the two-spin approximation (see Section 4) (Thomas et al., 1991; Kaluarachchi et al., 1991). At the time of this writing, however, the IRMA and MARDIGRAS approaches have not been compared to each other in a valid way to determine whether or not additional improvement in accuracy is gained for the additional computation time used in IRMA. In the absence of such a test, the less computationally demanding approach of iterative refinement at the distance level (MARDIGRAS) must be preferred to iterative refinement at the coordinate level (IRMA).

2.3.4. Direct Refinement of Three-Dimensional Structures. A given set of three-dimensional coordinates, upon application of the rigid-molecule approximation or some other motional model, predicts a particular pattern of NOESY intensities. If the structure is adjusted slightly, this predicted pattern changes, and will come into either worse or better agreement with the experimental data available. This is the conceptual basis for techniques of direct or "NOE-based refinement" of NMR structures against experimental data (Baleja et al., 1990; Bonvin et al., 1991b). The distinction from IRMA-type algorithms is that structures are not calculated *de novo* from a refined set of distance constraints at each of a discrete set of iterations, but are adjusted smoothly from a starting structure through coordinate space so as to optimize agreement with the data. Whether this increased agreement reflects improved structural accuracy depends on the appropriateness of the motional model used. For the implementation of such calculations, an evaluation of the gradient of NOE intensities with respect to coordinates is needed. The derivation of an analytical form for this gradient by Yip and Case (1989) has advanced work in this field dramatically, although other approximations are possible (Bonvin et al., 1991b).

NOE-based refinement by means of the analytical gradient approach (Yip and Case, 1989) was implemented in the widely-used commercial software package X-PLOR (Brünger, 1992). This implementation is thus accessible to practicing spectroscopists, although the CPU requirements are rather formidable at this point (Nilges et al., 1991). When this procedure was applied to the refinement of a 29-residue peptide, for which the calculations were tractable, improved agreement of calculated structures with the x-ray structure of the molecule was found (Nilges et al., 1991). NOE-based refinement as implemented in X-PLOR has also been used to refine the structures of an RNA tetraplex (Cheong and Moore, 1992), an isolated domain of *E. coli* 5S RNA (White et al., 1992), and a DNA dodecamer with a bound drug moiety (Schweitzer et al., 1994). NOE-based refinement is very flexible in that it can handle data at different mixing times and with different relaxation matrices (in both H_2O and D_2O, for example) in a natural way by imposing appropriately-weighted energetic penalties for the various datasets available (Brünger, 1992). This alone is a considerable advantage over iterative schemes. The NOE-based algorithm also can be modified in a straightforward way to refine structures against three-dimensional NOESY-NOESY data (Bonvin et al., 1991a), thus combining the improved information content of such spectra with a calculational tool that makes the most of it. At present, the most significant obstacle to the general acceptance of NOE-based refinement is the substantial demand placed on computer hardware. Intensive development of the calculational algorithms with the goal of overcoming this limitation is ongoing (Yip, 1993; Dellwo and Wand, 1993). A more fundamental limitation is that the calculation of NOE intensities from a coordinate set requires knowledge of (or assumptions regarding) the internal dynamics of the molecule; since the nature, time scale, and amplitude of internal motions in proteins are not yet clearly understood, the extent to which small disagreements between observed and calculated intensities may be due to an inadequate description of molecular dynamics is not known.

2.3.5. Summary. The general term "relaxation matrix refinement" covers a variety of techniques, which fall into the four general categories above. The differences may be clarified by superimposing the schemes used on the "standard procedure" for NMR structure determination summarized in Figure 1; such a comparison is shown in Figure 3. This figure emphasizes the fundamental difference between the procedure of Koehl and Lefèvre and all other schemes, in that fitting at the level of cross-relaxation rates does not depend on the application of the rigid-molecule approximation, and thus allows separation of the analysis of multispin effects and internal motion. This method is likely to be preferred in cases where the nature of internal motions is the question under study, since it supplies the best possible

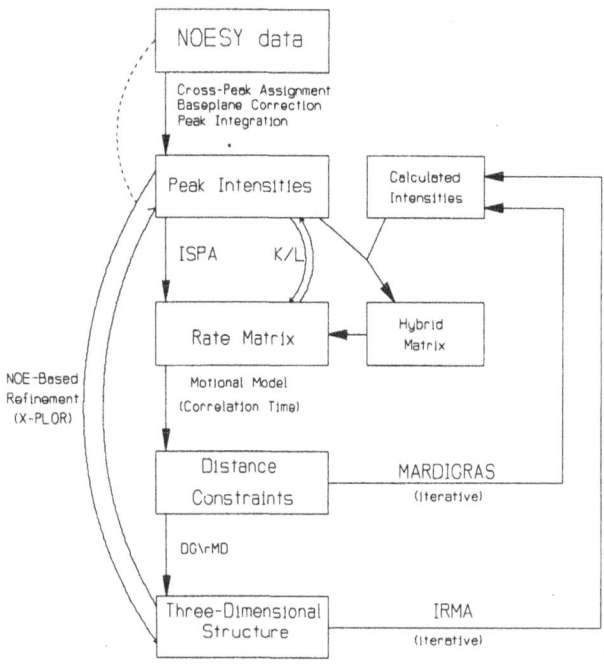

Figure 3. Relaxation matrix refinement. Schematic versions of the various matrix refinement schemes are superimposed on the scheme of Figure 1. Direct minimization procedures are indicated by bidirectional, curved arrows, while unidirectional arrows represent single steps in iterative schemes. K/L refers to the direct fitting of cross-relaxation rates demonstrated by Koehl and Lefèvre (1990). IRMA, MARDIGRAS, and X-PLOR are particular software packages that exemplify the algorithms described (see text).

set of "raw data" for such studies in the form of an optimized set of cross-relaxation rates. MARDIGRAS-type algorithms differ crucially from IRMA-type algorithms and NOE-based refinement in that the actual coordinates are not derived until after the refinement procedure produces an optimized set of distance constraints. This choice is a significant advantage in computational efficiency; whether it is a disadvantage in terms of the ultimate accuracy of refined structures remains to be seen. Direct refinement techniques, at whatever level, are inherently more flexible than iterative techniques since the algorithm is not "bottlenecked" by the back-calculation of a single rate matrix from hybrid NOE datasets, and thus can handle a variety of experimental data in a natural way. A final point has been emphasized by Jardetzky (1991); in the absence of algorithms to calculate NMR proton chemical shifts and linewidths from structural models, refinement against the "raw data" means refinement against an extracted set of peak intensities. The passage from a NOESY spectrum to such a dataset involves a number of nonlinear, error-prone steps including cross-peak assignment, baseplane flattening, accurate integration, and accounting for peak overlap.

What the variety of matrix-based refinement techniques has in common is the philosophy of taking advantage of the real information that is present in spin-diffusion contributions to NOE peaks. In curve c of Figure 2, for example, the observed cross peak is more intense because of the spin-diffusion contribution than the same cross peak would be if only the direct contribution (represented by the tangent line) were present. The magnitude of this intensity depends on the structure of the molecule, and can be taken advantage of to better define that structure. Such procedures thus depend fundamentally on the measurement of a network of connected NOEs so that the sources of multispin effects may be identified. In the next section, we describe a set of experimental techniques developed for the purpose of removing spin-diffusion contributions from cross-relaxation data sets so that the relationship of a single cross peak to a single distance constraint is conserved.

3. PRINCIPLES OF MAGNETIZATION EXCHANGE NETWORK EDITING

The feature common to all the techniques discussed above for overcoming the spin-diffusion problem is that experimental data are acquired in the presence of spin diffusion with the multispin effects accounted for at a later stage. This is a fundamental characteristic of any method relying on unperturbed NOE or ROE spectroscopy, since processes during the mixing delay are governed by the inherent multispin nature of the generalized Solomon equations. In order to acquire experimental data free of spin-diffusion contributions, the relaxation processes must be perturbed by suitable manipulations of the magnetization during the mixing time. In the past several years, a class of experiments known as magnetization exchange network editing (MENE) has been developed that applies such manipulations to prevent selected proton pairs from engaging in net magnetization transfer due to cross relaxation (Macura et al., 1992). In favorable cases, MENE experiments alleviate the spin-diffusion problem by allowing the direct contribution between two protons but forbidding the indirect contributions. This editing of the relaxation network allows the determination of the direct contribution and thus of the interproton distance. The fundamental advantage of such experiments is the extension of the range of applicability of the two-spin approximation or the initial-buildup rate approach to longer mixing times, thus allowing the determination of distances at longer range. The most serious disadvantage is that the extra information present in spin-diffusion contributions, which can be taken advantage of in the matrix-refinement approach, is not present in the spectrum. In this section, we discuss the classification, experimental nature, and utility of the MENE experiments that have appeared in the literature.

3.1. Classification and Epitome of Network Editing Sequences

3.1.1. Nonselective Sequences: Separation of Cross-Relaxation and Chemical Exchange. The NOESY pulse sequence is identical to that used for two-dimensional studies of chemical exchange rates (Jeener et al., 1979), since both processes lead to incoherent magnetization transfer during the mixing time. When chemical exchange is present, the mathematical formalism for NOESY is modified by replacing the relaxation matrix R with a generalized exchange matrix L. The elements of L are given by (Macura et al., 1994):

$$L_{ij} = K_{ij} - R_{ij} \tag{9}$$

in which K_{ij} is the rate of chemical exchange between spins i and j (for example, if i and j are the two delta protons of a tyrosine residue, then K_{ij} is the rate of ring flips). L (or L_{ij}) then replaces $-R$ (or $-R_{ij}$) in equations derived in the absence of exchange considerations. Analyses of exchange rates K_{ij} using two-dimensional spectroscopy are often confounded by cross-relaxation processes contributing to the peaks of interest.

The least selective MENE scheme eliminates all cross relaxation processes from the spectrum, leaving only the operation of chemical exchange processes ($L_{ij} = K_{ij}$). Exchange rates may be measured from such spectra free of confounding effects from cross-relaxation. This scheme was implemented as the pure exchange, or Clean EXSY (EXchange SpectroscopY) pulse sequence (Fejzo et al., 1990; Fejzo et al., 1991c). Experimentally, Clean EXSY is accomplished by alternating the magnetization vectors between the longitudinal (NOESY) and transverse (ROESY) frames such that the total NOESY delay is twice the total ROESY delay. Since, for large molecules, ROESY cross-relaxation is twice as fast and occurs with the opposite sign as NOESY cross-relaxation, processes in the two frames cancel, and no net cross-relaxation is observed. The most intuitive way to accomplish this effect would be to use the pulse sequence $(\tau_N - 90°_x - \tau_R - 90°_{-x})*n$, where τ_N and τ_R are short NOESY and ROESY delays, respectively, and the total mixing time is determined by the number of repeats n. To reduce certain types of artifacts, it is preferable to adopt a $(\tau_N - 90°_x - 90°_{-x} - 90°_{-x} - 90°_x)*n$ pattern, in which τ_N is equal to twice the 90° pulse width and all of the ROESY cross-relaxation takes place during the pulses (Fejzo et al., 1991c).

Macura et al. (1992) introduced a conceptually useful description of MENE experiments using graph theory. In this formalism, protons are represented by nodes, or circles, and allowed cross-relaxation processes are indicated by edges, or lines connecting two circles. Graphs for the various experimental schemes discussed in this section are shown in Figure 4 for five-spin relaxation networks. Panel A shows the fully-connected graph representing the NOESY experiment, for which all proton pairs in principle have finite cross-relaxation rates and, in accordance with the multispin Solomon equations, the longitudinal relaxation of every proton depends on the magnetization state of every other proton. (In practical cases, of course, many cross-relaxation rates will be vanishingly small; the graph formalism recognizes only that no rates are set identically to zero by the pulse sequence). Panel B, the completely disconnected graph (identically zero cross-relaxation rates are indicated by dotted lines), represents the Clean EXSY experiment. It is immediately suggested by this graph that each proton in the network undergoes longitudinal relaxation independent of the magnetization state of any other proton, that is, in single-exponential form with rate R_{ii}. In the absence of exchange, therefore, the Clean EXSY sequence allows measurement of diagonal elements of the relaxation matrix R by analysis of the time course of diagonal peaks. These matrix elements are not accessible from NOESY because of the strongly multiexponential character of the autorelaxation (Westler, W.M., Zhao, H., and Markley, J.L., unpublished results).

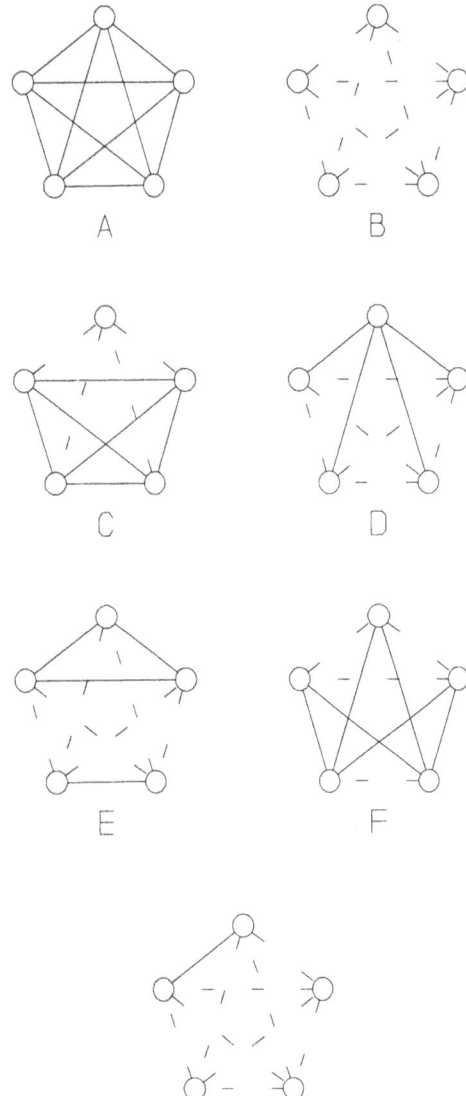

Figure 4. Graph represensions of network-editing experiments (Macura et al., 1992). Nodes represent protons in the relaxation network; lines represent finite cross-relaxation rates. Dotted lines indicate cross-relaxation rates set identically to zero by the experimental scheme. A, NOESY or ROESY; B, Clean EXSY; C, MINSY or XD-NOESY; D, S.NOESY; E, BD-NOESY or BD-ROESY; F, CBD-NOESY; G, Synchronous nutation or QUIET-NOESY.

3.1.2. Experiments Selective for a Single Resonance. If a spin k is known or suspected to mediate spin diffusion between two other spins i and j, the appropriate MENE scheme to study cross-relaxation between i and j is the removal of k from the relaxation network. The graph of this scheme is shown in Figure 4C, in which all cross-relaxation processes involving a single spin have been eliminated. The first experiment to accomplish this scheme used continuous decoupling of the spin to be removed over the course of the experiment (Olejniczak et al., 1986); this effectively cancels cross relaxation between the saturated spin and any other spin. Continuous irradiation of one spin, however, causes steady-state proton-proton NOEs to arise at all neighboring spins (see above). Because the steady-state NOE in macromolecules is large and negative, this approach leads to a significant loss of sensitivity. The sensitivity problem can be resolved by limiting the irradiation to the mixing time. This experiment, dubbed MINSY (Mixing Irradiation during a NOESY experiment), was demonstrated by Massefski and Redfield (1988) and was used to clarify cross-relaxation in a DNA fragment. Base-to-sugar NOEs in DNA are often badly perturbed by magnetization transfer within the sugar ring, and a systematic MINSY analysis was able to distinguish genuine from spin-diffusion NOEs and improve estimates of the glycosidic torsion angle (Massefski and Redfield, 1988).

An alternative to MINSY is to invert continuously or periodically the selected resonance during the mixing time using a selective pulse (Fejzo et al., 1991d). This technique is known as XD-NOESY (eXchange-Decoupled NOESY), since the initial application was to chemical exchange-mediated spin diffusion (Section 4.3). With a sufficiently short interval between inversion pulses, cross-relaxation involving the selected resonance before and after each inversion pulse will cancel, and no net magnetization transfer between the selected spin and the rest of the network will occur. The effects on spin-diffusion pathways are the same as for MINSY. This experiment may be generalized to decouple a number of individual resonances from the network by applying selective pulses sequentially at several frequencies. This feature of the experiment was exploited in a study of exchange-mediated spin diffusion (Fejzo et al., 1991d; Fejzo et al., 1991b). In this study, it was also possible to eliminate a number of exchange-mediated indirect NOEs between ring protons and protons on neighboring residues. Structures were calculated both with and without the incorrect restraints, and significant differences were found between the two resulting sets of structures (Fejzo et al., 1991b). This result provided evidence that exchange-mediated spin diffusion effects, unless corrected, can introduce inaccuracies in structures.

The complementary experiment to MINSY and XD-NOESY, in which only cross-relaxation to and from the selected resonance is allowed, was demonstrated experimentally as S.NOESY (Selective NOESY) (Fejzo et al., 1992). The graph of this experiment is shown in Figure 4D. Clearly, no indirect contributions to any of the observed cross peaks are possible. Thus, S.NOESY allows measurement of all cross-relaxation rates involving a selected proton in the absence of all spin diffusion. The S.NOESY mixing sequence uses alternation between NOESY and ROESY to cancel cross-relaxation and line-selective pulses to separate the selected resonance from the rest of the network. Magnetization alternates between the NOESY and ROESY frames as described above for the pure exchange experiment. The ROESY periods, however, are flanked by inversion pulses selective for the desired resonance. Thus, for only those proton pairs involving the selected resonance, the spins will be antiparallel during the ROESY periods with respect to their orientation during the NOESY periods; cross relaxation will take place with the same algebraic sign in both frames, and cross-peak intensity will build up over the course of the mixing time (Fejzo et al., 1992). The S.NOESY sequence as originally published is very insensitive, owing to off-resonance effects and to magnetization losses during both the spin lock periods and selective pulses. The off-resonance effects, however, potentially could be alleviated by application of the offset-compensation techniques described in the following section for the CBD-NOESY sequence. In S.NOESY, cross-relaxation is analyzed in the absence of all spin diffusion, but only a relatively small number of cross peaks can be analyzed in a single experiment. S.NOESY is thus more practical for answering specific questions about a cross-relaxation network in cases where spin diffusion is suspected than it is for deriving complete structural datasets.

The implementation of the schemes in Figures 4C and 4D depends on the ability to saturate or invert only a single resonance in the spectrum, without perturbing any other resonances. In macromolecules, this generally is not possible, since the saturation or inversion affects a number of resonances in the region of the selected frequency. For some purposes, this non-selectivity is advantageous. In the original work on MINSY, for example, it was found that irradiation at a single frequency was sufficient to saturate the bulk of the compact H2' region, thus allowing simultaneous analysis of several residues (Massefski and Redfield, 1988). This effect, however, somewhat complicates the interpretation of the results, since more than one spin in the network is perturbed. In the case of S. NOESY, for example, spin diffusion pathways mediated by a resonance that is accidentally within the bandwidth of the selective pulse will not be eliminated from the cross-peaks of interest. In any case, a clean separation of affected from non-affected resonances is desirable, because spins that are neither cleanly inverted or saturated nor unaffected will contribute to the relaxation network in complicated and unpredictable ways. There is some reason to prefer the XD-NOESY scheme to MINSY, since the effect of continuous-wave irradiation is characterized in the frequency domain by a (sin x)/x profile with several sidelobes, whereas shaped pulses can be used to obtain a relatively clean inversion profile (Freeman, 1991).

3.1.3. Division of the Relaxation Network Into Two Large Fragments. The XD-NOESY and S.NOESY experiments may be modified by the replacement of pulses tailored to invert a single resonance with pulses tailored to invert a large region of the spectrum. The effect is to separate the spectrum, not into one resonance versus the remainder of the molecule, but into two large blocks. The resulting experiments, with graphs shown in Figure 4E and 4F, are known as BD-NOESY (Block-Decoupled NOESY) (Hoogstraten et al., 1993) and CBD-NOESY (Complementary Block-Decoupled NOESY) (Hoogstraten et al., 1995b), respectively. On the basis of the above discussion of selectivity, these experiments might appear to form a continuum with the XD-NOESY and S.NOESY techniques; however, the soft pulse waveforms necessary to invert a fraction of a ppm (e.g., Gaussian) and several ppm (e.g., hyperbolic secant or IBURP) are different enough in useful bandwidth (Freeman, 1991) that the two classes of experiment are clearly distinguished in practice. In proteins, the most advantageous scheme is to separate the relaxation network into aliphatic protons (upfield) and amide and aromatic protons (downfield); the transition region of a selective pulse tailored to invert either of these blocks then falls into the normally empty spectral region between 5.3 ppm and 6.3 ppm. Examination of the graphs in Figure 4 reveals that spin-diffusion contributions in BD-NOESY (Figure 4E) are greatly diminished but not eliminated, as pathways taking place completely within one block are still allowed. Residual spin-diffusion contributions may be diagnosed using either buildup curves or a ROESY experiment tailored to the same relaxation submatrix (BD-ROESY) (Hoogstraten et al., 1995c). For CBD-NOESY (Figure 4F), no two-step pathways that contribute to the peaks of interest are allowed, and only a single class of three-step pathways (the rare case in which each of the three steps connects two different blocks) is allowed. Spin diffusion is thus essentially eliminated from CBD-NOESY.

The efficacy of the BD-NOESY and CBD-NOESY sequences for improving distance accuracy in proteins has been demonstrated on a small protein, turkey ovomucoid third domain (OMTKY3). By using BD-NOESY to restrict cross relaxation to the spectral block containing the amide and aliphatic protons, cross relaxation in this region could be analyzed in the absence of spin diffusion contributions mediated by aliphatic protons. Comparative buildup analyses of BD-NOESY and NOESY spectra demonstrated that substantial spin-diffusion contributions in difficult geometries had been removed (Figure 5). BD-NOESY enables the measurement of cross-relaxation rates and interproton distances at longer range than possible with standard NOESY techniques (Hoogstraten et al., 1993). CBD-NOESY was used to study cross relaxation into methylene groups, normally a severe source of spin diffusion in protein NOESY spectra due to the short interproton distance. CBD-NOESY allows the independent measurement of pairs of amide-to-methylene cross-relaxation rates in the absence of cross relaxation between the geminal protons. In one case (Figure 6) the distance estimated from NOESY data was underestimated by approximately 1 Å (2.5 Å, whereas CBD-NOESY yielded a distance of 3.5 Å) (Hoogstraten et al., 1995b). This difference could cause the misclassification of the peak, and consequently an application of an inappropriately short upper distance bound, even when a conservative "strong, medium,

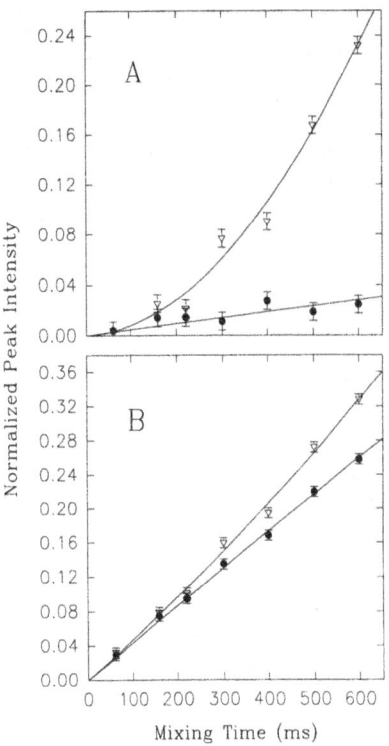

Figure 5. Buildup curves in OMTKY3 for NOESY (∇) and BD-NOESY ($\bullet$). (A) NOE intensity between the amide protons of Lys-29 and Thr-30 (K^{29} H^N/T^{30} H^N). (B) NOE intensity between the amide protons of Ser-44 and Asn-45 (S^{44} H^N/N^{45} H^N). For plotting and analysis, the peak intensities were normalized by reference to the corresponding diagonal peak intensities at the same mixing time (Macura et al., 1986) and corrected for diagonal peak overlap by assuming equal longitudinal relaxation rates for superimposed peaks. Curves shown are least squares fits to a second degree polynomial with zero constant term, forced to a straight line in the case of BD-NOESY. Error limits were derived by repeating a single experiment (BD-NOESY, 220 ms) seven times and determining the standard deviation of the normalized peak intensity. Panel A represents a long-range amide proton-amide proton distance badly contaminated by spin diffusion mediated by an aliphatic proton (K^{29} H^α) in NOESY, with the indirect contribution removed in BD-NOESY; panel B represents a short helical amide proton-amide proton distance for which the two experiments give similar results. Figure reproduced from Hoogstraten et al. (1993); used by permission.

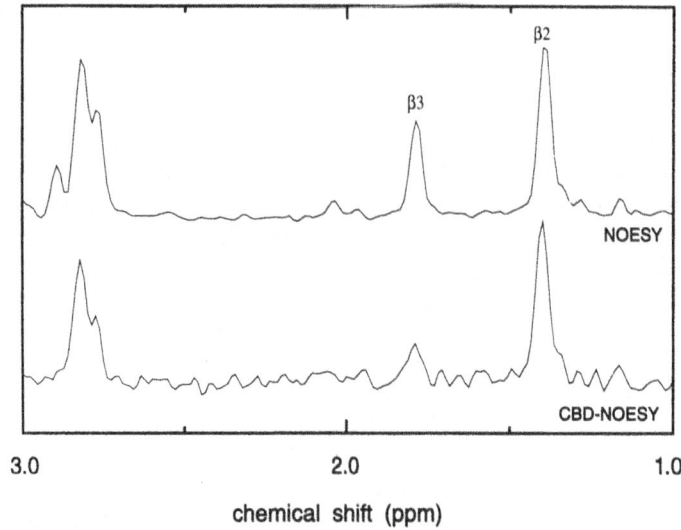

Figure 6. Slices through NOESY (top) and CBD-NOESY (bottom) spectra of OMTKY3 at the ω_2 frequency of C^{38} H^N (9.17 ppm). Spectra were acquired on a 3 mM sample of OMTKY3, 298K, pH 4.1, on a Bruker DMX-750 spectrometer. The NOESY mixing time was 180 ms; for CBD-NOESY, the NOESY delay was set to 19.3 ms, the ROESY delay was set to 12.5 ms, the precession delay τ_p to 35 μs, and the mixing sequence was repeated for N=4 times. The band-selective pulse was a 1.05 ms IBURP2 (Geen and Freeman, 1991) pulse applied to the aliphatic region of the spectrum. Z-gradient pulses of 1 ms duration were applied at the beginning and end of the mixing period,. 768 t_1 points of 80 transients each were acquired using States-TPPI (Marion et al., 1989) detection in the indirect dimension. Spectra were extended by 33% in t_1 using complex linear prediction; 3 Hz exponential broadening in t_2 and a cosine-squared window in t_1 were applied; and the spectra were zero-filled and transformed to a final matrix of 4096 by 1024 points. Figure reproduced from Hoogstraten et al. (1995b); used by permission.

weak" bounds protocol is applied, and was sufficient to cause an incorrect result for the side-chain conformation of this residue when used in structure calculations (Hoogstraten et al., 1995b). The magnitude of the potential improvement in distance accuracy is seen in Figure 7, which compares "actual" distances in a target structure (derived from a crystallographic model of OMTKY3) with distances derived from NOESY and CBD-NOESY spectra simulated from this structure. A very serious systematic underestimation of amide-methylene distances is observed in the NOESY simulations, particularly at the longer mixing time of interest for observing longer-range interactions; these inaccuracies are removed essentially completely by the MENE technique (Hoogstraten and Markley, in preparation).

A strategy combining the use of BD-NOESY and CBD-NOESY allows, in a small number of two-dimensional spectra, the analysis of every peak observed in NOESY in the presence of greatly reduced contributions from indirect transfer. Such studies are further facilitated by incorporating the MENE mixing sequences into a multidimensional heteronuclear-edited pulse sequence to improve peak dispersion, as is commonly done with standard NOESY experiments (Clore and Gronenborn, 1991). Figure 8 shows pulse sequences for incorporating the BD-NOESY (A) and CBD-NOESY (B) mixing sequences into a heteronuclear-edited three-dimensional NOESY-HSQC sequence, as implemented in a recent refinement of the OMTKY3 solution structure (Hoogstraten et al., 1995a). Other implementations of these schemes can be obtained by placing the indicated mixing sequences into other framework pulse sequences (two-dimensional NOESY, for example). These same pulse

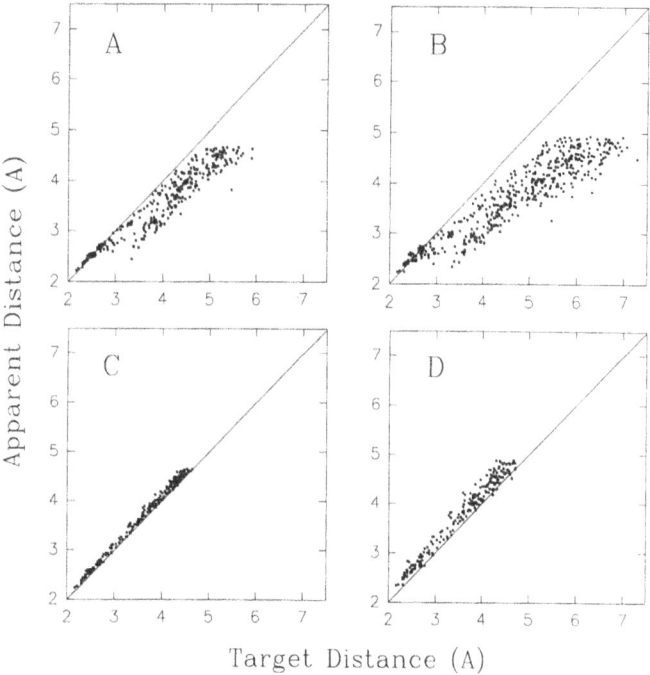

Figure 7. Apparent distance, as calculated by an ISPA calibration from simulated data, versus distance in the target structure for all backbone amide-β methylene NOEs with intensity greater than 0.1% of single-proton diagonal intensity at zero mixing time. A, 60 ms NOESY; B, 140 ms NOESY; C, 60 ms CBD-NOESY; D, 140 ms CBD-NOESY. Solid lines correspond to equal apparent and target distance (i.e., perfect accuracy). The target structure was the 1CHO crystal structure of OMTKY3 (Fujinaga et al., 1987), minimized in the X-PLOR force field (Hoogstraten and Markley, in preparation). Data were simulated using the CORMA algorithm of James and coworkers (Borgias et al., 1990; Liu et al., 1994) and calibrated with the average of the intensities for the 2.2 Å $d_{\alpha N}$ distances in the regular beta sheet regions of OMTKY3.

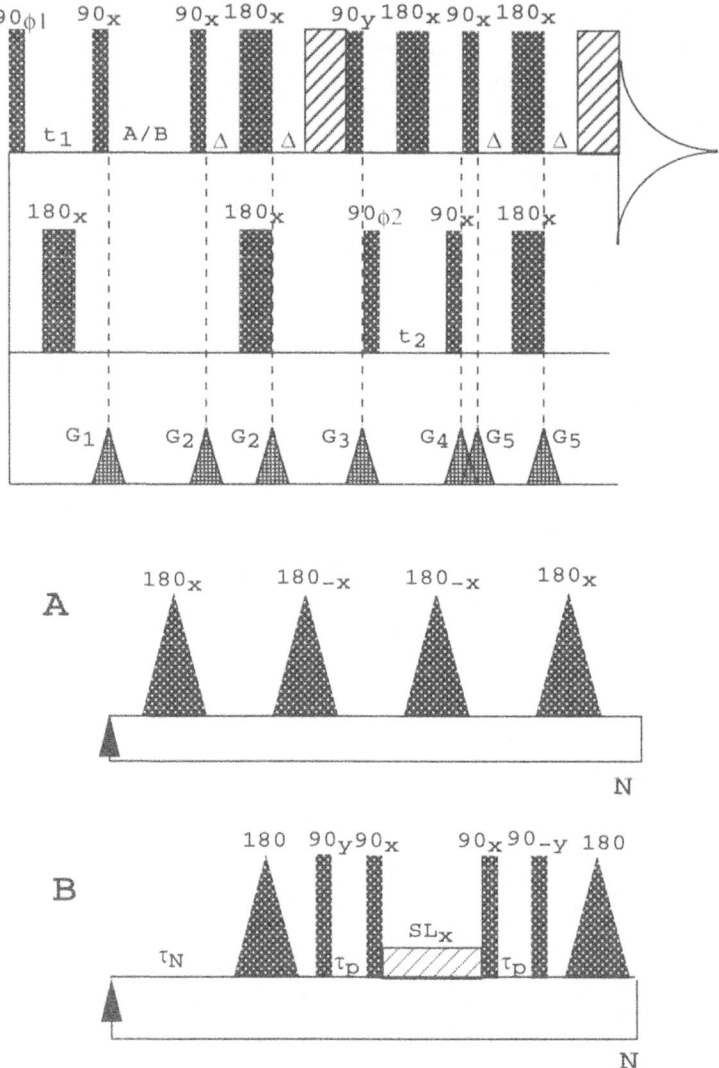

Figure 8. NMR pulse sequences for the BD-NOESY-HSQC (A) and CBD-NOESY-HSQC (B) experiments. The upper portion shows the ^{1}H RF pulses, ^{15}N RF pulses, and z-gradient pulses for the framework NOESY-HSQC sequence; network-editing mixing sequences are indicated separately (A, BD-NOESY; B, CBD-NOESY). Shaded triangles for RF pulses indicate low-power selective pulses. Diagonal shading indicates spin lock. Abbreviations: τ_N, longitudinal relaxation period; SL_x, transverse relaxation period (spin-locked); τ_p, precession delay; Δ, 1/4J delay for evolution of heteronuclear coupling. Phase cycle: $\phi1$: 0, π; $\phi2$: 0, 0, π, π; receiver: 0, π, π, 0. Figure reproduced from Hoogstraten et al. (1995a); used by permission.

sequences, using line-selective rather than band-selective pulses, yield the XD-NOESY and S.NOESY experiments. Recently, a version of BD-NOESY was developed in which the "inversion pulses" are spin-echo sequences that invert only magnetization attached to a given heteronucleus (^{15}N or ^{13}C); in labeled macromolecules, this experiment can divide the spin network into blocks consisting of protons that are attached to the given heteronucleus and protons that are not (Zolnai et al., 1995). In combination with differential isotopic labeling patterns, this type of experiment has promise for the analysis of intermolecular complexes by allowing the separation of intra- from inter-molecular cross-relaxation contributions.

3.1.4. Isolation of a Single Interaction. The opposite extreme from the indiscriminate magnetization transfer in NOESY experiments is the measurement of a single cross-relaxation rate in the absence of all multispin effects. This scheme, graphed in Figure 4G, was achieved by Bodenhausen and coworkers in the synchronous nutation (Boulat et al., 1992; Burghardt et al., 1993) and QUIET-NOESY (Quenching Undesirable Indirect External Trouble in NOESY) (Zwahlen et al., 1994) experiments. Both of these experiments are modifications of one-dimensional analogs of NOESY that give rapid recording of data for a single cross-peak. The physical description of processes during the mixing period, however, is identical to that for a standard NOESY experiment.

Synchronous nutation relies on the application of an RF field modulated in such a way that irradiation sidebands fall on the resonances of interest. A carefully adjusted field will modulate the magnetization of the two spins in concert with respect to each other but not to any other spin, thus allowing cross-relaxation to take place between these two spins without magnetization transfer to or from any other spin in the network (Boulat et al., 1992; Burghardt et al., 1993). This sequence is an effective way to isolate two spins; the experimental difficulties are substantial, however, and extraction of quantitative information may require the precise determination of several other parameters, including proton autorelaxation rates and multiple-quantum relaxation rates (Burghardt et al., 1993). Since this is a one-dimensional sequence, it is difficult to compare the sensitivity of synchronous nutation in a valid manner to NOESY; however, the results presented are not encouraging. Synchronous nutation was demonstrated on bovine pancreatic trypsin inhibitor (BPTI), a protein of very high solubility (Wüthrich, 1986); the concentration used was not stated. In the two published studies, one intraresidue α-β distance was measured as 2.3 Å, a second was qualitatively reported as corresponding to an unmeasurably small cross-relaxation rate (Burghardt et al., 1993), and an interresidue α-α distance of 2.0 Å was measured (Boulat et al., 1992). Since NOESY studies routinely analyze distances as long as 4.5 Å, representing a 56-fold smaller cross-relaxation rate than a 2.3 Å distance, in proteins much less soluble than BPTI, synchronous nutation seems to be too insensitive to reliably quantitate even moderately long distances in practical conditions.

The QUIET-NOESY experiment is a more recent attempt to overcome problems associated with synchronous nutation (Zwahlen et al., 1994). The mixing sequence in this case consists of a single selective inversion pulse adapted to invert both spins of interest; the experiment is thus a version of the scheme used in XD-NOESY or BD-NOESY but with a differently tuned inversion profile. This procedure effectively divides the relaxation network into two resonances (and the resonances with which they overlap) in one block and the remainder of the network in another; cross-relaxation within the larger block is not observed due to the selective nature of the excitation and observe pulses. In the experimental examples presented, a single inversion is sufficient to give a high degree of cancellation of multispin effects. In some cases, however, a single inversion may not be adequate to remove all spin-diffusion effects (Zolnai et al., 1995). The number of inversions that can be used is limited in this case by the considerable length of the individual inversion pulses (Zwahlen et al., 1994). In the initial report, QUIET-NOESY successfully distinguished between direct and spin-diffusion cross peaks in a DNA fragment; quantitative distance measurements were not reported (Zwahlen et al., 1994). This experiment thus forms an alternative to S.NOESY for the analysis of cross-relaxation in the absence of all spin-diffusion processes. Again, valid comparisons of QUIET-NOESY and S.NOESY in terms of sensitivity and efficiency are difficult because of the very different nature of the two experiments. It should be noted, however, that QUIET-NOESY fails to remove spin-diffusion contributions due to protons that overlap with either selective proton, whereas S.NOESY is vulnerable to residual effects only from protons overlapping with the resonance that is selectively inverted. For NOEs into methylene pairs, which often have similar chemical shifts and are difficult to separate with selective pulses, this limitation may be an important consideration.

3.2. Experimental Considerations

The two fundamental operations of network-editing pulse sequences, excluding sequences involving continuous-wave irradiation (MINSY and synchronous nutation), are alternation between cross-relaxation in the longitudinal and transverse frames and the selective inversion of a portion of the spectrum. Both of these procedures are somewhat problematic. For the pure exchange sequence, as discussed above, the best results are obtained if all ROESY relaxation takes place during the pulses and longitudinal delays are reduced to times on the order of a hard pulse. In sequences such as S.NOESY and CBD-NOESY, which combine soft pulses with NOESY/ROESY alternation, this procedure cannot be used because of the long duration of the selective pulses (milliseconds to tens of milliseconds), and NOESY and ROESY delays of macroscopic length must be used. For off-resonance spins, the transfers to and from a spin-locked condition cause a loss of magnetization due to a projection from the transverse plane to the spin-lock axis, which in the general case is tipped above or below the transverse plane, and back (Griesinger and Ernst, 1987). For spins offset by the spin-lock field strength in Hz (at the edges of the spectrum in a normal experimental setup), this process leads to a reduction in sensitivity by a factor of $(\frac{1}{2})^n$, where n is the number of NOESY-ROESY cycles, in an S.NOESY or CBD-NOESY spectrum. This rather serious spectral degradation can be overcome by the application of the offset-compensation procedure illustrated in scheme B of Figure 8. For CBD-NOESY (or S.NOESY), simple application of the high-power 90° offset-compensation pulses of Griesinger and Ernst (1987) would have no effect, as the selected coherences are aligned along the x-axis immediately following the $90°_y$ pulse. Therefore, a short precession delay (τ_p), adjusted so that resonances at the edges of the spectrum will precess by 45° during τ_p, is inserted prior to the ROESY sequence. The offset-compensation pulse then rotates the magnetization vectors approximately onto their spin-lock axes, minimizing the sensitivity loss; a second precession delay following the spin lock and second offset-compensation pulse allows the magnetization vectors to refocus along the x-axis and subsequently be restored to the longitudinal frame by the $90°_{-y}$ pulse (Hoogstraten et al., 1995b). This procedure is critical for adequate sensitivity to be obtained in CBD-NOESY spectra.

A wide variety of line- and band-selective shaped pulses has been described in the literature. This field has been reviewed recently in depth (Freeman, 1991; Kessler et al., 1991; Geen and Freeman, 1991; Emsley, 1994). Network-editing experiments place great demands on selective pulses because of the large number of pulses applied; imperfections that are acceptable when an inversion pulse is applied once in a pulse sequence may cause the failure of an XD-NOESY or BD-NOESY experiment that employs a train of several dozen such pulses over the course of the mixing time; because signals within the bandwidth of the soft pulse are not interpreted in these experiments, the key consideration is minimal perturbation of spins outside the bandwidth. For S.NOESY and CBD-NOESY, cross peaks involving resonances within the pulse bandwidth are interpreted; thus, complete inversion within the bandwidth, as well as minimal pulse length for the obtained bandwidth, are also critical. For all experiments, the soft pulse inversion profile must be as square as possible; that is, the size of the "transition region" that is neither cleanly inverted nor unaffected must be minimized. A sample pulse profile showing this transition region for a hyperbolic secant band-selective pulse is shown in Figure 9. Careful tuning of the adjustable parameters used in generating the soft pulse and optimization of the soft pulse hardware itself can have a

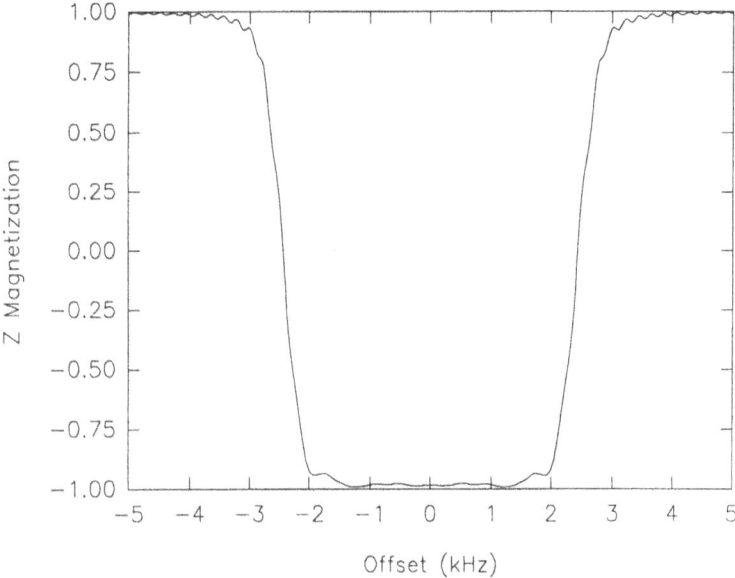

Figure 9. Example inversion profile for a shaped pulse. The z-magnetization following the soft pulse is plotted as a function of resonance offset calculated for a hyperbolic secant shaped pulse (Silver et al., 1985) with phase factor $\mu=15$ truncated at 15% using a pulse width of 5 ms at a power of 1600 Hz. The central region or pulse bandwidth is cleanly inverted, whereas the fringes of the displayed region are unaffected. The wings of the pulse bandwidth, for which final z-magnetization is near neither $+1$ nor -1, are the transition regions for this pulse. Calculations, which ignored relaxation effects, were performed with PASCAL code kindly provided by Dr. William M. Westler.

great effect on the quality of network-editing spectra. Such tuning is often done either in simulation, using numerical solutions of the Bloch equations that neglect relaxation, or on small-molecule test samples that have very different relaxation characteristics from macromolecules. Relaxation during the pulse can have a profound effect on the profile obtained in macromolecules, however, and these effects may be of importance in the selection or adjustment of the pulse used (Norris et al., 1991; Hajduk et al., 1993).

A final choice made in the setup of a network-editing pulse sequence is the number of inversions (for XD-NOESY, BD-NOESY, or QUIET-NOESY) or NOESY-ROESY cycles (for S.NOESY or CBD-NOESY). Since the application of any pulse, and particularly any selective pulse, involves a loss of some magnetization, a smaller number of these inversions or cycles will improve sensitivity. At some point, however, multispin effects in the experimental delays themselves will become significant and some spin-diffusion contributions may reappear. Often, these problems may be diagnosed by the presence or absence of cross peaks corresponding to the putatively cancelled cross-relaxation pathways. In applying the heteronuclear spin-echo edited version of BD-NOESY with a 300 ms mixing time to human ubiquitin, Zolnai et al. (1995) found that a single inversion was sufficient to reduce such cross peaks to 12% of their value in NOESY, while four inversions rendered them undetectably small. These authors have also provided a full theoretical account of processes occurring during the mixing time of a network-editing experiment and of the factors that affect the quality of the edited result (Zolnai et al., 1995).

The potential sensitivity of MENE experiments as compared to the parent NOESY sequence also merits discussion. It might be thought that the cancellation of some cross-relaxation pathways and concomitant zeroing of the associated cross peaks would increase the intensity of non-cancelled direct cross peaks, because less magnetization is "drained" to other pathways. In fact, network-editing experiments are always less sensitive than NOESY for all peaks. This is true because, whereas MENE sequences set cross-relaxation rates to zero on average, the corresponding dipole-dipole interactions are not decoupled in an instantaneous sense, and the "deselected" protons therefore still contribute to the autorelaxation of the resonances of interest, that is, the diagonal elements of the matrix R are the same for the edited experiment as for NOESY. Since the inversion of a portion of the spectrum converts cross-relaxation processes, which normally are conservative of total magnetization, into dissipative processes, the spectral lines will decay faster in the MENE spectrum than in the comparable NOESY experiment (Zolnai et al., 1995). The magnitude of these effects varies with the particular relaxation network and experimental scheme, and may be assessed in advance by matrix-type simulations of normal and edited spectra using suitably modified versions of programs such as CORMA (Liu et al., 1994). Additional losses occur, as discussed above, due to the effects of the shaped inversions and (for sequences involving spin locks) due to the enhancement of autorelaxation in the transverse frame in macromolecules. This loss of sensitivity is opposed by the ability of network-editing experiments to give reliable data at longer mixing times, for which sensitivity is improved. The above discussion refers to the signal-to-noise for the direct cross-relaxation components; the cancellation of multispin effects will also naturally lead to a loss of "signal" compared to NOESY for those peaks that contain spin-diffusion contributions.

3.3. Relationship to Other Techniques

Given the extensive history of techniques designed to deal with spin diffusion (described in Section 2), the question arises whether optimal strategies may be constructed by combining MENE pulse sequences described in this Section with other techniques. Buildup curves, for example, can diagnose residual spin-diffusion contributions in BD-NOESY spectra. By removing spin-diffusion contributions, network editing extends the range of validity of the initial-rate approach, and accurate measurements of very small cross-relaxation rates that would be completely masked by indirect contributions in NOESY may be obtained using buildup series of network editing data (Hoogstraten et al., 1993). The situation is more advantageous still if normalization on the diagonal (or, more practically, on the direct peak in a three-dimensional heteronuclear-edited spectrum) is possible; in this case, the sources of both negative curvature (autorelaxation) and positive curvature (spin diffusion) are eliminated, and buildup curves may be linear to very long mixing times indeed. This allows the analysis of each cross-peak at a series of mixing times in the region of its particular sensitivity optimum.

It is not clear, by contrast, that the application of network-editing sequences to random fractionally deuterated molecules would lead to improved results. This is because the quantitative analysis of network-editing data is based on a large number of identical relaxation networks, whereas in random deuteration a statistical distribution of deuteration patterns is obtained. In some cases, network editing may be useful to analyze or remove remaining spin diffusion contributions in specifically-deuterated samples; the sequential application of two techniques to simplify the relaxation network may allow a very full analysis. Neither of these possibilities has yet been investigated experimentally.

An intriguing possibility is the application of matrix-refinement techniques to network-editing data. A first step toward this was taken in the modification of the MARDIGRAS program to analyze BD-NOESY or CBD-NOESY data (Liu et al., 1994). This program, however, can be used to interpret data from a single network-editing scheme (amide/aromatic BD-NOESY, for example) but not to analyze simultaneously data from different network-editing schemes or from network editing and NOESY. This is not a fault in the particular code but a fundamental feature of the iterative refinement scheme; this algorithm necessarily involves a step in which a relaxation matrix is calculated using full matrix analysis from a single spectral matrix. Since each network-editing scheme gives data consistent with a different relaxation matrix (in that different elements have been set to zero), this algorithm cannot be generalized in an obvious way to handle data from multiple network-editing schemes.

In the broadest sense, the presence of a cross peak in a NOESY spectrum and its absence or diminution in a network-editing experiment are independent pieces of experimental data, and it is reasonable to require that a final set of calculated results (a refined structure, in most cases) be consistent with both of them. This concept can be implemented only awkwardly if at all in IRMA- or MARDIGRAS-type iterative schemes, but can be implemented in a natural way in direct refinement of either the cross-relaxation matrix or the three-dimensional structure. Such an algorithm would allow the information on multispin pathways present in the full NOESY matrix to be used, but constrained by the better determinations of specific matrix elements provided by network editing.

The above sections describe a spectrum of possible modifications to the basic NOESY scheme, ranging from analysis of the entire matrix with a single spin removed to isolation of a single interaction. The selection of a particular network-editing experiment involves a tradeoff between information and editing; as more selective sequences are considered, fewer cross peaks with correspondingly fewer spin diffusion contributions are observed. The combination of BD-NOESY and CBD-NOESY allows the analysis of all cross peaks present in NOESY in a small number of experiments, and thus is useful for the identification of those structural regions in which spin diffusion is occurring, as well for as the analysis of large numbers of NOEs for structure calculations. Remaining ambiguities, or individual rates of particular interest, may be investigated with sequences such as S.NOESY or QUIET-NOESY that remove all spin diffusion. XD-NOESY and MINSY are useful when particular resonances are strongly suspected of mediating spin diffusion, as in the case of aromatic rings in slow exchange (Section 4.3) or base-to-sugar NOEs in DNA. All of these experiments are less sensitive than the nonselective analogs, but many, nonetheless, have been used successfully to analyze quite weak interactions. The use of matrix-refinement techniques to combine the high information content of NOESY spectra with the well-defined measurements in network-editing spectra holds promise for improved analysis and structural definition.

4. EFFECTS OF MORE ACCURATE DISTANCES ON MACROMOLECULAR STRUCTURES

Overwhelming evidence, reviewed above, shows that appropriate experimental or calculational techniques improve the accuracy of measured cross-relaxation rates and/or distances by overcoming multispin effects. This in itself does not rigorously demonstrate that the application of these techniques in the context of an NMR structure determination will result in improved accuracy for the resultant structural models. Indeed, some arguments against this hypothesis exist in the common assumption that the great number of NOE constraints generally obtained will overcome the ill effects of any particular inaccurate distance constraint (James, 1994). On the other hand, the systematic nature of distance errors due to multispin effects (Figure 7) may mean that converged but inaccurate structures can be obtained in the presence of spin diffusion; alternatively, the underestimation of particular distances may lead to the exclusion of certain regions of conformation space from representation in the final set of structures, that is, the refinement of the structural model to an unjustified level of apparent precision (as measured, for example, by positional root-mean-squared deviation or rmsd). These issues can be addressed by side-by-side structure calculations from "more accurate" and "less accurate" distance constraint sets. Such calculations, in particular using data derived or simulated from an assumed structure to facilitate evaluation of absolute structural accuracy, have been extensively used to test the magnitude of the effects of errors introduced by the two-spin approximation independent of any particular scheme for overcoming them. These studies have recently led to contradictory and somewhat controversial results (Kominos et al., 1992; Clore et al., 1993; Zhao and Jardetzky, 1994) and are not discussed in detail here. Instead, we concentrate on efforts to quantify the magnitude of improvement seen when a particular experimental or calculational tool for overcoming multispin effects is applied.

4.1. Matrix-Based Refinement

The two major variants of iterative relaxation matrix refinement, one operating at the level of distances and one at the level of coordinates, have been tested in comparison with the simple ISPA approach by comparative structure calculations from an assumed (i.e., perfectly known) structure. Thomas et al. (1991) tested the effect of refinement at the level of distances using the MARDIGRAS algorithm using simulated data from the crystallographic structure of the small protein bovine pancreatic trypsin inhibitor (BPTI). Simulated data were interpreted according to "conservative ISPA," (broad distance ranges), "restrictive ISPA," (tighter distance ranges including explicit lower distance bound constraints), and with distances and bounds assigned by the MARDIGRAS algorithm (see Section 2.3.2). The application of tighter distance bounds under ISPA was found to improve structural precision and accuracy according to global statistics, but to introduce some local distortions, apparently due to distance constraint errors. Distance refinement with MARDIGRAS produced highly precise and accurate structural models in which the observed local distortions were removed, and was concluded to lead to a real improvement in the quality of structures for this protein with a minimal investment of computer time. The ability of MARDIGRAS to increase the number of distance constraints derived by allowing the use of longer mixing times was not tested in this work. Iterative refinement of three-dimensional structures (IRMA-type; see Section 2.3.3) was tested by Gorenstein and coworkers (Kaluarachchi et al., 1991) using their MORASS (Multiple Overhauser Relaxation AnalysiS and Simulation) code, incorporating restrained molecular dynamics for structure determination, to calculate structures from simulated data corresponding to a DNA dodecamer. In this work, relaxation matrix calculations were performed both with the same number of distances as the ISPA analysis, to demonstrate the effects of constraint accuracy, and with an increased number of constraints made accessible by the algorithm. It was found that, at the levels of global statistics, helical parameters, and backbone torsion angles, the MORASS calculations were considerably superior to the ISPA-derived constraints in accurately reproducing the features of the target structure. In particular, the MORASS calculations were able to distinguish interesting variations in nucleotide conformation across the dodecamer, whereas the ISPA calculations had insufficient resolution to allow this analysis. The use of a greater number of constraints within MORASS allowed a further improvement in structural definition. The differences between the unimproved and matrix-refined structures were more marked in this study than in the MARDIGRAS/BPTI calculations discussed above, but this is presumably due to the greater difficulty in defining nucleic acid structures with approximate NOE data due to the relative scarcity of interproton distances that constrain the phosphodiester backbone (Wijmenga et al., 1993). As discussed above, it is not known whether the less computation-intensive MARDIGRAS-type algorithm could have achieved a similar refinement of the dodecamer structure.

The direct refinement of three-dimensional structures (Section 2.3.4) has been tested by Nilges et al. (1991) using the implementation in X-PLOR to refine the solution structure of the 29-residue peptide CMTI-I. In this case, the algorithm was tested using experimental data for the molecule of interest, rather than data simulated from an assumed coordinate set. This procedure has the advantage of testing the method under realistic experimental conditions, but the disadvantage that there is no absolute standard by which the accuracy of the derived structures may be assessed and compared. In this case, the matrix-refined structures were compared to the high-resolution crystallographic coordinates for the same molecule, and the refinement procedure was found to lead to structures closer to the crystal coordinates than the original, ISPA-based NMR models. This does indeed indicate an increase in accuracy, assuming no real differences between crystallographic and solution conformations and neglecting possible inaccuracies in the crystallographic determination. The primary barrier preventing the application of this procedure to much larger systems is the significant CPU requirement, as noted above. Again, the magnitude of improvement achieved with this very computation-intensive procedure was not compared to that achievable with the much less demanding MARDIGRAS algorithm; however, the direct refinement was able to handle data from both H_2O and D_2O, as well as a variety of mixing times, in a natural way.

4.2. Noesy and Roesy Buildup Curves

When adequate sensitivity can be obtained in ROESY spectra, the simultaneous analysis of NOESY and ROESY buildup curves can determine cross-relaxation rates with excellent precision and accuracy, since multispin effects are detected by the different time dependence in the two regimes (Section 2.2.4) (Fejzo et al., 1989). Under the rigid-molecule approximation, these rates can be converted into precise interproton distances. This approach has been applied to the refinement of the solution structure of OMTKY3 by estimating a fraction of the distance constraints in this molecule from such a buildup analysis, and the application of appropriately-tightened bounds about the new distance estimates (Krezel et al., 1994). The resulting structures were found to be considerably more precise than unimproved structures at both the dihedral angle and global-coordinate level, apparently due to the use of tighter distance constraints allowed by the improved analysis. Again, the effects on structural accuracy were assessed by reference to crystallographic results; in this case, the application of the NOESY/ROESY method was found to cause the structures to converge toward a set of eight crystal structures of ovomucoid third domains according to global positional rmsd.

4.3. Magnetization Exchange Network Editing

As discussed above (Section 3.1.2) network-editing experiments designed to remove one or a few resonances from the relaxation network, such as XD-NOESY or MINSY, can be very helpful in unraveling situations in which a particular resonance is strongly suspected of mediating spin-diffusion contributions. Such a situation arises when side-chain aromatic rings rotate at a rate such that the pairs of delta and epsilon resonances are in slow exchange on the chemical-shift time scale. Since such exchange will usually be fast relative to cross relaxation, the ring rotation provides an efficient pathway for multistep magnetization transfer across long distances through chemical exchange in series with cross relaxation. This case is found in OMTKY3 at low temperature, under which conditions Tyrosine 31 is in slow exchange and a number of protons give rise to apparent NOEs arising from exchange-mediated spin diffusion (Fejzo et al., 1991b). These incorrect NOEs were identified, and the corresponding cross-relaxation rates analyzed with the error removed, by the use of XD-NOESY to systematically remove the tyrosine ring protons from the relaxation network. Resulting structure calculations with and without the spin-diffusion errors showed differences of up to 2 Å in the immediate vicinity of the tyrosine ring and up to 1 Å in other regions of the protein, indicating that these particular errors had caused a significant structural perturbation that the XD-NOESY study was able to remove (Fejzo et al., 1991b).

More recently, the effect of the combined BD-NOESY/CBD-NOESY approach to improving large numbers of distance estimates (Section 3.1.3) has been tested for the ability to improve solution models of OMTKY3. These studies have been performed both from data simulated from a given model, to allow assessments of absolute accuracy, and in an experimental refinement of the OMTKY3 structure. In the simulation studies (Hoogstraten and Markley, in preparation), the precision and accuracy of the distance constraints were varied independently: the accuracy by using either exact distances, distances derived by an ISPA analysis of simulated NOESY data, or distances derived from simulated MENE data, and the precision by assigning upper and lower bounds to each of the distance sets according to a number of conservative or restrictive protocols. When global statistics were examined, it was found, interestingly, that increasingly accurate restraints led to less accurate structures according to global positional rmsd, but more accurate structures according to an analysis of dihedral angle distributions. The "rmsd from target" metric, however, was strongly correlated to a global contraction or expansion of the structures, which is poorly determined by short-range NMR data, and was, therefore, argued to be a problematic measure of the quality of NMR-derived solution models, whereas the dihedral angle analysis gave an intuitively satisfying correlation of input with output accuracy and was independent of global structural superpositions. By this dihedral angle analysis, experimentally-achievable MENE data was found to lead to a substantial increase in the quality of calculated structures for this globular protein; in addition, in cases for which more accurate MENE data were available for some but not all distances, the most appropriate protocol was found to be the application of tight distance bounds to the MENE-derived distances but more conservative bounds to unimproved distances. These insights were applied to an experimental refinement of the solution structure of OMTKY3, in which 212 of the 655 distance constraints observed for this molecule were improved in accuracy by the application of three-dimensional heteronu-clear-edited BD-NOESY-HSQC and CBD-NOESY-HSQC spectroscopy (Hoogstraten et al., 1995a). Structures were calculated both from the original "crude" dataset and an "improved" dataset containing the MENE-derived data. The resulting sets of structures were found to differ significantly, with the differences centering in a hinging of the conformation of a reverse turn in a beta sheet, with backbone displacements of up to 2 Å at the tip of the turn, and in the expansion of explored conformational space in the active site for proteinase inhibition (Figure 10). The changes in the beta turn upon application of the MENE analysis were found to cause the structure in this region to be less similar to two crystal structures of OMTKY3 in complex with protease, but more similar to the crystal structure of the closely-related silver pheasant ovomucoid third domain (OMSVP3) in uncomplexed form; the improved structures were therefore concluded to be a more accurate representation of the solution conformation of free OMTKY3. The changes in the active site were interpreted as the restriction of the crude structures by the spin-diffusion contaminated NOESY data to an unjustified level of apparent precision, re-emphasizing the inadequacy of measures of precision such as rmsd to assess the accuracy of solution structures (James, 1994; Zhao and Jardetzky, 1994). The application of the MENE approach outlined in Section 3.1.3 is thus seen to allow measurable improvements in the accuracies of structural models with a moderate input of experimental time.

In summary, a number of the approaches to improving the accuracy of cross-relaxa-tion rates and interproton distances discussed in Sections 2 and 3 have been tested, using simulations, experimental refinements, or both, to determine their effectiveness in improving the derived solution structural models. In many cases, the changes in the resulting structures

Figure 10. Average structures of OMTKY3 from NOESY and MENE data. The polypeptide backbones for residues 7-56 of unminimized average structures calculated from standard NOESY data (blue) and NOESY supplemented with BD-NOESY and CBD-NOESY data (red) are compared. For calculation of average structures, structures within a set were superimposed by reference to the backbone atoms for residues 7 through 56; the two average structures were superimposed in the same way for comparison. The beta turn in which the two structure sets significantly diverged is at the top of the figure; the changes in the reactive-site loop involve an expansion rather than a displacement of the set of structures (see text) and are not clearly visible in the perspective shown here. The widths of the ribbons are proportional to the backbone pairwise rmsd for the appropriate set of structures. Each calculation was based on fifty converged coordinate sets. The figure was prepared using the INSIGHT-II 2.3.5 software package (Biosym Technologies, Inc., San Diego). Figure reproduced from Hoogstraten et al. (1995a); used by permission. A color representaion of this figure can be found facing p. 112.

have been significant; when simulation studies have allowed the assessment of the true structural accuracy, the structures resulting from attempts to overcome multispin effects were more accurate than those calculated from unimproved NOESY "data." Particularly in the case of the various matrix-refinement schemes, which vary greatly in CPU requirements, comparisons among methods to identify those most effective at removing structural distortions in a given situation are still lacking. The two broad approaches of network editing and matrix refinement may prove to be complementary, in that each approach removes different classes of spin-diffusion errors, and the application of the two in combination may yield better structures than either alone. Attempts to combine these approaches, by developing matrix-refinement procedures that handle MENE data either alone or in combination with NOESY data, are a subject of current research in our laboratory.

5. CONCLUSION

The common procedure by which NOE intensities are translated directly into distance constraints, whether after explicit measurement or classification into ranges, has been very successful in allowing the efficient calculation of three-dimensional macromolecular solution structures, but has tended to obscure the fact that the fundamental physical parameters under study in NOE spectroscopy are the cross-relaxation rates R_{ij}, which are related both to the structure and to the dynamical properties of the molecule. The interpretation in terms of distances alone is facilitated by the favorable error-propagation properties of the sixth-root dependence of distances on cross-relaxation rates (Section 2.1); however, the interpretation of NOE data in terms of cross-relaxation rates does not depend upon any given motional model, and unambiguous cross-relaxation rates can be calculated if multispin effects and the (not trivial) problems of spectral interpretation and quantification are overcome. The procedures described in this Chapter for overcoming multispin effects have opened the possibility of deriving reasonably accurate cross-relaxation rates in macromolecules; in some cases, these rates will contain new information on the nature of the internal dynamics in the system of interest (Koehl and Lefèvre, 1990).

Even when it is desired to interpret NOE studies in terms of average single-conformer distances, however, the complete neglect of multispin effects for structure determination is not justified. The use of very broad distance ranges does not guarantee that distance underestimation due to severe spin-diffusion will not cause the distance bounds to be inaccurate in the sense of not containing the "true average" distance. A variety of studies, summarized in Section 4, have concluded that the neglect of multispin effects can indeed cause distortions in calculated structures, and a variety of calculational and experimental procedures have been shown to be viable tools for removing these errors. Magnetization exchange network editing is a particularly flexible tool, since different experiments may be applied in cases where individual resonances are suspected as sources of multispin effects, where individual interactions are of particular interest, or where nonselective improvement of large numbers of distances is desired. As discussed in Section 3.3, it is anticipated that the most accurate distances possible will be derived by a combined treatment of several types of data (MENE as well as unedited NOESY) with the next generation of matrix-refinement schemes. As useful as algorithms such as MARDIGRAS have proven, the most flexible way to take a variety of types of data into account will likely be some type of direct refinement. The best-developed current techniques are for direct refinement of three-dimensional structures; however, direct refinement of the cross-relaxation matrix seems to have untapped potential, particularly in cases in which the nature of internal motions is the question being investigated. Network-editing datasets, by "pinning" some elements of the cross-relaxation matrix very precisely, can be expected to dramatically improve the convergence of those algorithms optimized to take best advantage of them. With MENE pulse sequences now available that span the spectrum of editing schemes diagrammed in Figure 4, the next round of advances in network-editing applications is likely to come from the development of calculational tools that take advantage of the improved data, rather than in the development of additional, radically novel pulse sequences.

ACKNOWLEDGMENTS

The authors are grateful to Dr. William M. Westler (NMRFAM) and Prof. Slobodan Macura (Mayo Foundation) for stimulating discussions. The CORMA software used to derive Figure 7 was supplied by the laboratory of Prof. Thomas L. James (UCSF). This work was carried out at the National Magnetic Resonance Facility at Madison with support from NIH grants RR02301 and GM35976. Equipment in the NMR Facility was purchased with funds from the NIH National Center for Research Resources, Biomedical Research Technology Program (grant RR02301), the University of Wisconsin, the NSF Biological Instrumentation Program (grant DMB-8415048), the NIH Shared Instrumentation Program (grants RR02781 and RR08438), the NSF Academic Infrastructure Program (grant BIR-0214394), and the U.S. Department of Agriculture. C.G.H. was supported by a Predoctoral Fellowship from the Howard Hughes Medical Institute.

REFERENCES

Anil Kumar, Wagner, G., Ernst, R.R. and Wüthrich, K. (1981). Buildup rates of the nuclear Overhauser effect measured by two-dimensional proton magnetic resonance spectroscopy : Implications for studies of protein conformation. *J. Am. Chem. Soc. 103*, 3654-3658.

Baleja, J.D., Moult, J. and Sykes, B.D. (1990). Distance measurement and structure refinement with NOE data. *J. Magn. Reson. 87*, 375-384.

Baleja, J.D. and Sykes, B.D. (1991). Correlation time adjustment factors for NOE-based structure refinement. *J. Magn. Reson. 91*, 624-629.

Bax, A. and Davis, D.G. (1985). Practical aspects of two-dimensional transverse NOE spectroscopy. *J. Magn. Reson. 63*, 207-213.

Bloembergen, N., Purcell, E.M. and Pound, R.V. (1948). Relaxation effects in nuclear magnetic resonance absorption. *Phys. Rev. 73*, 679-712.

Boelens, R., Koning, T.M.G. and Kaptein, R. (1988). Determination of biomolecular structures from proton-proton NOE's using a relaxation matrix approach. *J. Mol. Struct. 173*, 299-311.

Boelens, R., Koning, T.M.G., van der Marel, G.A., van Boom, J.H. and Kaptein, R. (1989a). Iterative procedure for structure determination from proton-proton NOEs using a full relaxation matrix approach. Application to a DNA octamer. *J. Magn. Reson. 82*, 290-308.

Boelens, R., Vuister, G.W., Koning, T.M.G. and Kaptein, R. (1989b). Observation of spin diffusion in biomolecules by three-dimensional NOE-NOE spectroscopy. *J. Am. Chem. Soc. 111*, 8525-8526.

Bonvin, A.M.J.J., Boelens, R. and Kaptein, R. (1991a). Direct structure refinement using 3D NOE-NOE spectra of biomolecules. *J. Magn. Reson. 95*, 626-631.

Bonvin, A.M.J.J., Boelens, R. and Kaptein, R. (1991b). Direct NOE refinement of biomolecular structures using 2D NMR data. *J. Biomolec. NMR 105*, 305-309.

Bonvin, A.M.J.J., Vis, H., Breg, J.N., Burgering, M.J.M., Boelens, R. and Kaptein, R. (1994). NMR solution structure of the Arc repressor using relaxation matrix calculations. *J. Mol. Biol. 236*, 328-341.

Borgias, B.A., Gochin, M., Kerwood, D.J. and James, T.L. (1990). Relaxation matrix analysis of 2D NMR data. *Prog. NMR Spect. 22*, 83-100.

Borgias, B.A. and James, T.L. (1990). MARDIGRAS—A procedure for matrix analysis of relaxation for discerning geometry of an aqueous structure. *J. Magn. Reson. 87*, 475-487.

Bothner-By, A.A., Stephens, R.L. and Lee, J. (1984). Structure determination of a tetrasaccharide: Transient nuclear Overhauser effects in the rotating frame. *J. Am. Chem. Soc. 106*, 811-813.

Boulat, B., Burghardt, I. and Bodenhausen, G. (1992). Measurement of Overhauser effects in magnetic resonance of proteins by synchronous nutation. *J. Am. Chem. Soc. 114*, 10679.

Brünger, A.T. 1992. *X-PLOR Version 3.1: A system for x-ray crystallography and NMR*. New Haven: Yale University Press.

Brünger, A.T. and Nilges, M. (1993). Computational challenges for macromolecular structure determination by X-ray crystallography and solution NMR-spectroscopy. *Quart. Rev. Biophys. 26*, 49-125.

Burghardt, I., Konrat, R., Boulat, B., Vincent, S.J.F. and Bodenhausen, G. (1993). Measurement of cross relaxation between two selected nuclei by synchronous nutation of magnetization in nuclear magnetic resonance. *J. Chem. Phys. 98*, 1721-1736.

Case, D.A., Dyson, H.J. and Wright, P.E. (1994). Use of chemical shifts and coupling constants in nuclear magnetic resonance structural studies on peptides and proteins. *Methods Enzymol. 239*, 392-416.

Cheong, C. and Moore, P.B. (1992). Solution structure of an unusually stable RNA tetraplex containing G- and U-quartet structures. *Biochemistry 31*, 8406-8414.

Clore, G.M., Robien, M.A. and Gronenborn, A.M. (1993). Exploring the limits of precision and accuracy of protein structures determined by nuclear magnetic resonance spectroscopy. *J. Mol. Biol. 231*, 82-102.

Clore, G.M. and Gronenborn, A.M. (1991). Structures of larger proteins in solution: Three- and four-dimensional heteronuclear NMR spectroscopy. *Science 252*, 1390-1399.

Davis, J.H., Bradley, E.K., Miljanich, G.P., Nadasdi, L., Ramachandran, J. and Basus, V.J. (1993). Solution structure of Ω-conotoxin GVIA using 2-D NMR spectroscopy and relaxation matrix analysis. *Biochemistry 32*, 7396-7405.

Dellwo, M.J. and Wand, J. (1993). Computationally efficient gradients for relaxation matrix-based structure refinement including the accommodation of internal motions. *J. Biomolec. NMR 3*, 205-214.

Edmondson, S., Khan, N., Shriver, J., Zdunek, J. and Gräslund, A. (1991). The solution structure of motilin from NMR distance constraints, distance geometry, molecular dynamics, and an iterative full relaxation matrix refinement. *Biochemistry 30*, 11271-11279.

Emsley, L. (1994). Selective pulses and their applications to assignment and structure determination in nuclear magnetic resonance. *Methods Enzymol. 239*, 207-246.

Farmer, B.T.,II, Macura, S. and Brown, L.R. (1987). Relay artifacts in ROESY spectra. *J. Magn. Reson. 72*, 347-352.

Farmer, B.T.,II, Macura, S. and Brown, L.R. (1988). The effect of molecular motion on cross relaxation in the laboratory and rotating frames. *J. Magn. Reson. 80*, 1-22.

Fejzo, J., Zolnai, Zs., Macura, S. and Markley, J.L. (1989). Analysis of laboratory-frame and rotating-frame cross-relaxation buildup rates from macromolecules. *J. Magn. Reson. 82*, 518-528.

Fejzo, J., Westler, W.M., Macura, S. and Markley, J.L. (1990). Elimination of cross-relaxation effects from two-dimensional chemical-exchange spectra of macromolecules. *J. Am. Chem. Soc. 112*, 2574-2577.

Fejzo, J., Krezel, A.M., Westler, W.M., Macura, S. and Markley, J.L. (1991a). Refinement of the NMR solution structure of a protein to remove distortions arising from neglect of internal motion. *Biochemistry 30*, 3807-3811.

Fejzo, J., Krezel, A.M., Westler, W.M., Macura, S. and Markley, J.L. (1991b). Direct cross-relaxation NOESY (D.NOESY). A method for removing spin-diffusion cross peaks from two-dimensional NOE spectra of macromolecules. *J. Magn. Reson. 92*, 651-657.

Fejzo, J., Westler, W.M., Macura, S. and Markley, J.L. (1991c). Strategies for eliminating unwanted cross-relaxation and coherence-transfer effects from two-dimensional chemical-exchange spectra. *J. Magn. Reson. 92*, 20-29.

Fejzo, J., Westler, W.M., Macura, S. and Markley, J.L. (1991d). Elimination of chemical-exchange-mediated spin diffusion from exchange spectra of macromolecules. Exchange-Decoupled NOESY (XD-NOESY). *J. Magn. Reson. 92*, 195-202.

Fejzo, J., Westler, W.M., Markley, J.L. and Macura, S. (1992). Complete elimination of spin diffusion from selected resonances in two-dimensional cross-relaxation spectra of macromolecules by a novel pulse sequence (SNOESY). *J. Am. Chem. Soc. 114*, 1523-1524.

Freeman, R. (1991). Selective Excitation in high-resolution NMR. *Chem. Rev. 91*, 1397-1412.

Fujinaga, M., Sielecki, A.R., Read, R.J., Ardelt, W., Laskowski, M.,Jr. and James, M.N.G. (1987). Crystal and molecular structures of the complex of α-chymotrypsin with its inhibitor turkey ovomucoid third domain at 1.8 Å resolution. *J. Mol. Biol. 195*, 397-418.

Geen, H. and Freeman, R. (1991). Band-selective radiofrequency pulses. *J. Magn. Reson. 93*, 93-141.

Griesinger, C. and Ernst, R.R. (1987). Frequency offset effects and their elimination in NMR rotating-frame cross-relaxation spectroscopy. *J. Magn. Reson. 75*, 261-271.

Grzesiek, S., Anglister, J., Ren, H. and Bax, A. (1993). ^{13}C line narrowing by ^{2}H decoupling in ^{2}H/^{13}C/^{15}N-enriched proteins. Application to triple resonance 4D J connectivity of sequential amides. *J. Am. Chem. Soc. 115*, 4369-4370.

Grzesiek, S., Wingfield, P., Stahl, S., Kaufman, J.D. and Bax, A. (1995). Four-dimensional ^{15}N-separated NOESY of slowly tumbling perdeuterated ^{15}N-enriched proteins. Application to HIV-1 NEF. *J. Am. Chem. Soc. 117*, 9594-9595.

Habazettl, J., Ross, A., Oschkinat, H. and Holak, T.A. (1992a). Secondary NOE pathways in 2D NOESY spectra of proteins estimated from homonuclear three-dimensional NOE-NOE nuclear magnetic resonance spectroscopy. *J. Magn. Reson. 97*, 511-521.

Habazettl, J., Schleicher, M., Otlewski, J. and Holak, T.A. (1992b). Homonuclear three-dimensional NOE-NOE nuclear magnetic resonance spectra for structure determination of proteins in solution. *J. Mol. Biol. 228*, 156-169.

Hajduk, P.J., Horita, D.A. and Lerner, L.E. (1993). Theoretical analysis of relaxation during shaped pulses. I. The effects of short T_1 and T_2. *J. Magn. Reson. A 103*, 40-52.

Holak, T.A., Habazettl, J., Oschkinat, H. and Otlewski, J. (1991). Structures of proteins in solution derived from homonuclear three-dimensional NOE-NOE nuclear magnetic resonance spectroscopy. High-resolution structure of squash trypsin inhibitor. *J. Am. Chem. Soc. 113*, 3196-3198.

Hoogstraten, C.G., Westler, W.M., Macura, S. and Markley, J.L. (1993). Improved measurement of longer proton-proton distances in proteins by relaxation network editing. *J. Magn. Reson. B 102*, 232-235.

Hoogstraten, C.G., Choe, S., Westler, W.M. and Markley, J.L. (1995a). Comparison of the accuracy of protein solution structures derived from conventional and network-edited NOESY data. *Protein Science* (In press).

Hoogstraten, C.G., Westler, W.M., Macura, S. and Markley, J.L. (1995b). NOE measurements in the absence of spin diffusion: Application to methylene groups in proteins and effects on local structural parameters. *J. Am. Chem. Soc. 117*, 5610-5611.

Hoogstraten, C.G., Westler, W.M., Mooberry, E.S., Macura, S. and Markley, J.L. (1995c). Analysis of transverse cross relaxation within a simplified relaxation submatrix: BD-ROESY. *J. Magn. Reson. B 109*, 76-79.

Hyberts, S.G. and Wagner, G. (1989). Taylor transformation of 2D NMR τ_m series from time dimension to polynomial dimension. *J. Magn. Reson. 81*, 418-422.

James, T.L. (1994). Assessment of quality of derived macromolecular structures. *Methods Enzymol. 239*, 416-439.

James, T.L. and Basus, V.J. (1991). Generation of high-resolution protein structures in solution from multidimensional NMR. *Ann. Rev. Phys. Chem. 42*, 501-542.

Jardetzky, O. 1991. Interpretation of NMR data in terms of protein structure: Summary of a round table discussion. In: Hoch, J.C., Poulsen, F.M. and Redfield, C., eds. *Computational aspects of the study of biological macromolecules by nuclear magnetic resonance spectroscopy*. New York: Plenum Press. pp 375-389.

Jeener, J., Meier, B.H., Bachmann, P. and Ernst, R.R. (1979). Investigation of exchange processes by two-dimensional NMR spectroscopy. *J. Chem. Phys. 71*, 4546-4553.

Kalk, A. and Berendsen, H.J.C. (1976). Proton magnetic relaxation and spin diffusion in proteins. *J. Magn. Reson. 24*, 343-366.

Kaluarachchi, K., Meadows, R.P. and Gorenstein, D.G. (1991). How accurately can oligonucloeotide structures be determined from the hybrid relaxation rate matrix/NOESY distance restrained molecular dynamics approach? *Biochemistry 30*, 8785-8797.

Keepers, J.W. and James, T.L. (1984). A theoretical study of distance determinations from NMR. Two-dimensional nuclear Overhauser effect spectra. *J. Magn. Reson. 57*, 404-426.

Kessler, H., Mronga, S. and Gemmecker, G. (1991). Multi-dimensional NMR experiments using selective pulses. *Magn. Reson. Chem. 29*, 527-557.

Koehl, P. and Lefèvre, J.F. (1990). The reconstruction of the relaxation matrix from an incomplete set of nuclear Overhauser effects. *J. Magn. Reson. 86*, 565-583.

Kominos, D., Suri, A.K., Kitchen, D.B., Bassolino, D. and Levy, R.M. (1992). Simulating the effect of the two-spin approximation on the generation of protein structures from NOE data. *J. Magn. Reson. 97*, 398-410.

Koning, T.M.G., Boelens, R., van der Marel, G.A., van Boom, J.H. and Kaptein, R. (1991). Structure determination of a DNA octamer in solution by NMR spectroscopy. Effect of fast local motions. *Biochemistry 30*, 3787-3797.

Krezel, A.M., Darba, P., Robertson, A.D., Fejzo, J., Macura, S. and Markley, J.L. (1994). Solution structure of turkey ovomucoid third domain as determined from nuclear magnetic resonance data. *J. Mol. Biol. 242*, 203-214.

LeMaster, D.M. (1990). Uniform and selective deuteration in two-dimensional NMR of proteins. *Annu. Rev. Biophys. Biophys. Chem. 19*, 243-266.

LeMaster, D.M. (1994). Isotope labeling in solution protein assignment and structural analysis. *Prog. NMR Spect. 26*, 371-419.

LeMaster, D.M. and Richards, F.M. (1988). NMR sequential assignment of Escherichia coli thioredoxin utilizing random fractional deuteriation. *Biochemistry 27*, 142-150.

Liu, H., Thomas, P.D. and James, T.L. (1992). Averaging of cross-relaxation rates and distances for methyl, methylene, and aromatic ring protons due to motion or overlap. Extraction of accurate distances iteratively via relaxation matrix analysis of 2D NOE spectra. *J. Magn. Reson. 98*, 163-175.

Liu, H., Anil Kumar, Borgias, B.A., Thomas, P.D. and James, T.L. 1994. *CORMA 5.0/MARDIGRAS 3.0 (UCSF), University of California, San Francisco.*

Macura, S., Farmer, B.T.,II and Brown, L.R. (1986). An improved method for the determination of cross-relaxation rates from NOE data. *J. Magn. Reson. 70*, 493-499.

Macura, S., Fejzo, J., Hoogstraten, C.G., Westler, W.M. and Markley, J.L. (1992). Topological editing of cross-relaxation networks. *Israel J. Chem. 32*, 245-256.

Macura, S. (1994). Evaluation of errors in 2D exchange spectroscopy. *J. Magn. Reson. B 104*, 168-171.

Macura, S., Westler, W.M. and Markley, J.L. (1994). Two-dimensional exchange spectroscopy of proteins. *Methods Enzymol. 239*, 106-144.

Macura, S. and Ernst, R.R. (1980). Elucidation of cross relaxation in liquids by two-dimensional N.M.R. spectroscopy. *Molec. Phys. 41*, 95-117.

Madrid, M., Llinás, E. and Llinás, M. (1991). Model-independent refinement of interproton distances generated from ^{1}H NMR Overhauser intensities. *J. Magn. Reson. 93*, 329-346.

Majumdar, A. and Hosur, R.V. (1989). . *Biochem. Biophys. Res. Comm. 159*, 886-892.

Marion, D., Ikura, M., Tschudin, R. and Bax, A. (1989). Rapid recording of 2D NMR spectra without phase cycling. Application to the study of hydrogen exchange in proteins. *J. Magn. Reson. 85*, 393-399.

Markley, J.L., Putter, I. and Jardetzky, O. (1968). High-resolution nuclear magnetic resonance spectra of selectivity deuterated Staphylococcal nuclease. *Science 161*, 1249-1251.

Markley, J.L. and Kainosho, M. 1993. Stable isotope labelling and resonance assignments in larger proteins. In: Roberts, G.C.K., ed. *NMR of macromolecules: A practical approach.* Oxford: Oxford University Press. pp 101-152.

Massefski, W.,Jr. and Redfield, A.G. (1988). Elimination of multiple-step spin diffusion effects in two-dimensional NOE spectroscopy of nucleic acids. *J. Magn. Reson. 78*, 150-155.

Mujeeb, A., Kerwin, S.M., Kenyon, G.L. and James, T.L. (1993). Solution structure of a conserved DNA sequence from the HIV-1 genome: Restrained molecular dynamics simulation with distance and torsion angle restraints derived from two-dimensional NMR spectra. *Biochemistry 32*, 13419-13431.

Neuhaus, D. and Williamson, M. 1989. *The nuclear Overhauser effect in structural and conformational analysis.* New York: VCH.

Nikonowicz, E.P., Meadows, R.P., Fagan, P. and Gorenstein, D.G. (1991). NMR structural refinement of a tandem GA mismatched decamer d(CCAAGATTGG)$_2$ via the hybrid matrix procedure. *Biochemistry 30*, 1323-1334.

Nilges, M., Habazettl, J., Brünger, A.T. and Holak, T.A. (1991). Relaxation matrix refinement of the solution structure of squash trypsin inhibitor. *J. Mol. Biol. 29*, 499-510.

Noggle, J.H. and Schirmer, R.E. 1971. *The nuclear Overhauser effect: Chemical applications.* New York: Academic Press.

Norris, D.G., Lüdemann, H. and Leibfritz, D. (1991). An analysis of the effects of short T$_2$ values on the hyperbolic-secant pulse. *J. Magn. Reson. 92*, 94-101.

Olejniczak, E.T., Gampe, R.T.,Jr. and Fesik, S.W. (1986). Accounting for spin diffusion in the analysis of 2D NOE data. *J. Magn. Reson. 67*, 28-41.

Pachter, R., Arrowsmith, C.H. and Jardetzky, O. (1992). The effect of selective deuteration on magnetization transfer in larger proteins. *J. Biomolec. NMR 2*, 183-194.

Press, W.H., Flannery, B.P., Teukolsky, S.A. and Vetterling, W.T. 1988. *Numerical recipes in C.* Cambridge: Cambridge University Press.

Schmitz, U., Sethson, I., Egan, W.M. and James, T.L. (1992). Solution structure of a DNA octamer containing the Pribnow box via restrained molecular dynamics simulation with distance and torsion angle constraints derived from two-dimensional nuclear magnetic resonance spectral fitting. *J. Mol. Biol. 227*, 510-531.

Schweitzer, B.I., Mikita, T., Kellogg, G.W., Gardner, K.H. and Beardsley, G.P. (1994). Solution structure of a DNA dodecamer containing the anti-neoplastic agent arabinosylcytosine: Combined use of NMR, restrained molecular dynamics, and full relaxation matrix refinement. *Biochemistry 33*, 11460-11475.

Silver, M.S., Joseph, R.I. and Hoult, D.I. (1985). Selective spin inversion in nuclear magnetic resonance and coherent optics through an exact solution of the Bloch-Riccati equation. *Phys. Rev. A 31*, 2753-2755.

Solomon, I. (1955). Relaxation processes in a system of two spins. *Phys. Rev. 99*, 559-565.

Thomas, P.D., Basus, V.J. and James, T.L. (1991). Protein solution structure determination using distances from two-dimensional nuclear Overhauser effect experiments: Effect of approximations on the accuracy of derived structures. *Proc. Natl. Acad. Sci. USA 88*, 1237-1241.

Tsang, P., Wright, P.E. and Rance, M. (1990). Specific deuteration strategy for enhancing direct nuclear Overhauser effects in high molecular weight complexes. *J. Am. Chem. Soc. 112*, 8183-8185.

Venters, R.A., Metzler, W.J., Spicer, L.D., Mueller, L. and Farmer, B.T.,II (1995). Use of $^{1}H_N$-$^{1}H_N$ NOEs to determine protein global folds in perdeuterated proteins. *J. Am. Chem. Soc. 117*, 9592-9593.

Wagner, G. (1990). NMR investigations of protein structure. *Prog. NMR Spect. 22*, 101-139.

White, S.A., Nilges, M., Huang, A., Brünger, A.T. and Moore, P.B. (1992). NMR analysis of helix I from the 5S RNA of Escherichia coli. *Biochemistry 31*, 1610-1621.

Wijmenga, S.S., Mooren, M.M.W. and Hilbers, C.W. 1993. NMR of nucleic acids; from spectrum to structure. In: Roberts, G.C.K., ed. *NMR of macromolecules: A practical approach*. Oxford: Oxford Univ. Press. pp 217-288.

Wüthrich, K. 1986. *NMR of proteins and nucleic acids*. New York: Wiley.

Yamazaki, T., Lee, W., Arrowsmith, C.H., Muhandiram, D.R. and Kay, L.E. (1994a). A suite of triple resonance NMR experiments for the backbone assignment of ^{15}N, ^{13}C, ^{2}H labeled proteins with high sensitivity. *J. Am. Chem. Soc. 116*, 11655-11666.

Yamazaki, T., Lee, W., Revington, M, Dahlquist, F.W., Arrowsmith, C.H. and Kay, L.E. (1994b). An HNCA pulse scheme for the backbone assignment of ^{15}N,^{13}C,^{2}H-labeled proteins: Application to a 37-kDa Trp repressor-DNA complex. *J. Am. Chem. Soc. 116*, 6464-6465.

Yip, P. and Case, D.A. (1989). A new method for refinement of macromolecular structures based on nuclear Overhauser effect spectra. *J. Magn. Reson. 83*, 643-648.

Yip, P.F. (1993). A computationally efficient method for evaluating the gradient of 2D NOESY intensities. *J. Biomolec. NMR 3*, 361-365.

Zhao, D. and Jardetzky, O. (1994). An assessment of the precision and accuracy of protein structures determined by NMR: Dependence on distance errors. *J. Mol. Biol. 239*, 601-607.

Zolnai, Zs., Juranic, N., Markley, J.L. and Macura, S. (1995). Magnetization exchange network editing: Mathematical principles and experimental demonstration. *J. Chem. Phys.* (In press).

Zwahlen, C., Vincent, S.J.F., Di Bari, L., Levitt, M.H. and Bodenhausen, G. (1994). Quenching spin diffusion in selective measurements of transient Overhauser effects in nuclear magnetic resonance. Applications to oligonucleotides. *J. Am. Chem. Soc. 116*, 362-368.

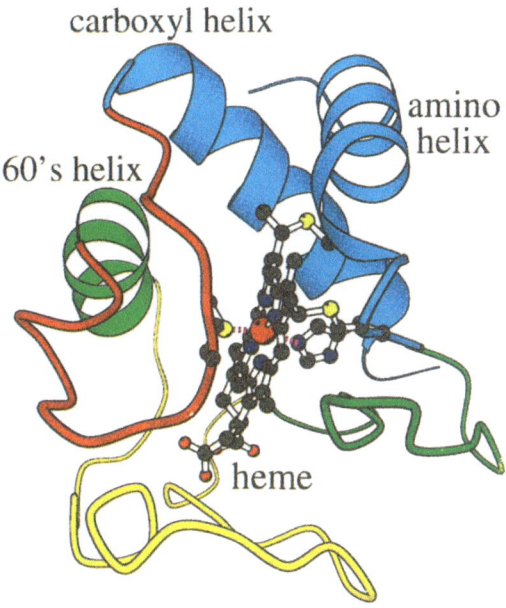

Figure 1.1. The cooperative substructural units that compose the cytochrome c protein, obtained from the native state hydrogen exchange experiment. The spectral color code keys to the free energy of the unfolding reactions (in order of red, yellow, green, blue).

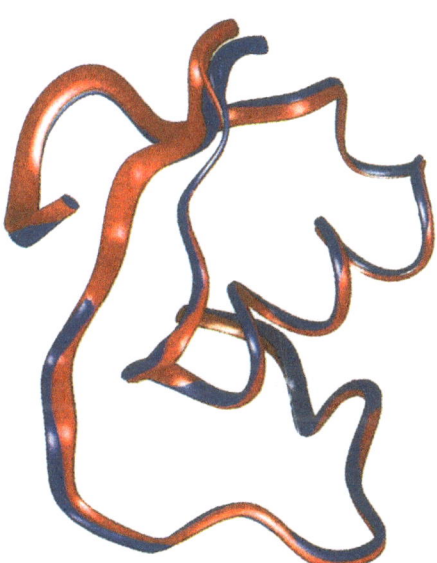

Figure 8.10. Average structures of OMTKY3 from NOESY and MENE data. The polypeptide backbones for residues 7-56 of unminimized average structures calculated from standard NOESY data (blue) and NOESY supplemented with BD-NOESY and CBD-NOESY data (red) are compared. For calculation of average structures, structures within a set were superimposed by reference to the backbone atoms for residues 7 through 56; the two average structures were superimposed in the same way for comparison. The beta turn in which the two structure sets significantly diverged is at the top of the figure; the changes in the reactive-site loop involve an expansion rather than a displacement of the set of structures (see text) and are not clearly visible in the perspective shown here. The widths of the ribbons are proportional to the backbone pairwise rmsd for the appropriate set of structures. Each calculation was based on fifty converged coordinate sets. The figure was prepared using the INSIGHT-II 2.3.5 software package (Biosym Technologies, Inc., San Diego). Figure reproduced from Hoogstraten et al. (1995a); used by permission.

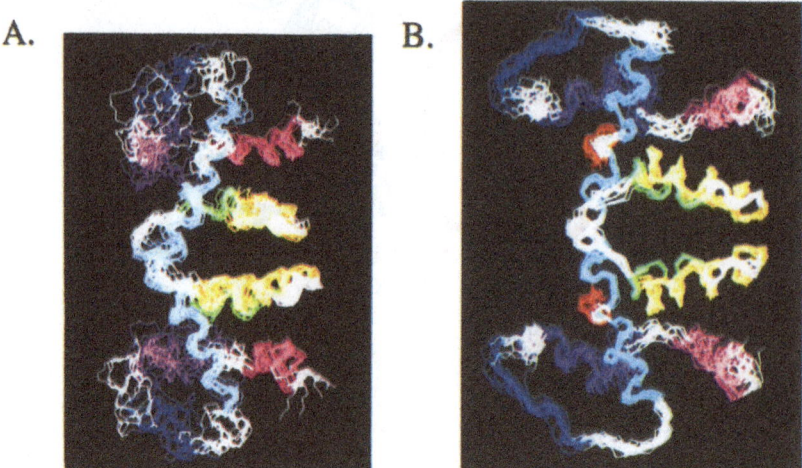

Figure 15.1. A family of 10 solution structures of (A) aporepressor and (B) holorepressor obtained from calculations using the XPLOR program: only the backbone atoms are displayed. The 6 heleces (A,B,C,D,E,F) are displayed in yellow, green, light blue, dark blue, purple, and pink, respectively. The L-trp ligands are indicated in red.

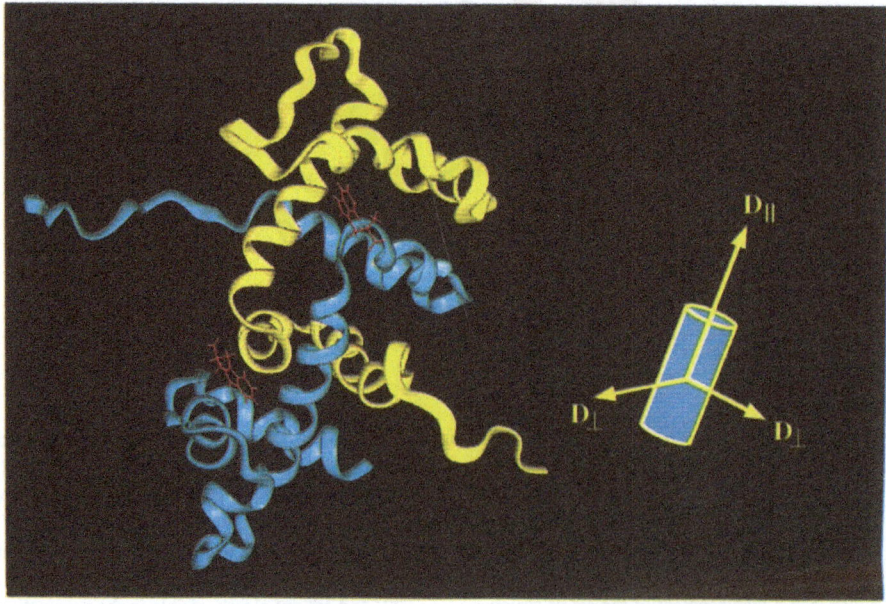

Figure 15.3. Ribbon diagram of one of the *trp*-repressor solution structures (Zhao et al., 1993). The axes of the axially symmetric rotational diffusion tensor (D∥ and D) are given schematically in a cylindrical diagram. The unique axis of the diffusion tensor is oriented closely along the two C helices of the dimeric molecule.

9

PROTEIN DYNAMICS

From the Native to the Unfolded State and Back Again

Martin Karplus, Amedeo Caflisch, Andrej Šali, and Eugene Shakhnovich

Department of Chemistry
Harvard University
12 Oxford Street, Cambridge, Massachusetts 02138

ABSTRACT

Simulations to study protein unfolding and folding were performed. The unfolding simulations make use of molecular dynamics and treat an atomic model of barnase in aqueous solvent. The cooperative nature of the unfolding transition and the important role of water are described. The folding calculation are based on a bead model of the protein on a cubic lattice. It is shown for the 27-mer model that a large energy gap between the lowest energy (native) state and the excited states is a necessary and sufficient condition for fast folding.

I. INTRODUCTION

The dynamics of proteins includes a wide range of length and time scales (Karplus & Shakhnovich, 1992). Figure 1 shows a schematic diagram of the energy of a protein as a function of a configurational coordinate, such as the radius of gyration. There is the native state, which includes fluctuations of up to 2 Å in length on a time scale for most fundamental events in the picosecond to nanosecond range. Many molecular dynamics simulations have explored this region of conformational space (Brooks *et al.*, 1988), which is of importance for protein function. A second region involves the transition from the native state to a compact globule, which may be an intermediate of the "molten globule" type that is of great current interest (Ptitsyn, 1992). Here the length scale is in the range 2 to 5 Å and the time scale of the motions involved in the nanosecond to microsecond range. Finally, there are the vast number of configurations of the denatured (coil) state and the folding process from the coil to the organized compact globule. In this region the length scales are on the order of 10 to 20 Å and the time scales are in the microsecond to millisecond range. This paper briefly reviews some recent studies that explore the native to globule transition in barnase (Caflisch & Karplus, 1994a,b) and makes use of a lattice model to study the folding process from the denatured to the organized globule state (Sali *et al.*, 1994a,b).

Dynamics and the Problem of Recognition in Biological Macromolecules, edited by Jardetzky and Lefèvre
Plenum Press, New York, 1996

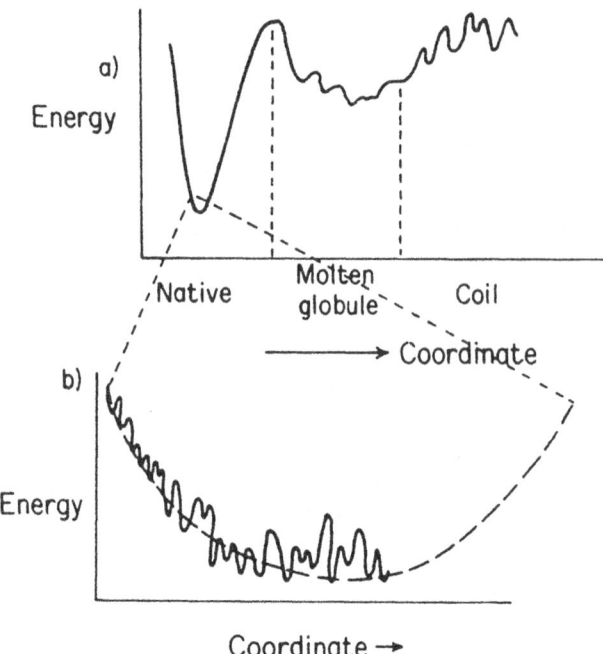

Figure 1. Schematic representation of the configuration space of a protein giving the energy as a function of a configurational coordinate: (a) complete space; (b) enlarged view in the vicinity of the native state.

II. BARNASE DENATURATION

Molecular dynamics simulations of the initial stages of unfolding of barnase (a 110 amino acid ribonuclease from *Bacillus amyloliquefaciens*) have been made at high temperature in the presence of water (Caflisch & Karplus, 1994a,b). This protein is a particularly good system for such studies because transition states and pathways of barnase folding and unfolding have been investigated by protein engineering and NMR hydrogen-exchange trapping experiments (Fersht, 1993). The rate determining step for both folding and unfolding involves the crossing of a free energy barrier near the native state (Serrano *et al.*, 1992a,b). Figure 2 shows a schematic drawing of the barnase structural elements. The present simulations provide a detailed mechanism for solvent denaturation of the secondary structure and the hydrophobic cores.

The barnase denaturation simulations used a deformable boundary potential (Serrano *et al.*, 1992b; Brooks & Karplus, 1983) and standard molecular dynamics methodology (Brooks *et al.*, 1988). The system consisted of 1091 protein atoms and 3003 water molecules in a sphere of 30 Å radius. Two denaturation simulations were performed at 600 K (A600, 120 ps; R600, 230 ps) and a 300 K control trajectory was run for 250 ps. R600 was recently continued to 250 ps; in addition, a third simulation at 600 K (200 ps) was performed and analyzed (Caflisch & Karplus, to be submitted). The denaturation process is similar in the three 600 K simulations. The 600 K temperature was used to speed up the unfolding transition. An increase in the unfolding rate by many orders of magnitude should result since the activation energy is expected to be large; the activation free energy for unfolding is 20 kcal/mol (Matouschek *et al.*, 1990).

The radius of gyration (R_g) and heavy-atom root-mean-square deviation (RMSD) from the x-ray structure as a function of time are given in Fig. 3. In A600 and R600, R_g starts

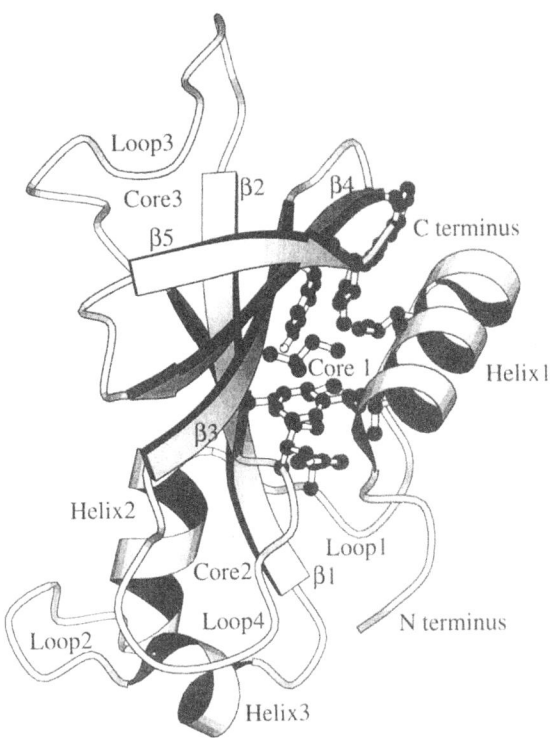

Figure 2. Schematic picture of the backbone of barnase, emphasizing the secondary structural elements; sidechains of hydrophobic core₁ are plotted in a ball and stick representation. The structural elements include the following residues: N-terminus (1-5), helix₁ (6-18), loop₁ (19-25), helix₂ (26-34), loop₂ (35-40), helix₃ (41-45), type II ß-turn (46-49), strand₁ (50-55), loop₃ (56-69), strand₂ (70-75), loop₄ (76-84), strand₃ (85-90), type I ß-turn (91-94), strand₄ (95-100), type III' ß-turn (101-104), strand₅ (105-108), C-terminus (109-110).

to increase after 30 ps, while the RMSD increases immediately. Both R_g and RMSD then increase over most of the simulation. However, the increase is not uniform; e.g. in A600, R_g is nearly constant for 20 ps between 45 and 65 ps; this may be indicative of an intermediate (Mark & van Gunsteren, 1992). In the control simulation at 300 K, R_g shows a very small increase (the average R_g is 13.7 Å, relative to the x-ray value of 13.6 Å); the RMSD from the x-ray structure is 1.9 Å (the mainchain atom RMSD is 1.5 Å) during the last 50 ps.

In the A600 simulation (120 ps) and the first half (115 ps) of the R600 simulation, there are similar structural changes. The N-terminus, loop₁ and loop₂ begin to unfold during the first 30 ps. This is followed by partial denaturation of the hydrophobic cores; core₂ denatures relatively rapidly, followed by core₁, core₃ and loop₃ in both 600 K simulations. The solvation of hydrophobic core₁ is coupled with a large distortion of helix₁ and of the edge strands of the ß-sheet. Both helix₁ and helix₂ lose about half of the native α-helical hydrogen bonds; helix₃ unfolds after about 20 ps. In the ß-sheet, about half of the native interstrand hydrogen bonds have disappeared after 100 ps; in R600 the ß-sheet is fully solvated after 150 ps. During the last 50 ps of R600, the mainchain still shows essentially the same overall fold as in the native structure, although the polypeptide chain is almost fully solvated and all of the secondary structure is lost except for the last two turns of helix₁.

Core₁, which is an important stabilizing element of barnase (Fersht, 1993; Serrano *et al.*, 1992), is formed by the packing of helix₁ against the β-sheet and is centered around

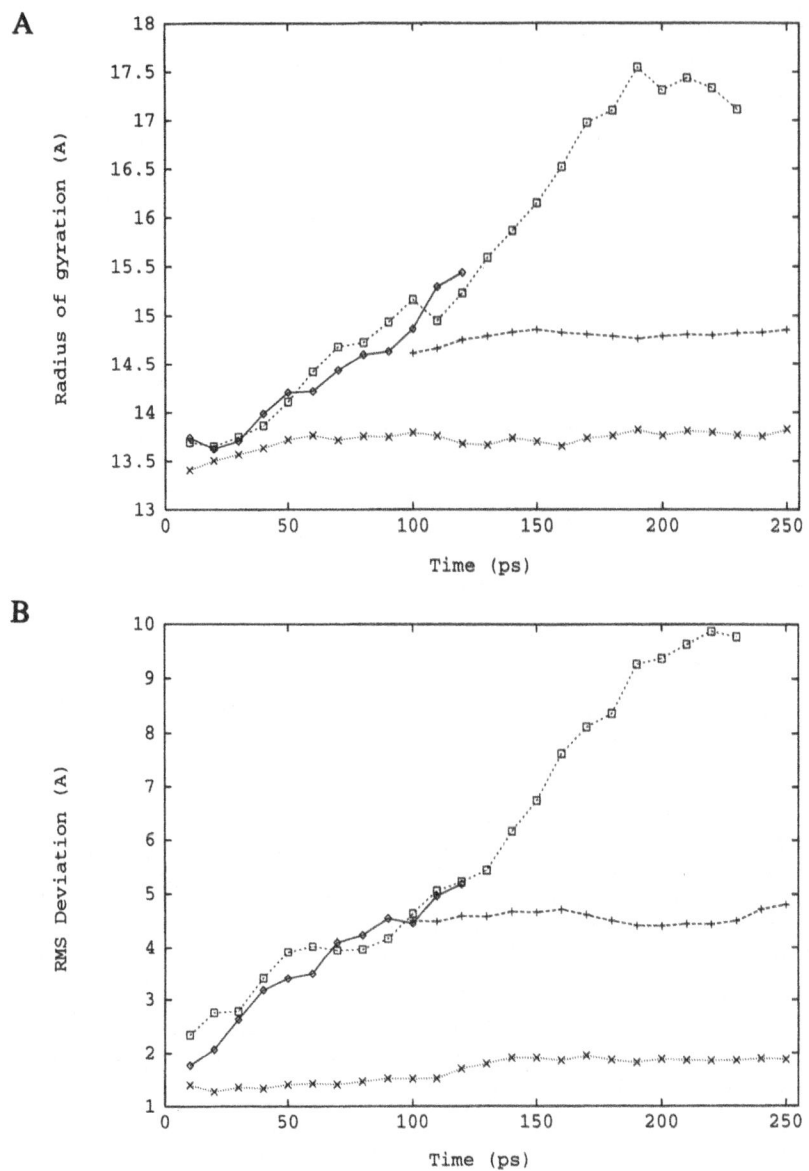

Figure 3. (a) R_g as a function of simulation time averaged over 10-ps intervals. (b) RMSD from the x-ray structure as a function of simulation time averaged over 10-ps intervals: (——) A600, (- - - - -) B300, (- - -) R600, (·····) control run at 300 K.

the sidechain of Ile 88 (Fig. 2). Figure 4 shows the time dependence in A600 of the solvent accessible surface area of the sidechains of $core_1$ and the number of water molecules in the core; similar behavior is seen during the first half of the R600 trajectory. Increase in accessible surface area and water penetration are nearly simultaneous and begin at about 35 ps. Sixteen water molecules have penetrated at 82 ps; this falls to 10 between 89-98 ps and increases to 17 in the period from 111 to 120 ps. Many of the solvating waters make hydrogen bonds to waters outside the core. The accessibility of $core_1$ to water is coupled with the relative motion of $helix_1$ and the ß-sheet (see Fig. 5); i.e., they begin to move apart at about 30 ps and their separation is continuous during the 30-80 ps period; between 80 and 100 ps

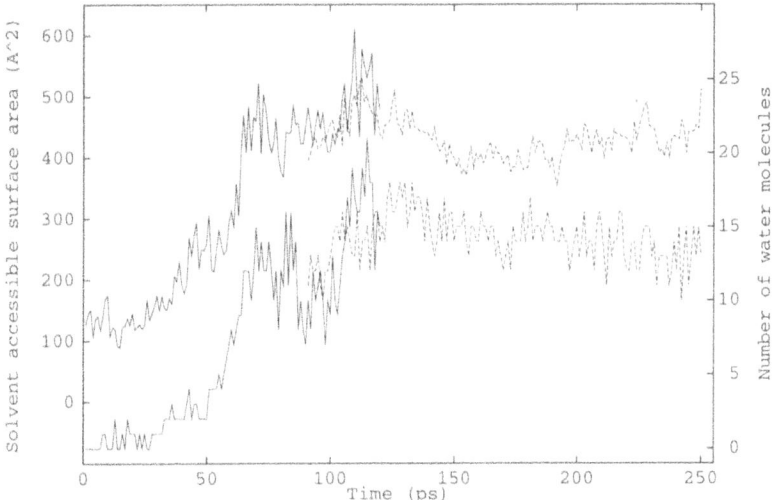

Figure 4. Solvent accessible surface area and number of water molecules for core$_1$ as a function of time. (——) A600; (- - - -) B300. For the exposed surface area, shown in the upper curve with the scale in Å^2 on the left, the Lee and Richards algorithm (CHARMM implementation) and a probe sphere of 1.4 Å radius were utilized. For the number of water molecules, shown in the lower curve with the scale on the right, those within 7 Å of the center of the core (the instantaneous center of geometry of the carbon atoms of the sidechains of residues Phe 7, Val 10, Ala 11, Leu 14, Leu 20, Tyr 24, Ala 74, Ile 76, Ile 88, Tyr 90, Trp 94, Ile 96, Ile 109) were included.

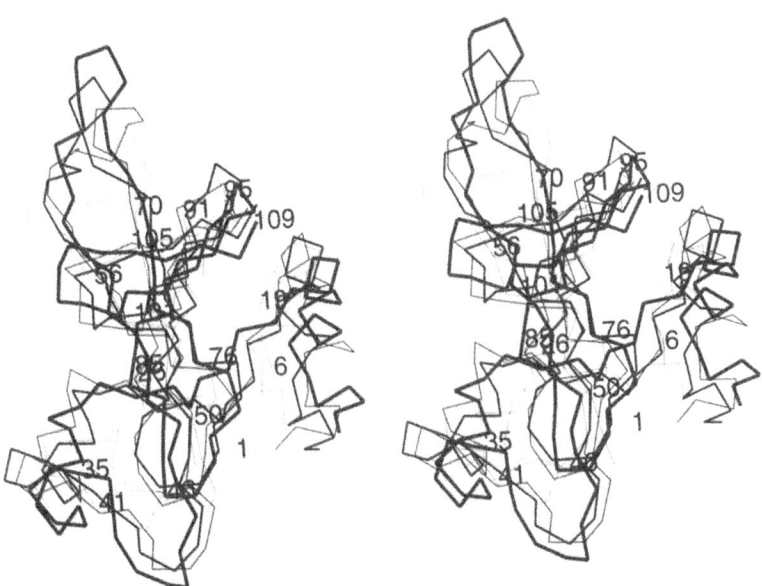

Figure 5. Stereo view of the barnase C$_\alpha$ atoms to illustrate the relative helix$_1$/ß-sheet motion during A600. 1 ps (thin line and labels), 70 ps (medium line), 90 ps (thicker line), 120 ps (thick lines).

A.

B.

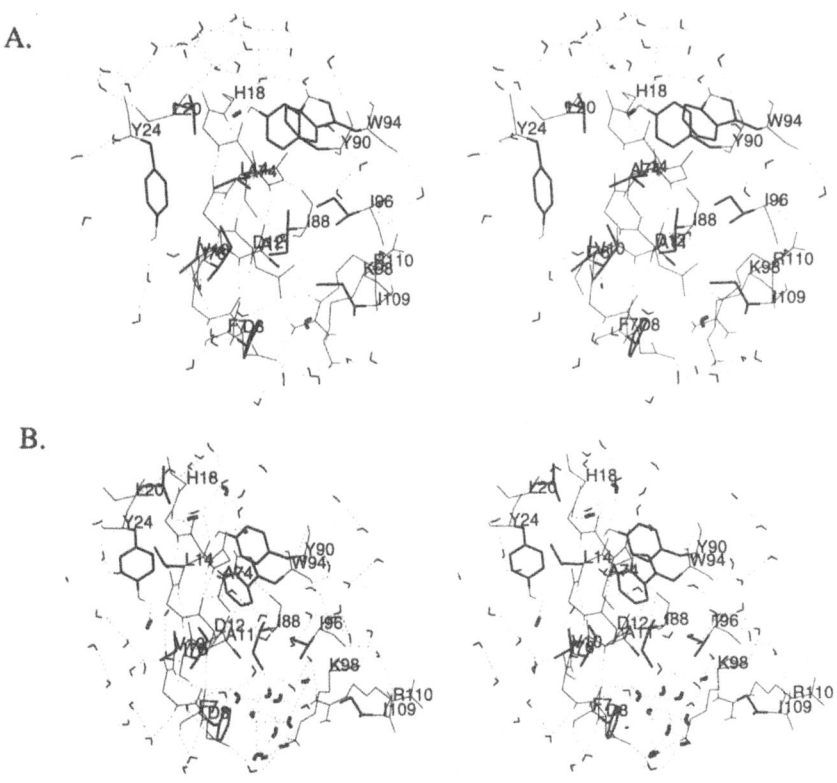

Figure 6. Stereo views of A600 dynamics: (a) 9 ps, (b) 39 ps; (on opposite page) (c) 69 ps, (d) 115 ps. Hydrophobic sidechains of core₁ are shown as thick lines; α-helix backbone (residues 7-18), Asp 8, Asp 12, Lys 98, and Arg 110 sidechains are shown as thin lines. Water molecules within 12 Å of the center of geometry of core₁ are included. Hydrogen bonds are dotted (acceptor-hydrogen distance smaller than 2.5 Å; no angular criterion). Waters discussed in text are shown with thick lines.

there is a small closing movement in accord with the decrease of water in the core, followed by expansion for the remainder of the simulation. During the B300 simulation the number of water molecules in core₁ and the solvent accessible surface area of its sidechains are nearly constant (see Fig. 4); the average number of water molecules is 14. This steady state of solvation of the core is correlated with the nearly constant number of hydrogen bonds in helix₁ and the central part of the ß-sheet (strands 2-3 and 3-4).

A number of detailed results of the present analysis are of interest because they may play a role in protein denaturation, in general. The polar OH and NH groups of tyrosines and tryptophan sidechains, respectively, play an important role in the penetration of water (from the top part of core₁), while the motion of a lysine sidechain helps the water molecules reaching the center of core₁ from the bottom part (Fig. 6a-b). Clusters of hydrogen bonded water molecules surrounding hydrophobic sidechains participate in hydrogen bonds with polar groups of the backbone and/or of the sidechains (Fig. 6c-d). The β-sheet disruption starts near the irregular element (β-bulge at residues 53-54) and at the edges (strands 1 and 5); it is promoted by an increase in the twist and an influx of water molecules, some of which insert between adjacent strands and participate in hydrogen bonds as both donors and acceptors with the mainchain polar groups (Fig. 7). Water molecules act mainly as hydrogen

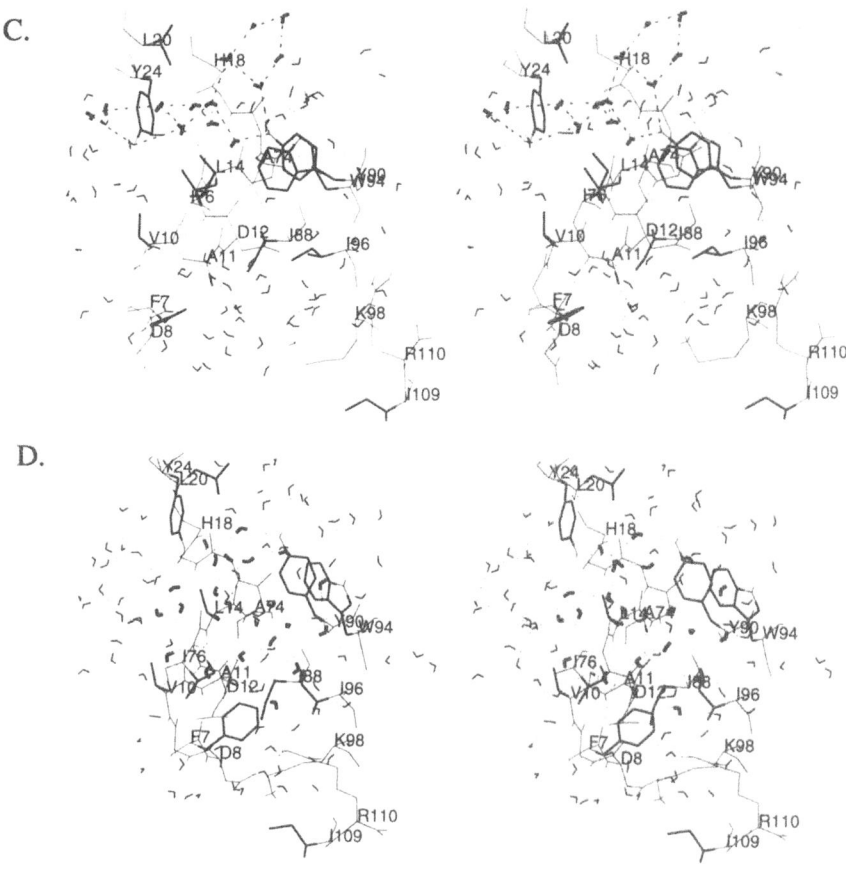

Figure 6. (Continued.)

bonding donors in the initial phase of solvation of the main helix; sometimes they insert and replace the helical hydrogen bond, as previously found in the fragment studies (Tirado-Rives & Jorgensen, 1991; Tobias & Brooks, 1991).

The present results suggest a possible mechanism for the solvation of hydrophobic cores and for the dissolution of secondary structural elements in protein denaturation. Testing of this mechanism is a challenge for experimentalists. Techniques that make use of photo-CIDNP (Kaptein *et al.*, 1978), NOE's of water interactions with specific residues (Otting & Wüthrich, 1989) and chemical markers (Ghelis, 1980) are possible approaches.

III. LATTICE MODEL FOR FOLDING AND ITS IMPLICATIONS

The essential question is how a polypeptide chain is able to fold rapidly, in ms to s, to the native state despite the very large number of conformations that exist for the denatured chain (Levinthal paradox) (Levinthal, 1969). To approach this problem, we use a simplified model that consists of a 27-bead self-avoiding chain on a cubic lattice (Fig. 8). The native (lowest-energy) state can be determined exactly (Shakhnovich *et al.*, 1991) and a survey of the folding behavior of many sequences is possible (Sali *et al.*, 1994a,b). In addition, the full phase space density of the system can be obtained and the thermodynamic properties can be

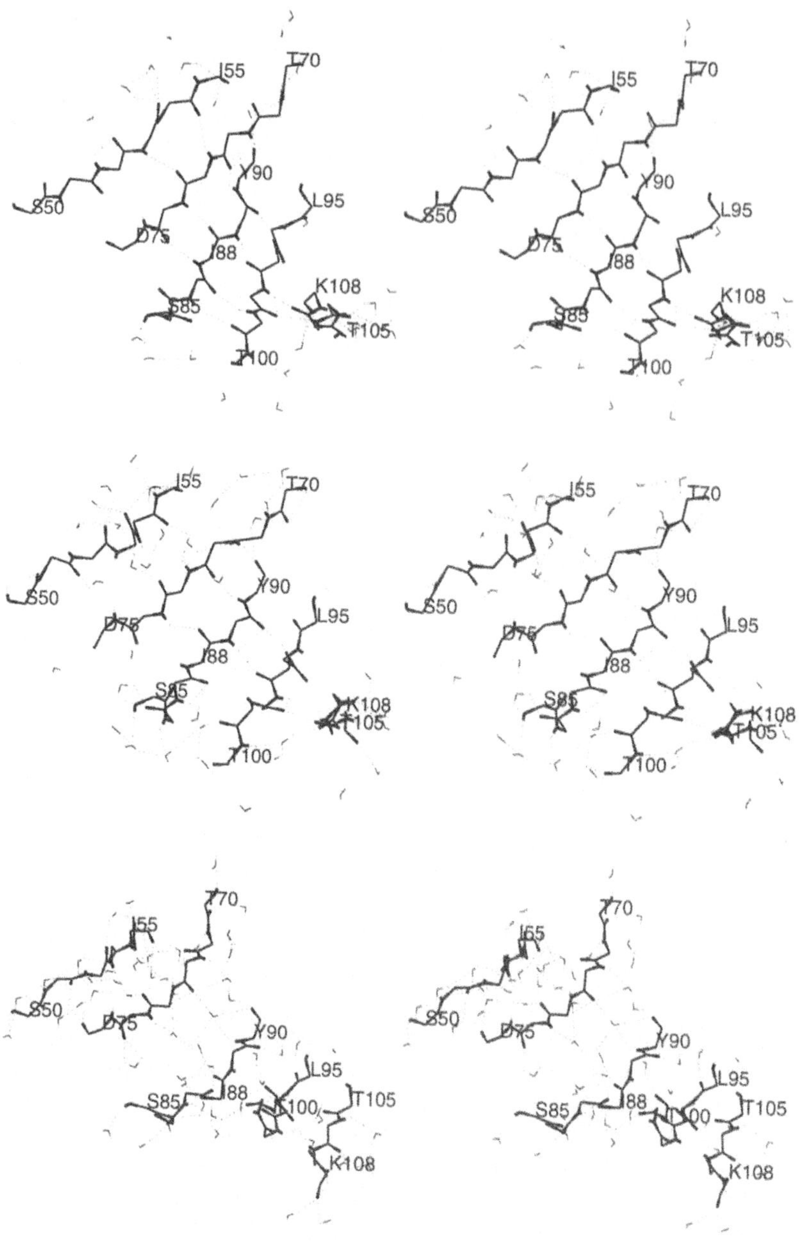

Figure 7. Stereo views of water penetration into the ß-sheet during R600. Mainchain N and O atoms are thick, hydrogens are thin and hydrogen bonds are dotted; water molecules within 3 Å of any mainchain atom of the ß-sheet are shown. Top, 1 ps; middle, 60 ps; bottom 150 ps.

calculated as a function of the folding reaction coordinate, which is defined as the fraction of native contacts (out of a total of 28). The model is sufficiently complex that the resolution of the Levinthal paradox is required for folding; i.e., some sequences find the native state in only ~10^7 Monte Carlo (MC) steps even though there are ~10^{16} conformations. Since the lattice model does not include the amino acid sidechains, the process considered here may

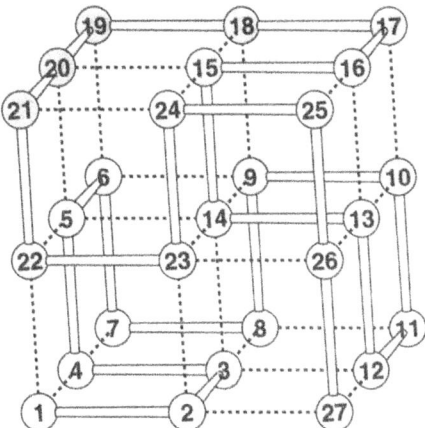

Figure 8. Lattice model of protein folding. An example of a compact self-avoiding structure of a chain of 27 monomers (filled numbered circles) with 28 contacts (dashed lines). The total energy of a conformation is the sum of contact energies: $E = \Sigma_{i<j} \Delta(r_i, r_j)B_{ij}$, where r_i are the positions of monomers i, B_{ij} are the contact energies for pairs of monomers i, j, and $\Delta(r_i, r_j)$ is 1 if monomers i and j are in contact and is 0 otherwise; two monomers are in contact if they are not successive in sequence and at unit distance from each other. The values of the B_{ij} are obtained from a Gaussian distribution with a mean B_0 and standard deviation σ_B. The parameter B_0 is an overall attractive term that emulates the hydrophobic effect observed in globular proteins. The *native* conformation is the compact self-avoiding chain with the lowest energy.

correspond to the folding of real proteins to the molten globule stage; i.e., the molten globule, if it has a defined fold (Ptitsyn, 1992), is the native state in the present model.

In the first part of the analysis, 200 sequences with random interactions were generated and subjected to MC folding simulations (Sali *et al.*, 1994a). Of these, 30 chains found the known native state in a short time. These chains correspond to actual protein sequences in the present model; the remaining sequences, which do not fold, do not correspond to protein sequences and serve as controls. The 30 folding sequences were analyzed and compared with the non-folding sequences. Several suggested mechanisms for resolving the Levinthal paradox do not apply to the present model; i.e., the features assumed to be responsible for rapid folding are found to be the same for the folding and non-folding sequences. These include a high number of short versus long range contacts in the native state (Wetlaufer, 1973), a high content of secondary structure in the native state (Kim & Baldwin, 1990), a strong correlation between the native contact map and the interaction parameters (Go & Abe, 1981), and the existence of a high number of low energy states with near-native conformations (Shakhnovich *et al.*, 1991). Moreover, there is no repetitive trapping of the non-folding sequences in the same local minimum, so that the native state cannot be a metastable state (Honeycutt & Thirumalai, 1992). The only significant difference is that the native state is at a pronounced energy minimum. As can be seen in Fig. 9, there is a large energy gap between the native and the "excited" states in the sequences that fold; no such gap is present in the nonfolding sequences. This energy gap is the necessary and sufficient condition for a sequence to fold rapidly in the present model.

The time history of the folding process (Fig. 10) shows that there is a rapid collapse in $\sim 10^4$ MC steps to a semi-compact random globule; i.e., the number of contacts increases, the energy decreases, while the fraction of native contacts remains below 0.3. In this way, the total of $\sim 10^{16}$ random-coil conformations is reduced to $\sim 10^{10}$ random semi-compact globule states. The fast collapse results from a large energy gradient and the presence of

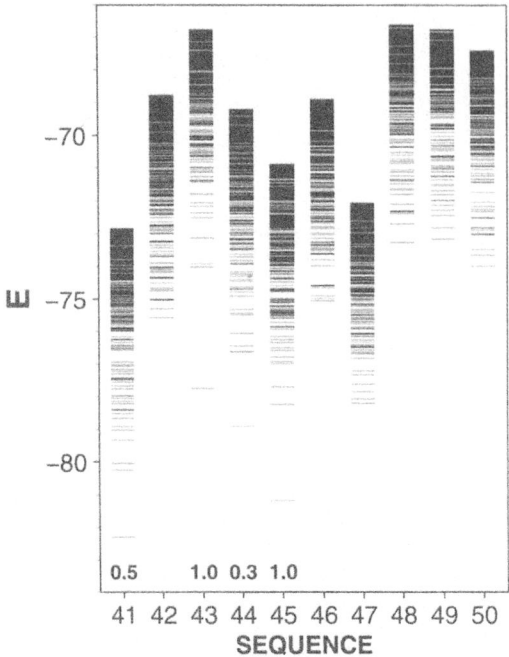

Figure 9. Energy spectra for 10 folding and nonfolding random sequences. The energies of the 400 lowest compact self-avoiding conformations are shown. The native state corresponds to the bottom bar. The numbers below the spectra show the probabilities that the corresponding sequence will fold under the conditions of the simulation (6); if no number is given, the probability is 0.

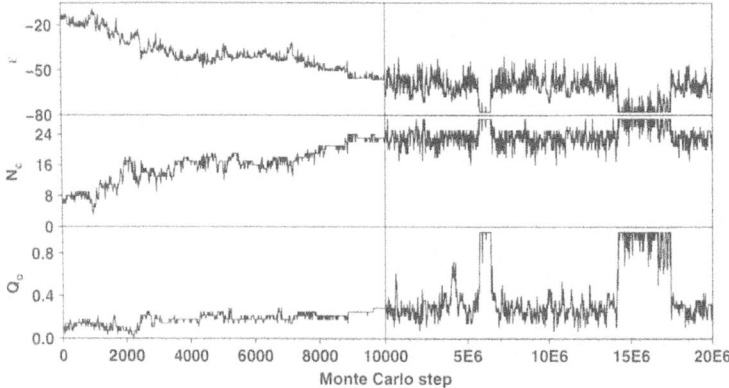

Figure 10. Typical trajectory for a folding sequence (T = 1.3). Energy, ε (in units of k_BT); the number of contacts, N_c; fraction of the number of contacts in common with the native state, Q_o. The instantaneous values of these quantities are plotted every 10 MC steps in the first part of the trajectory ($\leq$ 10000 MC steps) and every 20000 steps in the subsequent part. The folding trajectory starts with a random-coil conformation and consists of local MC moves of one or two successive monomers that preserve bond lengths and avoid multiple occupancy of the lattice sites (6).

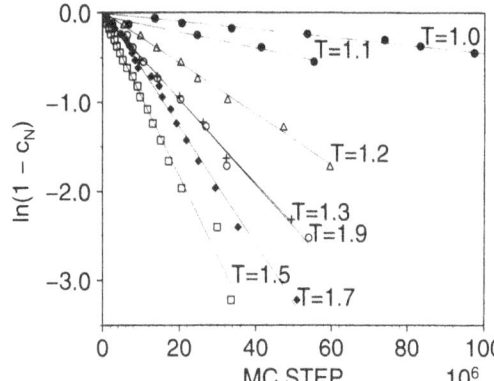

Figure 11. Distribution of the mean first passage times for a sample folding sequence. The distribution of the mean first passage times is shown at the temperatures indicated on the plot. The lines are the linear least-squares fits to the points. 100 independent folding trials were done to obtain the points shown.

many empty lattice points. In the second, rate limiting, stage the chain searches for one of the $\sim10^3$ transition states. The transition region consists of all states from which the chain folds rapidly to the native state. The transition states are structurally similar to the native state, with 23 to 26 of the native contacts. The mean first passage time, τ, for finding any of the n states among the total of N states by a random search which explores r states per unit of time is $N / (n \times r)$. For the present model, $\tau \sim 10^{10}/(10^3 \times 1) = 10^7$, identical to the observed time scale (Fig. 10). This indicates that the rate limiting stage in folding consists of a random search for a transition state in the semi-compact part of the phase space; i.e., a folding "pathway" is not involved in finding the native state. In the third stage, the chain rapidly (within $\sim10^5$ MC steps) attains the native conformation from any one of the transition states.

To examine the kinetics of the folding process, we write the unimolecular rate expression; i.e.,

$$\frac{dC_N}{dt} = k\, C_D$$

where C_N is the probability of finding the unique native state at time t, C_D is the probability of the denatured states (all other states; i.e., $C_D = 1-C_N$) and k is the rate coefficient. Since the folding process takes place on a complex multiminimum surface, it would be possible that k would not be the usual rate constant, but a function of time (Bai & Fayer, 1989). To determine the kinetic behavior, we show in Fig. 11 a plot of $\ln(1-C_N) = -kt$ versus the number of MC steps, which correspond to the time in the lattice model. The probability of finding the native state at time t is estimated from 100 independent folding simulations at several temperatures. It is calculated from the distribution of first passage times (i.e., when the native

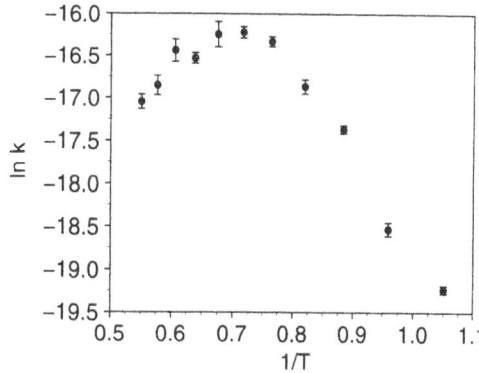

Figure 12. Temperature dependence of the folding rate constant for a sample folding sequence. This Arrhenius plot shows the rate constants at different temperatures that were obtained from Figure 11. The standard error of this estimate was found by the jacknife test.

state is reached for the first time). It is evident from the figure that the coefficient k is indeed a rate constant; i.e., it is independent of time. To examine the dependence of the folding rate on temperature, ln k is plotted versus 1 in Fig. 12. This plot shows a strong deviation from standard Arrhenius type behavior. For low T (0.90 < 1/T < 0.65) there is an increase in the folding rate, as expected from an activation energy limited process. However, at higher T (0.65 < 1/T < 0.50) the rate decreases with increasing temperature as expected from an activation entropy dominated process. This is in accord with analysis of the configuration space of the systems; i.e., at low temperatures, an increase in temperature makes it easier to get over the energy barriers which are rate limiting. However, at higher temperatures, a larger portion of the configuration space becomes accessible, which results in a slowing down of the folding process.

The pronounced energy minimum is the necessary condition for the folding of a 27-mer on a lattice because it guarantees that the native state is stable above the critical temperature, where the rearrangements with local unfolding required in the rate limiting stage are energetically possible. It is also a sufficient condition because a random search of compact globules with random structures can rapidly find a transition state that folds to the stable native state in a short time. While a non-folding sequence may also fold slowly to its native state above the critical temperature, it would not be stable and, therefore, could not correspond to a real protein.

The size of the search for the native state is greatly reduced when the chain is semi-compact as it is in real proteins (Dill, 1985). Nevertheless, in the 27-mer model as in real proteins, a random search of a collapsed globule cannot find the native state in the observed time. It is the existence of a transition region, consisting of a large number of states, that reduces the search time to realistic values when combined with the search of the random compact globule. Extrapolation of the folding time to real proteins suggests that the three-stage random search mechanism could be effective for small proteins (Fig. 13).

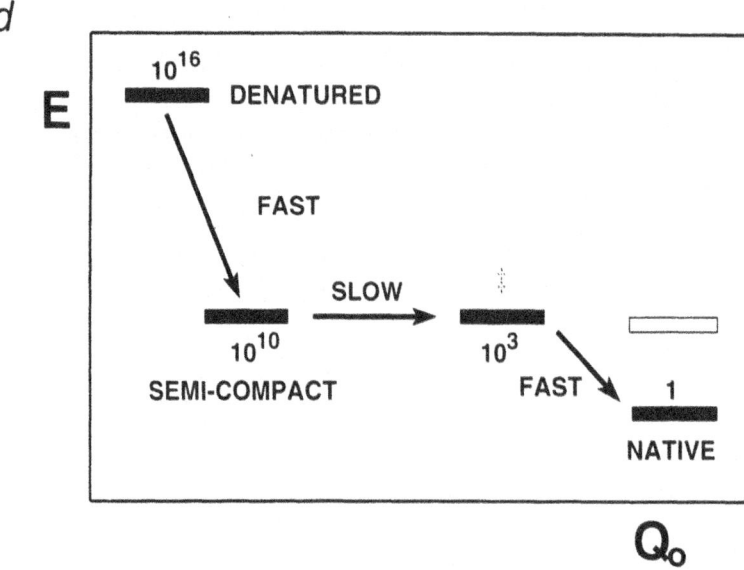

Figure 13. Three-stage random search model of folding. The numbers indicate the geometrically possible conformations obtained from the density of states that was determined as a function of energy and reaction coordinate. The empty rectangle indicates the energy of the native state of a nonfolding random sequence. The other states of a nonfolding sequence appear at the same positions as the corresponding states of a folding random sequence.

However, the mechanism breaks down for long chains because the folding time increases exponentially with chain length; i.e., the number of semi-compact states increases faster than the number of the transition states. Thus, a modification of the present mechanism is required for larger proteins.

It is likely that proteins existing early in evolution were small enough to fold according to the three-stage random search mechanism (Di Iorio *et al.*, 1993). Since the pre-biotic and early biotic environment was hot, unusually thermostable proteins were required, such as those found in the most primitive bacteria that live at temperatures as high as 105° (Stetter, 1992). If so, the stability condition required a native state that corresponded to a low energy minimum for which the folding problem was solved simultaneously. As the evolution progressed, longer proteins evolved. These proteins had to fold on the same time scale. One way of achieving this is by evolving proteins with sequences that have a larger difference between the native and non-native contact energies than the random folding sequences of the present model. Such sequences would have an even more pronounced global minimum in their potential surfaces, in line with the "consistency principle" (Go & Abe, 1984) and the "principle of minimal frustration" (Bryngelson & Wolynes, 1987). It is also in accord with the existence of a nucleus for folding (Wetlaufer, 1973; Taketomi & Go, 1975), or the early appearance of secondary structural elements (Karplus & Weaver, 1994), neither of which are found in the present model. As a result, the collapse would not result in random semi-compact globules and the very favorable native contacts would lead more directly to a native-like molten globule state. Actual proteins could use an intermediate mechanism that might vary with the external conditions.

The results of this study may have implications for the prediction of the structure of a protein from its amino acid sequence. The success of the three-stage random search in finding the pronounced global minimum on a potential surface suggests, at least for small proteins, that the bottleneck in structure prediction may be the derivation of a suitable potential function rather than the design of folding algorithms.

ACKNOWLEDGMENTS

Most of the material in this manuscript is based on a lecture at the 27th Jerusalem Symposium (May 1994) and it was taken mainly from Caflisch & Karplus (1994b) and Šali *et al.* (1994a,b). A. Caflisch was supported by a grant from the Schweizerischer National-fonds. A. Šali is a Fellow of the Jane Coffin Childs Memorial Fund for Medical Research. Support was also provided by a David and Lucille Packard Fellowship (E.S.), and by a grant from the National Science Foundation and a gift from Molecular Simulations Inc. (M.K.).

REFERENCES

Bai, Y. S., and Fayer, M. D. (1989) *Phys. Rev. B 39*, 11066.

Brooks, C.L. III, Karplus, M. (1983) *J. Chem. Phys. 79*, 6312.

Brooks III, C. L., Karplus, M., and Pettitt, B. M. (1988) *Proteins: A Theoretical Perspective of Dynamics, Structure, & Thermodynamics*, Adv. Chem. Phys. *LXXI* (John Wiley & Sons).

Bryngelson, J.D., and Wolynes, P. G. (1987) *Proc. Natl. Acad. Sci. USA 84*, 7524.

Caflisch, A., and Karplus, M. (1994a) in *The Protein Folding Problem and Tertiary Structure Prediciton*, K. Merz, Jr. and S. Le Grand, Editors (Birkhäuser, Boston, MA), 193.

Caflisch, A., and Karplus, M. (1994b) *Proc. Natl. Acad. Sci. USA 91*, 1746.

Di Iorio, E. E., Yu. W., Calonder, C., Winterhalter, K. H., De Sanctis, G., Falcioni, G., Ascoli, F., Giardina, B., and Brunori, M. (1993) *Proc. Natl. Acad. Sci. USA 90*,, 2025.

Dill, K. A. (1985) *Biochemistry 24*, 1501.

Fersht, A.R. (1993) *FEBS letters 325*, 5.

Ghelis, C. (1980) *Biophysical J. 32*, 503.

Go, N., and Abe, H. (1981) *Biopolymers 20*, 1013.

Go, N., and Abe, H. (1984) *Adv. Biophysics 18*, 149.

Honeycutt, J. D., and Thirumalai, D. (1992) *Biopolymers 32*, 695.

Kim, P., and Baldwin, R. (1990) *Ann. Rev. Biochem. 59*, 631.

Kaptein, R., Dijkstra, K., and Nicolay, K. (1978) *Nature 274*, 293.

Karplus, M. and Shakhnovich, E. (1992) in *Protein Folding,* T. Creighton, editor (W.H. Freeman & Sons), 127.

Karplus, M., and Weaver, D. L. (1994) *Protein Science 3*, 650.

Levinthal, C. (1969) in *Mossbauer Spectroscopy in Biological Systems, Proceedings of a Meeting held at Allerton House, Monticello, Illinois,* Debrunner, P., Tsibris, J. C. M. and Münck, E., editors (University of Illinois Press, Urbana), 22.

Mark, A.E., and van Gunsteren, W. F. (1992) *Biochemistry 31*, 7745.

Matouschek, A., Kellis, J.T. Jr., Serrano, L., Bycroft ,M., and Fersht A.R. (1990) *Nature 346*, 440.

Otting, G., and Wüthrich, K. (1989) *J. Am. Chem. Soc. 111*, 1871.

Ptitsyn O.B. (1992) in *Protein Folding*, Creighton T.E., ed. (W.H. Freeman, New York) 243.

Šali, A., Shakhnovich, E., and Karplus, M. (1994a) *J. Mol. Biol. 235*, 1614. Šali, A., Shakhnovich, E., and Karplus, M. (1994b) *Nature 369*, 248.

Serrano, L., Matouschek, A., and Fersht, A. R. (1992a) *J. Mol. Biol. 224*, 805.

Serrano, L., Matouschek, A., and Fersht, A.R. (1992b) *J. Mol. Biol. 224*, 347.

Shakhnovich, E., Farztdinov, G., Gutin, A. M., and Karplus, M. (1991) *Phys. Rev. Lett. 67*, 1665.

Stetter, K. O. (1992) in *Frontiers of Life*, Trân Thanh Vân, J. K., Mounolou, J. C., Schneider, J., and McKay, C., editors (Editions Frontières, Gif-sur-Yvette, France), 195.

Taketomi, H., and Go, N. (1975) *Intl. Journal Peptide Prot. Res. 7*, 445.

Tirado-Rives, J. and Jorgensen, W. L. (1991) *Biochemistry 30*, 3864.

Tobias, D. J. and Brooks, C. L. III, (1991) *Biochemistry 30*, 6059.

Wetlaufer, D. B. (1973) *Proc. Natl. Acad. Sci. USA 70*, 697.

10

INSIGHTS INTO PROTEIN DYNAMICS BY NMR TECHNIQUES

Lorna J. Smith and Christopher M. Dobson[*†]

Oxford Centre for Molecular Sciences and
New Chemistry Laboratory
University of Oxford
South Parks Road, Oxford OX1 3QT England

ABSTRACT

NMR techniques can give insight into a wide variety of motional events that occur in proteins over a range of time scales. Using data from studies of hen lysozyme and human interleukin-4 performed in our laboratory, we demonstrate here two NMR methods which can be used to give residue specific quantitative data about both main chain and side chain dynamics, namely coupling constant measurements and relaxation studies. The experimental data are compared with crystallographic data and the possible functional significance of experimentally observed protein motions in binding, activity and folding is also discussed.

It is now well established that proteins are not rigid systems but that they undergo a wide variety of internal motions, some of which have important functional significance (Karplus & McCammon, 1981; Brooks et al., 1988; Williams, 1993). Since it first became possible, some 20 years ago, to resolve the individual resonances of the NMR spectra of proteins, NMR spectroscopy has played an important role in providing insight into these internal motions (Wüthrich & Wagner, 1978; Jardetzky, 1981; Campbell et al., 1985; Dobson & Karplus, 1986) through studies, for example, of the rotational motion of the aromatic side chains of tyrosine and phenylalanine residues (Campbell et al., 1975; Dobson et al., 1975; Wüthrich & Wagner, 1975), the hydrogen exchange rates of individual amide and labile side chain protons within a protein (Roder et al., 1985; Wand et al., 1986; Radford et al., 1992) and the rate of *cis-trans* isomerism of prolyl peptide bonds (Evans et al., 1987; Kördel et al., 1990). Recently, the scope of NMR studies of protein dynamics has increased still further with the development of heteronuclear NMR techniques which can provide information about the relaxation behaviour of both the backbone and side chains of each residue in the

---

[*] Contribution from the Oxford Centre for Molecular Sciences which is supported by the U.K. Engineering and Physical Sciences Research Council, the Biotechnology and Biological Sciences Research Council and the Medical Research Council.

[†] To whom correspondence should be addressed.

Dynamics and the Problem of Recognition in Biological Macromolecules, edited by Jardetzky and Lefèvre
Plenum Press, New York, 1996

127

protein (Wagner, 1993). Heteronuclear techniques are also enabling the extension of studies of protein dynamics to partly folded non-native states including folding intermediates (van Mierlo et al., 1993; Redfield et al., 1994a).

The utility of NMR in characterising dynamical properties arises from the nature of the resonance phenomenon. Resonance techniques are distinct from other structural methods such as diffraction or optical/vibrational spectroscopy in their timescales. The interaction of, for example, X-rays or infrared radiation with matter occurs on a timescale of less than about 10^{-18}s. Within such a time, significant movement of atoms cannot take place. Any motional information from such methods will therefore be indirect, being manifest in a superimposition of images. NMR, however, is directly sensitive to motions on a timescale of ca 10^{-12} to 10^2s. In the short timescale range, this is a consequence of nuclear relaxation properties that depend on fluctuations that occur near to the Larmor (resonant) frequency (typically ca $10^9 s^{-1}$). On a slower timescale, the influence of molecular dynamics is through chemical exchange effects which transfer nuclei between different chemical or physical environments. These can be observed through lineshape effects, magnetisation transfer phenomena and the averaging of NMR parameters. The breadth of the timescale range is such that virtually all events likely to be of significance for chemical or biological process can in principle be examined.

EVIDENCE FOR PROTEIN DYNAMICS

In this article we concentrate on two areas where NMR can give residue specific quantitative data about dynamics of proteins, namely coupling constant measurements and relaxation studies. We use illustrations from the results of NMR studies of hen lysozyme and human interleukin-4 (IL-4) (structures shown in Figure 1) performed in our laboratory, together with other work that has been reported in the literature, including comparisons with crystallographic data. We also discuss the possible significance of protein dynamics in protein function and folding.

If there is mobility in the protein structure, either involving librations around a single conformation or interconversion between populations of distinct multiple conformations, then the observed values of NMR parameters such as chemical shifts and coupling constants will be a population weighted mean of the values expected for the individual conformers, as long as the rate of interconversion between the conformers is fast on the NMR time scale. Fast exchange for homonuclear coupling constants corresponds, for example, to a rate of interconversion between the contributing conformers that is greater than approximately 20 Hz (Smith et al., 1991). If such averaged NMR parameters can be recognised they can provide direct evidence for motional processes occurring in the protein in solution. Recognition and interpretation of motional averaged NMR parameters is simplest in the case of spin-spin coupling constants, the most commonly measured values being $^3J_{HN\alpha}$ and $^3J_{\alpha\beta}$. These can be related to the main chain ϕ and side chain χ_1 angles respectively via the Karplus relationship (Karplus, 1959), although because of the degeneracy of this relationship measurement of a single coupling constant value does not provide a unique torsion angle solution. Recent developments in heteronuclear NMR techniques have been increasing, however, the range of protein coupling constants that can be measured (Wagner, 1990) giving the potential for almost all the torsion angles in proteins to be probed.

The structure of the enzyme hen lysozyme has been determined in solution by NMR techniques (Smith et al., 1993) and detailed comparisons with crystal structures of the protein have shown the close similarity of the structure of this protein in solution and in crystals. For lysozyme we have measured both $^3J_{HN\alpha}$ (for 106 residues) and $^3J_{\alpha\beta}$ (for 57 residues) coupling constants (Smith et al., 1991). These values have been compared with those

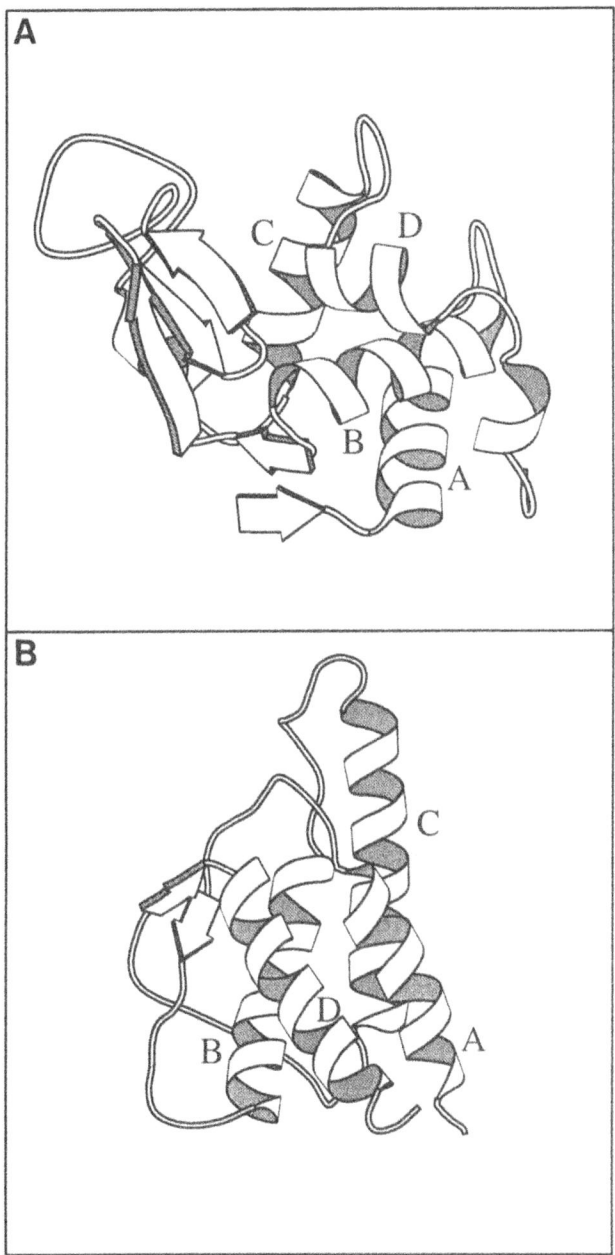

Figure 1. Schematic diagrams showing the structures of hen lysozyme (A) and human interleukin-4 (B). In each case the four helices are labelled A-D. The diagrams were generated using the program MOLSCRIPT (Kraulis, 1991).

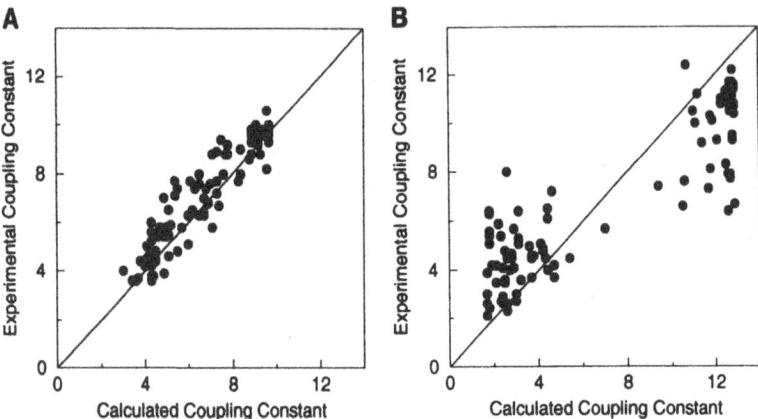

Figure 2. Comparison of the experimental coupling constants for hen lysozyme with those calculated from the angles in the tetragonal type 2 crystal structure (Handoll, 1985) of the protein using the Karplus relationship. In (A) the main chain $^3J_{HN\alpha}$ coupling constants are shown and in (B) the side chain $^3J_{\alpha\beta}$ coupling constants. Data from Smith et al. (1991). All values are in Hz.

predicted from crystal structure ϕ and χ_1 angles using the Karplus equation. For $^3J_{HN\alpha}$ there is found to be a close agreement between the experimental and calculated values (RMSD 0.88 Hz) (Figure 2A). This shows the close similarity between the backbone conformations of the protein in solution and in crystals, and suggests that any large scale motions of the backbone in solution are very limited (Smith et al., 1991). Similar results have been found for other proteins including BPTI (Pardi et al., 1984), barnase (Ludvigsen et al., 1991) and staphylococcal nuclease (Vuister & Bax, 1993).

For $^3J_{\alpha\beta}$ (Figure 2B) it is immediately evident that the agreement between the experimental and predicted values for hen lysozyme is less good. For residues where the side chain adopts a single staggered conformation one would expect for residues with two β protons, two small $^3J_{\alpha\beta}$ values or one large and one small value. For many residues this is indeed observed (see examples in Table 1A) and there is good agreement with the values predicted from the crystallographic χ_1 angles, although for residues with a small coupling constant the experimental value is generally larger than the calculated one, while for residues with a large coupling constant the experimental value is usually smaller than the calculated one. These differences reflect small fluctuations about the average torsion angle (Dobson & Karplus, 1986). For residues where multiple χ_1 conformations are adopted in solution, the observed $^3J_{\alpha\beta}$ values will be averaged and in general will both lie in the range 6-8 Hz. This motional averaging of the $^3J_{\alpha\beta}$ values is observed for 16 of the 57 residues of lysozyme studied (see Examples in Table 1B) (Smith et al., 1991). All of these residues, with one exception Val 99, are on the protein surface and, as the crystal χ_1 angles for these residues are usually those for specific staggered rotamers, there is a poor agreement between the calculated and experimental coupling constant values.

An important question is whether there is evidence that similar motions are occurring for the protein in the crystalline state. In order to examine this, the various crystal structures of lysozyme were examined. In the tetragonal type 2 crystal structure, for example, 50% of the residues with averaged $^3J_{\alpha\beta}$ coupling constants in solution were found to have γ atom temperature factors in the crystal that are greater than 30Å^2 (compared with 7% of residues that adopt single χ_1 conformations). Although none of the crystal structures reported for hen lysozyme resolve multiple side chain conformations, comparison of the χ_1 conformations of structures of different crystal forms reveal different structures for some side chains. For

Table 1. Examples of $^3J_{\alpha\beta}$ coupling constants for hen lysozyme and the corresponding χ_1 angles in the tetragonal type 2 and triclinic crystal structures of the protein (data from Smith et al., 1991)

Residue	$^3J_{\alpha\beta}$ (Hz)[‡]	χ_1 tetragonal	χ_1 triclinic
A) Residues that occupy a single staggered χ_1 conformation in solution.			
Cys 6	3.5, 11.5	−68.2°	−67.5°
Tyr 20	2.3, 11.7	−172.8°	179.7°
Asn 39	10.8, 4.5	−172.3°	−172.5°
Cys 64	2.7, 4.6	63.6°	67.3°
Trp 123	2.9, 10.6	−69.0°	−68.9°
B) Residues that occupy multiple χ_1 conformations in solution.			
Glu 7	6.7, 6.4	−177.6°	−65.7°
Arg 45	6.9, 6.7	−175.1°	−66.4°
Val 99	6.3	80.3°	171.1°
Asp 101	6.6, 5.6	−151.4°	−97.9°
Arg 125	7.9, 6.1	−52.9°	166.4°

[‡]In each case the coupling constant for the β proton whose resonance has the high chemical shift is listed followed by that for the β proton with the lower chemical shift.

example, comparison of tetragonal and triclinic structures of lysozyme shows that 9 of the 16 residues with motionally averaged $^3J_{\alpha\beta}$ values are found to adopt significantly different conformations in the two crystal forms (see examples in Table 1B) (Smith et al., 1991).

The result for hen lysozyme that 55% of the residues studied with a solvent accessibility greater than 60% adopt multiple χ_1 conformations in solution is similar to those found for other proteins including BPTI (Nagayama & Wüthrich, 1981), FK506 binding protein (Xu et al., 1992) and ribonuclease A (Rico et al., 1993). For BPTI and ribonuclease A there are high resolution crystal structures available containing discrete multiple conformations for certain side chains. It is interesting that for ribonuclease A, 14 of the 30 residues with averaged $^3J_{\alpha\beta}$ values adopt multiple conformations in the crystal (Rico et al., 1993) while for BPTI one of the residues with averaged $^3J_{\alpha\beta}$ values, Asp 52, is modelled in multiple conformations in the structure of crystal form III of the protein (Berndt et al., 1992). Hence, side chain mobility in solution can be reflected in disorder or multiple conformations in crystal structures although in general the mobility in solution appears more extensive than analysis of a single crystal structure would suggest.

The study of the relaxation behaviour of uniformly ^{15}N labelled proteins provides a powerful technique for characterising main chain dynamics through probing the motion of the ^{15}N-^1H vectors (Wagner, 1993). In general ^{15}N T_1 and T_2 relaxation rates and ^1H-^{15}N NOE values are measured and the data analysed using the model free approach of Lipari and Szabo (1982), order parameters (S^2) being extracted for the individual amide groups in the sequence. These order parameters give a measure of the degree of spatial restriction of the ^1H-^{15}N bond vector, very restricted fluctuations on a fast picosecond timescale (i.e. involving motions faster than the rotational correlation time of the molecule) giving a value close to 1.0. In some cases the data cannot be analysed assuming only a single fast motion, and an extended model which considers additional fast motions is required [e.g. that of Clore et al. (1990a) which considers two fast motions, one on a slightly slower timescale than the other]. Yet slower timescale motions can also be detected by these relaxation techniques; these result in exchange terms being required in the analysis of the experimental data.

The ^1H-^{15}N order parameters resulting from these ^{15}N relaxation experiments performed for two proteins, hen lysozyme (Buck et al., 1994) and human IL-4 (Redfield et al.,

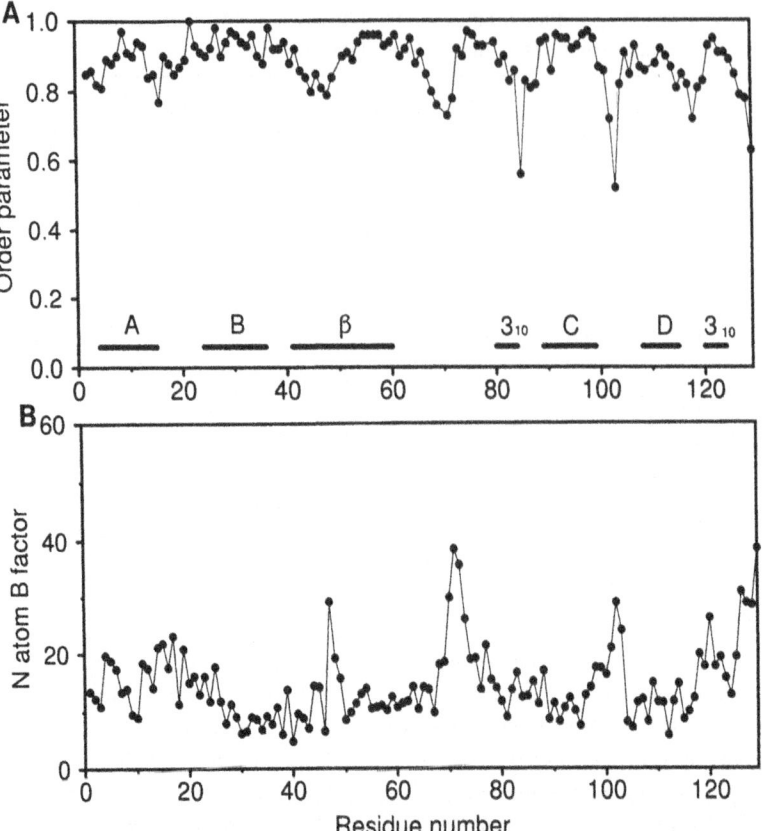

Figure 3. Main chain ^1H-^{15}N order parameters (S^2) (A) (Buck et al., 1994) and crystallographic temperature factors (in $Å^2$) for the main chain nitrogen atoms (B) (Handoll, 1985) of hen lysozyme. The values are plotted versus sequence number and the positions of the secondary structure regions in the protein are indicated.

1992), in our laboratory are shown in Figures 3A and 4A. For hen lysozyme only two residues, Ser 85 and Asn 103, have order parameters considerably less than 0.7; there is therefore no evidence for mobile loop regions or rapid hinge bending motions in this protein. For IL-4, in contrast, there are a significant number of residues with low order parameter values. The structure of this protein, determined initially by 3D heteronuclear NMR techniques (Smith et al., 1992; Powers et al., 1992), is a four helix bundle with an up-up-down-down connectivity, giving two long loops that run the length of the molecule (Figure 1B). As shown by the order parameters for IL-4 in Figure 4A, high order parameters (average 0.88) are seen throughout the four helices, but lower order parameters (0.3-0.8) are observed for the terminal regions of the sequence and the long AB and CD loops, except in the region where they are linked to each other by a short β-sheet. These regions with low order parameters are amongst those where differences have been observed within the ensemble of experimentally determined NMR structures for IL-4 structures, indicating that in these cases the disorder across the family of structures reflects the presence of internal motions and not merely the lack of NMR restraints (Redfield et al., 1994b).

From these results for lysozyme and IL-4, together with those from ^{15}N relaxation studies on a range of other proteins including staphylococcal nuclease (Kay et al., 1989),

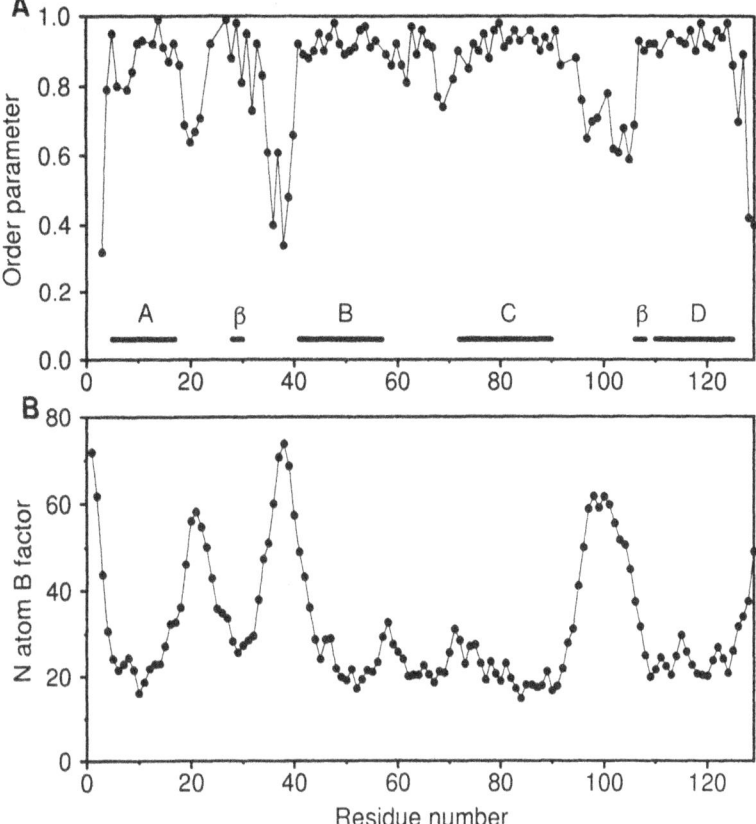

Figure 4. Main chain ^{1}H-^{15}N order parameters (S^2) (A) (Redfield et al., 1992) and crystallographic temperature factors (in Å^2) for the main chain nitrogen atoms (B) (Wlodawer et al., 1992) of human interleukin-4. The values are plotted versus sequence number and the positions of the secondary structure regions in the protein are indicated.

interleukin-1β (Clore et al., 1990b), calmodulin (Barbato et al., 1992) and thioredoxin (Stone et al., 1993), an overall picture is emerging that for most regions of a protein fast fluctuations of the main chain are of limited magnitude, reflecting the fact that close packing restrains the backbone into a single well defined conformation. It is only where this close packing is lost, for example in surface loops or at the termini, that the possibilty for fast motions arises.

There has been considerable interest in attempts to assess the relationship between main chain ^{15}N order parameters and crystallographic B factors (Powers et al., 1993). In some cases, such as IL-4 (see Figure 4B), there is a high degree of correlation, regions with low order parameters in solution exhibiting high N atom temperature factors in crystal structures of the protein (Smith et al., 1994). In other proteins, however, there are residues where there is little or no correlation between these two parameters, reflecting the fact that, for example, lattice contacts can restrict the dynamics of regions in crystals that are mobile in solution and crystallographic B factors can include contributions from disorder resulting from mobility on a slower timescale than that reflected in order parameters calculated in the conventional way (Powers et al., 1983).

The dynamics of side chains can also be probed by ^{15}N relaxation studies for those residues where appropriate nuclei exist; these are the indole NH of tryptophan residues, the NH$_2$ of asparagine and glutamine residues and the NεH of arginine residues. In the case of hen lysozyme the results of these experiments show that there is a wide variety of motional behaviour for different side chains, the experimental order parameters ranging from 0.2-0.9 (Buck et al., 1994). Of particular interest, however, in agreement with the results of the $^3J\alpha\beta$ coupling constant analysis, is that a clear correlation between the side chain order parameters and solvent accessibility has been identified (Figure 5), residues with relatively buried side chains having high order parameters while those with exposed side chains have much lower values of S^2. A comparison of the side chain order parameters for lysozyme with molecular dynamics simulations of the protein has provided models for the motions giving rise to the observed side chain order parameters. These range from limited fluctuations in χ potential energy wells ($S^2 > 0.7$) to conformational disorder arising from transitions between potential energy wells for two or more χ torsion angles ($S^2 < 0.3$) (Smith et al., 1995).

^{15}N relaxation studies have also be applied successfully to non-native states of proteins. Here, in contrast with native proteins where most of the residues have high main chain order parameters, considerable regions are found with lower order parameters reflecting the loss of close packing within the structure. Examples include a derivative of BPTI containing only one of the three disulphide bridges found in the native protein (van Mierlo et al., 1993), and a low pH state of human IL-4 (Redfield et al., 1994a). In the case of the partly folded form of IL-4 generated at low pH, there is a dramatic reduction in the main chain order parameters for residues 68-77 from an average of 0.86 in the high pH form to an average of 0.53 in the low pH form (Figure 6). These residues comprise the end of the BC loop and the start of the C helix, and the reduction in order parameters reflects a local unfolding transition in this region. The rest of the helical hydrophobic core of the molecule stays highly ordered in this state but it is surrounded by substantially less structured regions of the polypeptide chain.

SIGNIFICANCE OF PROTEIN DYNAMICS

Combining the insights that NMR techniques have provided about a range of dynamic processes in proteins enables us to build up a picture of the mobility of a protein structure. In the close packed interior of a native folded protein, motions of both the main chain and side chains are limited giving rise, for example, to high order parameters and coupling constants that are characteristic of a single well defined conformation. However, even here some motions have been identified such as the flipping of aromatic rings (Campbell et al., 1975; Dobson et al., 1975; Wüthrich & Wagner, 1975) and the exchange of amide protons (Roder et al., 1985; Wand et al., 1986; Radford et al., 1992) or buried water molecules (Otting et al., 1991), showing the presence of dynamic fluctuations and the ability of the protein core to relax. At the protein

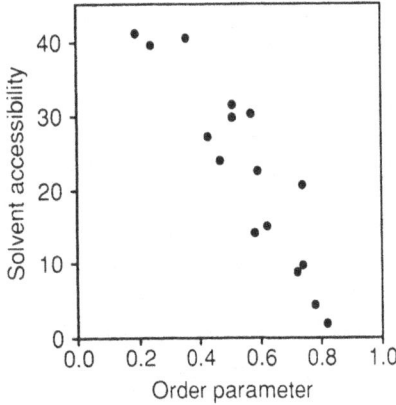

Figure 5. Comparison of the side chain ^{1}H-^{15}N order parameters (S^2) for the asparagine and glutamine residues in hen lysozyme with the solvent accessibility of the side chain nitrogen atoms (in $\mathring{A}^2$). Data from Buck et al. (1994).

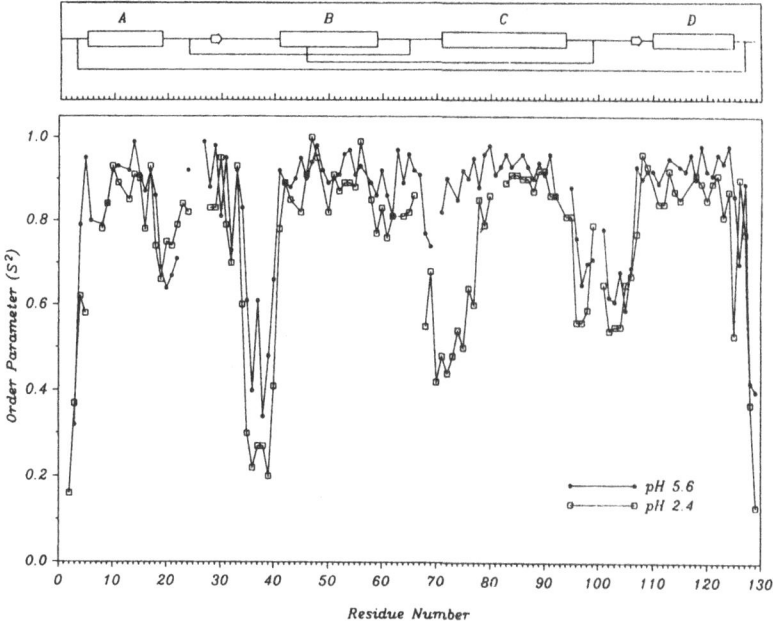

Figure 6. The calculated main chain ^{1}H-^{15}N order parameters (S^2) of human interleukin-4 at pH 5.6 and pH 2.4. The secondary structure of the protein is shown schematically above the diagram. Note the significant reduction of the order parameters for residues at the end of the BC loop and the start of the C helix on lowering pH. Diagram from (Redfield et al., 1994b).

surface the loss of complete close packing gives possibilities for far more extensive mobility. The backbone, especially in regions not involved in secondary structure, may have a considerable mobility and many side chains adopt multiple conformations. Still greater mobility has been observed in structural studies of linker regions connecting domains in large proteins and in peptides and partly folded or unfolded proteins where the loss of close packing is more extensive (Perham, 1991; Dobson, 1992). These dynamic characteristics of proteins, rather than being an exceptional property, can be understood in the light of the nature of the forces within the molecules. The protein chain is held into a globular fold in the main by weak noncovalent forces between the different residues. The majority of these are nonspecific van der Waals' interactions which can therefore readily allow and adapt to small changes in the structure such as those caused by dihedral angle fluctuations.

Having recognised the range and extent of dynamic processes occurring in proteins, it is important to consider their significance. It appears that, rather than being a property that decreases their functional effectiveness, in many cases biology may have actually exploited the dynamic nature of proteins (Dobson, 1993; Williams, 1993). Let us first consider regions involved in binding either to another protein, for example a receptor, or to a small molecule such as an enzyme substrate. Many of these binding regions are found to be disordered, presumable reflecting the fact that they are located on surfaces. Nevertheless, the mobility of such regions does give rise to speculation as to its possible value in the recognition process. The mobility of a binding site may enable the regions concerned to carry out a dynamic search for the required site on the binding partner so speeding up recognition and increasing adaptability. The flexibility then helps the complementarity of the surfaces to be maximised (induced fit) so improving the strength of binding. Both these features have been suggested to be important in the binding of the loop in the inhibitor eglin *c* to the active site of proteinases such as elastase and chymotrypsin (Hyberts et al., 1992). In the case of enzymes, such as dihydrofolate reductase which has a mobile loop involved in its active site, the mobility can also allow favourable binding to be maintained as the substrate structure changes along the reaction pathway (Li et al., 1992). The mobility of binding sites has another important thermodynamic consequence. Generally on binding much of the mobility is lost leading to an unfavourable entropy change. In some cases this is thought to be important for reducing the free energy of molecular association so providing reversibility of binding without loss of specificity (Searle et al., 1992).

If we consider multidomain proteins with flexible linkers between the domains, the mobility here gives motional independence to the different domains enabling the protein to perform multiple functions. An example of this is seen in urokinase-type plasminogen activator (u-PA), a multidomain protein where a long linker peptide joins a catalytic (serine protease) domain to a receptor binding (EGF-Kringle) domain (Nowak et al., 1993). This protein has a molecular weight of 48kD but NMR studies have shown that its resonances have linewidths typical of a significantly smaller protein. This is due to extensive motion about the linker region resulting in the domains exhibiting motions that are largely independent of each other. The flexibility of the linker peptide enables u-PA to be located on a receptor (via the EGF-Kringle) yet to search for its macromolecular substrate and bind to it in a manner appropriate for catalysis by the protease domain (Nowak et al., 1993).

As well as having important functional roles, the mobility of the polypeptide chain is undoubtedly of crucial importance in protein folding. The flexibility enables the polypeptide chain to explore a range of conformational states and hence to find the pathway where the barriers to structural transitions are lowered. A considerable degree of mobility is even observed in partly folded states that contain a significant amount of persistent structure such as the compact molten globules. For examples, hydrogen exchange studies of the α-lactalbumin molten globule (Baum et al., 1989) show that there is a high degree of persistent native-like secondary structure but the majority of side chains are disordered and there are few specific tertiary interactions resulting in broad NMR resonances, a reduced chemical shift dispersion, and the loss of the majority of native long range NOE's from the NMR spectra (Alexandrescu et al., 1993). These results in conjunction with kinetic studies of the refolding process show that it is only in the final steps of the folding pathway that the well ordered characteristics of the core of native proteins emerge (Radford & Dobson, 1995). This is important as it is recognised that it is often when close packed structure develops, and significant barriers to structural changes build up, that at least some of the slow steps in folding pathways might occur.

The dynamic characteristics of proteins described in this article have important consequences for the interpretation of structural data from NMR or X-ray diffraction techniques. When exposed regions, such as loops, linkers and binding surfaces, are being considered, crystal structures where there is generally only a single conformation, perhaps with high B factors, must be used with caution. The fact that these are often the regions which are of the greatest functional interest, highlights the importance of further detailed dynamics studies of proteins both by NMR and by other complementary techniques. In this regard comparison of NMR data with results from theoretical MD simulations appears to be particularly promising and has the potential to considerably increase our understanding of the dynamic properties of proteins (Dobson & Karplus, 1986; Chandrasekhar et al., 1992; Eriksson et al., 1993; Smith et al., 1995).

ACKNOWLEDGMENTS

This is a contribution from the Oxford Centre for Molecular Sciences which is supported by the U.K. Engineering and Physical Sciences Research Council, the Biotechnology and Biological Sciences Research Council and the Medical Research Council. The research of C.M.D. is supported in part by an International Research Scholars award from the Howard Hughes Medical Institute. L.J.S. is a Royal Society University Research Fellow.

REFERENCES

Alexandrescu, A. T., Evans, P. A., Pitkeathly, M., Baum, J. and Dobson, C. M. (1993) Biochemistry *32*, 1707-1718.

Barbato, G., Ikura, M., Kay, L. E., Pastor, R. W. and Bax, A. (1992) Biochemistry *31*, 5269-5278 .

Baum, J., Dobson, C. M., Evans, P. A. and Hanley, C. (1989) Biochemistry *28*, 7-13.

Berndt, K. D., Güntert, P., Orbons, L. P. M. and Wüthrich, K. (1992) J. Mol. Biol. *227*, 757-775.

Brooks, C. L. I., Karplus, M. and Pettitt, B. M. (1988) Adv. Chem. Phys. *71*, 1-249.

Buck, M., Boyd, J., Redfield, C., MacKenzie, D. A., Jeenes, D. J.,Archer, D. B. and Dobson, C. M. (1994) Biochemistry, *34* 4041-4055

Campbell, I. D., Dobson, C. M. and Williamson, R. J. P. (1975) Proc. Royal Soc. London, Series B *189*, 503-509.

Campbell, I. D., Dobson, C. M. and Williams, R. J. P. (1985) Biochem. J. *231*, 1-10.

Chandrasekhar, I., Clore, G. M., Szabo, A., Gronenborn, A. M. and Brooks, B. R. (1992) J. Mol. Biol. *226*, 239-250.

Clore, G. M., Szabo, A., Bax, A., Kay, L. E., Driscoll, P. C. and Gronenborn, A. M. (1990a) J. Amer. Chem. Soc. *112*, 4989-4991.

Clore, G. M., Driscoll, P. C., Wingfield, P. T. and Gronenborn, A. M. (1990b) Biochemistry *29*, 7387-7401.

Dobson, C. M. (1992) Cur. Opin. Struc. Biol. *2*, 6-12.

Dobson, C. M. (1993) Current Biology *3*, 530-532.

Dobson, C. M. and Karplus, M. (1986) Methods in Enzymol. *131*, 362-389.

Dobson, C. M., Moore, G. R. and Williamson, R. J. P. (1975) FEBS Lett. *51*, 60-65 .

Eriksson, M. A. L., Berglund, H., Härd, T. and Nilsson, L. (1993) Proteins *17*, 375-390.

Evans, P. A., Dobson, C. M., Kautz, R. A., Hatfull, G. and Fox, R. O. (1987) Nature *329*, 266-268.

Handoll, H. H. G. (1985) D. Phil. thesis, University of Oxford.

Hyberts, S. G., Goldberg, M. S., Havel, T. F. and Wagner, G. (1992) Protein Science *1*, 736-751.

Jardetzky, O. (1981) Acc. Chem. Res. *14*, 291-298.

Karplus, M. (1959) J. Chem. Phys. *30*, 11-15.

Karplus, M. and McCammon, J. A. (1981) CRC Crit. Rev. Biochem. *9*, 293-349.

Kay, L. E., Torchia, D. A. and Bax, A. (1989) Biochemistry *28*, 8972-8979.

Kördel, J., Forsen, S., Drakenberg, T. and Chazin, W. J. (1990) Biochemistry *29*, 4400-4409.

Kraulis, P. (1991) J. Appl. Crystall. *24*, 946-950.

Li, L., Falzone, C. J., Wright, P. E. and Benkovic, S. J. (1992) Biochemistry *31*, 7826-7833.

Lipari, G. and Szabo, A. (1982) J. Amer. Chem. Soc. *104*, 4546-4559.

Ludvigsen, S., Anderson, K. V. and Poulsen, F. M. (1991) J. Mol. Biol. *217*, 731-736 .

Nagayama, K. and Wüthrich, K. (1981) Eur. J. Biochem. *115*, 653-657.

Nowak, U. K., Li, X., Teuten, A. J., Smith, R. A. G. and Dobson, C. M. (1993) Biochemistry *32*, 298-309.

Otting, G. and Wüthrich, K. (1989) J. Amer. Chem. Soc. *111*, 1871-1875.

Pardi, A., Billeter, M. and Wüthrich, K. (1984) J. Mol. Biol. *180*, 741-751.

Perham, R. N. (1991) Biochemistry *30*, 8501-8512.

Powers, R., Garrett, D. S., March, C. J., Frieden, E. A., Gronenborn, A. M. and Clore, G. M. (1993) Science *256*, 1673-1677.

Powers, R., Clore, G. M., Garrett, D. S. and Gronenborn, A. M. (1993) J. Magn. Reson. Series B *101*, 325-327.

Radford, S. E. and Dobson, C. M. (1995) Phil. Trans. R. Soc. Lond. B *348*, 17-25.

Radford, S. E., Buck, M., Topping, K. D., Dobson, C. M. and Evans, P. A. (1992) Proteins *14*, 237-248.

Redfield, C., Boyd, J., Smith, L. J., Smith, R. A. G. and Dobson, C. M. (1992) Biochemistry *31*, 10431-10437.

Redfield, C., Smith, R. A. G. and Dobson, C. M. (1994a) Nature Struc. Biol. *1*, 23-29.

Redfield, C., Smith, L. J., Boyd, J., Lawrence, G. M. P., Edwards, R. G., Gershater, C. J., Smith, R. A. G. and Dobson, C. M. (1994b) J. Mol. Biol. *238*, 23-41.

Rico, M., Santoro, J., Gonzalez, C., Bruix, M., Neira, J. L. and Nieto, J. L. (1993) Applied Magn. Reson. *4*, 385-415.

Roder, H., Wagner, G. and Wüthrich, K. (1985) Biochemistry *24*, 7396-7407.

Searle, M. S., Williams, D. H. and Gerhard, U. (1992) J. Amer. Chem. Soc. *114*, 10697-10704.

Smith, L. J., Sutcliffe, M. J., Redfield, C. and Dobson, C. M. (1991) Biochemistry *30,* 986-996.

Smith, L.J., Redfield, C., Boyd, J., Lawrence, G.M.P., Edwards, R.G., Smith, R. A. G. and Dobson, C.M. (1992) J. Mol. Biol. *224,* 900-904.

Smith, L. J., Sutcliffe, M. J., Redfield, C. and Dobson, C. M. (1993) J. Mol Biol. *229* 930-944.

Smith, L. J., Redfield, C., Smith, R. A. G., Dobson, C. M., Clore, G. M., Gronenborn, A. M., Walter, M. R., Nagabushan, T. L. and Wlodawer, A (1994) Nature Struc. Biol. *1,* 301-310.

Smith, L. J., Mark, A. E., Dobson, C. M. and van Gunsteren, W. F. (1995) Biochemistry 10918-10931.

Stone, M. J., Chandrasekhar, K., Holmgren, A., Wright, P. E. and Dyson, H. J. (1993) Biochemistry *32,* 426-435.

van Mierlo, C. M. P., Darby, N. J., Keeler, J., Neuhaus, D. and Creighton, T. E. (1993) J. Mol. Biol. *229,* 1125-1146.

Vuister, G. W. and Bax, A. (1993) J. Amer. Chem. Soc. *115,* 7772-7777.

Wagner, G. (1990) Prog. NMR Spec. *22,* 101-139.

Wagner, G. (1993) Cur. Opin. Struc. Biol. *3,* 748-754.

Wand, A. J., Roder, H. and Englander, S. W. (1986) Biochemistry *25,* 1107-1114.

Williams, R. J. P. (1993) Eur. Biophys. J. *21,* 393-401.

Wlodawer, A., Pavlovsky, A. and Gustchina, A. (1992) FEBS Lett., *309,* 59-64.

Wüthrich, K. and Wagner, G. (1975) FEBS Lett. *50,* 265-268.

Wüthrich, K. and Wagner, G. (1978) Trends Biochem. Sci. *3,* 227-230.

Xu, R. X., Olejniczak, E. T. and Fesik, S. W. (1992) FEBS Lett. *305,* 137-143.

11

HETERONUCLEAR RELAXATION AND THE EXPERIMENTAL DETERMINATION OF THE SPECTRAL DENSITY FUNCTION

K. T. Dayie,[1] G. Wagner,[1] and J.-F. Lefèvre[2]

[1] Committee on Higher Degrees in Biophysics, Harvard University, and
Department of Biological Chemistry and Molecular Pharmacology
Harvard Medical School
240 Longwood Avenue, Boston, Massachusetts 02115
[2] Groupe de Cancérogenèse et de Mutagenèse Moléculaire et Structurale
ESBS
Bd Sébastien Brant. 67400 Strasbourg-Illkirch Cedex, France

1. INTRODUCTION

Proteins have always been known as dynamic molecules. With the success of X-ray crystallography in solving protein structures over the past thirty years, the static aspects of proteins have been emphasized, primarily because this technique can only characterize relatively rigid and completely folded proteins. On the other hand, NMR can study partially folded proteins and can characterize well internal motions. Among the most fruitful techniques to characterize internal mobility in proteins are relaxation time measurements, and a large number of studies have focused on that aspect recently (see e.g. Wagner, 1993). Globular proteins must undergo a range of motions in space and time to modulate the stunning array of critical biological processes such as enhance the rate of transcription of DNA, transport electrons, maintain structural integrity, or modulate cellular immune responses (McCammon & Harvey, 1985; Brooks et al., 1988). Understanding such motion inside a protein might thus provide some insight into their behavior. For example, it may allow the elucidation of the potential energy surface on which these proteins move, it may provide some clues on how they fold from a linear chain to three dimensional structure. Finally, dynamics is important in the context of structure refinement. Mobility studies can allow one to detect erroneously too tightly constrained structures, and identify regions that are rigid but appear artifactually disordered in structure calculations due to incomplete data analysis.

Till now, the analysis of heteronuclear relaxation measurements have relied on *a priori* presumed models all based on a Markovian diffusive process, biasing the analysis as a result. An alternative that provides the raw dynamic information upon which motional models can be evaluated *a posteriori* is discussed in the context of select published data

Dynamics and the Problem of Recognition in Biological Macromolecules, edited by Jardetzky and Lefèvre
Plenum Press, New York, 1996

taken from the literature. The theoretical and experimental foundations, as well as various models of correlation functions are discussed as a backdrop to the use of spectral density mapping technique to probe picoseconds to millisecond motions in proteins.

2. THEORETICAL ASPECTS OF RELAXATION: A LOOK AT THE POWER OF THE SPECTRAL DENSITY FUNCTION

2.1. A Heteronuclear Spin System at Equilibrium

A typical solution sample of a 20 kDa protein of concentration 1 mM dissolved in 500 ml of buffer contains $\approx 10^{16}$ molecules performing a thermally-activated random walk. Let us consider a heteronuclear spin pair in such a system, made of an atom X covalently attached to a proton (X being a member of the set $\{^{13}C^{\alpha}, \,^{13}C^{\beta}, \,^{13}C', \,^{15}N\}$ and the proton of the corresponding set $\{^{1}H^{\alpha}, ^{1}H^{\beta}, ^{1}H^{H\alpha}, ^{1}H^{HN}\}$, for example). As the sample is placed in a superconducting magnet, bulk magnetization develops as a slight excess of spins aligning with the field: those that align with the field and thus have the lowest energy will be called α, the others against the field are called β (in the case of ^{15}N, the lowest energy state is β, resulting in a negative gyromagnetic ratio). Within a two-spin system, the Zeeman states combine to give a four level diagram (Fig. 1). The energy of each level is affected by the scalar coupling between the two nuclei. However, the variation in energy brought by this coupling is fairly weak compared to the energy difference between the various levels. At equilibrium, after a while in the magnetic field, each level is populated by the spin systems according to the Boltzmann law. In terms of spin operators, the density matrix (σ) can be written as the sum of Pauli spin matrices and therefore as sum of spin angular momentum operators, the z operators of the pair-wise nuclei involved (I stands for the proton and S for heteronucleus X and γ's are the magnetogyric ratios):

$$\sigma = \gamma_I \cdot I_z + \gamma_S \cdot S_z \tag{1}$$

2.2. Displacement of the Spin System from Its Equilibrium

The system of spins can be perturbed away from equilibrium by combination of radiofrequency pulses. These non equilibrium states may involve only changes in population by inversion of all transitions ($-I_z$, $-S_z$), selective inversion of one transition ($2I_z S_z$), or by polarization transfer ($\pm(\gamma_I / \gamma_S)S_z, \pm(\gamma_S / \gamma_I)I_z$). When one quantum coherences are induced, the magnetizations alternate between in phase (I_x and S_x) and anti phase ($2I_y S_z$ or $2I_z S_y$)

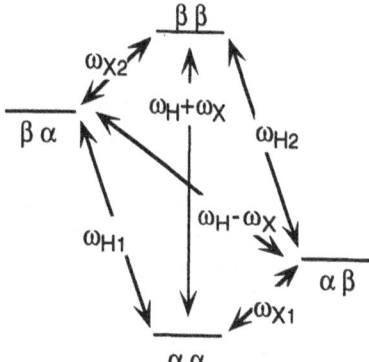

Figure 1. Four energy level diagram for a two spin system H - X. Simple quantum transitions are distinguished for the lower (ω_{H1}, ω_{X1}) and upper (ω_{H2}, ω_{X2}) frequency components of the doublets. ($\omega_H - \omega_X$) and ($\omega_H + \omega_X$) are the zero and double quantum transitions respectively. In the case of ^{15}N, the lower energy level being β, the energy levels should be permuted: $\alpha\alpha$ with $\alpha\beta$ and $\beta\alpha$ with $\beta\beta$.

states under the scalar coupling. Each state can be maintained by applying a spin locking B_1 field along the x or y axis. Zero and double quantum coherences are produced simultaneously when non selective pulses are used.

2.3. The Relaxation Perturbing Hamiltonian

The spin system relaxes back towards its equilibrium state via transitions between the four energy levels. The random transitions of spins from one level to another progressively destroy the coherences and redistribute the Boltzmann populations. Such transitions are induced by the fluctuation of the local magnetic field which, one assumes, mainly originates from the dipolar interactions (DD) with the neighboring nuclei and the anisotropy of the chemical shift (CSA) which is a property of the electronic environment of each spin-1/2 nucleus.

If one concentrates on the relaxation of the X nucleus, the magnetic field fluctuations comes from the angular dependence of the dipolar interaction with its attached protons and its own CSA, as the X-H vector tumbles. The perturbing Hamiltonians which correspond to the dipole-dipole and the chemical shift interactions are easily written in irreducible spherical tensor basis (a spherical tensor T(k,q) of rank k has 2k+1 components corresponding to q = -k, -k+1,..., k-1, k). In this basis, each dipole being a vector is represented by a first rank spherical tensor $T_\mu = T(1,\mu)$ with three components ($\mu = -1$, $\mu = 0$ and $\mu = +1$) which depend on the spherical polar coordinates (Θ and Φ). The magnetic field can also be represented by a spherical tensor of rank one with only one non-zero component (for $\mu = 0$). This representation is adopted for convenience in applying rotational transformations of tensors between different basis. For example, one particular interaction expressed in the molecule-fixed frame by the tensor T(k,q) is easily transformed into the laboratory frame with the same rank tensor but different components p by the simple expansion:

$$T(k,p) = \sum_{q=-k}^{k} D_{qp}^k(\Theta,\Phi)\, T(k,q) \tag{2}$$

where $D_{qp}^k(\Theta,\Phi)$ are the elements of the Wigner rotation matrix (Edmonds, 1957; Rose, 1957). In our case, the perturbing Hamiltonians being constructed from the coupling of two first rank tensors (I and S for the dipolar interaction and H and S for the chemical shift interaction) are second rank tensors and transform under rotation according to:

$$H_n = A_n \sum_{\mu=-1}^{1} \sum_{\mu'=-1}^{1} D_{q0}^k(\Omega_n(t))\, (-1)\, q\sqrt{2k+1} \begin{pmatrix} 1 & 1 & k \\ \mu & \mu' & -q \end{pmatrix} L_\mu K_{\mu'} \tag{3}$$

where $q = \mu + \mu'$ and k (the rank of the interaction tensor) = 2. The double sum on μ and μ' is equivalent to a sum on q from -2 to 2 and a sum on μ, replacing μ' by q-μ. Thus the Hamiltonian has the form of a second rank spherical tensor with 5 components. The index n refers to a particular type of interaction (dipolar or CSA). L_μ and $K_{\mu'}$ are the components of the first rank spherical tensors which are coupled in each Hamiltonian. The expression between the time varying Wigner matrix element and the product $L_\mu K_{\mu'}$ represents the Clebsch-Gordon coefficients which appear when two angular momentum are coupled. They are written here in the Wigner 3-j symbol (Zare, 1988). The coefficients A_n are, for DD and CSA interactions:

$$A_{DD} = -\frac{h\gamma_I\gamma_S}{4\pi r^3}, \qquad A_{CSA} = \frac{\gamma_I}{3}(\sigma_\parallel - \sigma_\perp) \tag{4}$$

Each Hamiltonian, H_n, can be written in a contracted notation where the spin interaction and the time dependence terms are factored out:

$$H_n = F_n(t) \cdot V_n \tag{5}$$

V_n is a second rank spherical tensor and $F_n(t)$ is the time varying rotation Wigner matrix. The components of this matrix are equivalent to the second rank spherical harmonics Y_{2q} $(\Omega_n(t))$. The total perturbing Hamiltonian is obtained by summing over all possible Hamiltonians (H_n):

$$H = \sum_n H_n = \sum_n F_n(t) \cdot V_n \tag{6}$$

2.4. The Relaxation Rate Constants

The evolution of the system of spin is inscribed in the kinetic equation of the density matrix σ, the master equation for relaxation (e.g. Abragam, 1961; Goldman, 1987):

$$\frac{d\sigma}{dt} = -i\,[H',\sigma] - \sum_n [(V_n)^\dagger, [V_n, (\sigma - \sigma_o)]] \int_0^{+\infty} <F_n(0) \cdot F_n^*(\tau)> \exp(i\omega_n\tau)d\tau \tag{7}$$

The first term gives the evolution under the scalar coupling Hamiltonian H' and the second one the relaxation due to the fluctuation of the dipole-dipole and the chemical shift interactions. For the sake of simplicity, the above equation is written for an autocorrelated relaxation mechanism. Cross correlation may occur between the chemical shift anisotropy and the dipolar interaction. In that case mixed terms containing V_n and $V_{n'}$ (with n = DD and n' = CSA) may appear in the spin interaction part of the equation, while corresponding mixed terms $<F_n(0) \cdot F_{n'}^*(\tau)>$ would occur in the dynamic part. This cross correlation will appear again in the final result below (Eq. 11).

There are four different angular frequencies ω_i which correspond to the four transitions which may occur: the one quantum transitions of X (at ω_X) and ^1H (at ω_H), the zero quantum (at $\omega_H - \omega_X$) and the double quantum (at $\omega_H + \omega_X$) transitions (Fig. 1). The stochastic dependence is now contained in the correlation function defined as:

$$G_n(\tau) = <F_n(0) \cdot F_n^*(\tau)> \tag{8}$$

and the Fourier transform which yields the spectral density function:

$$J_n(\omega) = \int_{-\infty}^{+\infty} G_n(\tau) \exp(i\omega\tau)d\tau = 2 \int_0^{+\infty} G_n(\tau) \exp(i\omega\tau)d\tau \tag{9}$$

so that Eq. 7 can be written as:

$$\frac{d\sigma}{dt} = -i\,[H',\sigma] - \frac{1}{2} \sum_n [(V_n)^\dagger,[V_n,(\sigma - \sigma_o)]]\,j_n(\omega_n) \tag{10}$$

Equivalently, in the spherical tensor operator notation, the time evolution of the expectation value of any spin operator Q relevant to relaxation (for example the transverse and longitudinal nitrogen magnetizations N_x and N_z) during the mixing time is readily deduced from the trace relation $<Q> = \text{Tr} \{\sigma Q\}$. Evaluating the resulting master equation in the rotating frame leads to the familiar expression for the relaxation rate for operator Q (here, we follow familiar notation, Abragam, 1961, and Peng & Wagner, 1994):

$$R_Q(Q_m) = \frac{5}{3h^2} \sum_{nn'} A_n \cdot A_{n'} \sum_{q=2}^{-2} \sum_{\mu=1}^{-1} \begin{pmatrix} 1 & 1 & 2 \\ \mu & q-\mu & -q \end{pmatrix}^2$$

$$\times J_{nn'}(\mu\omega_L^n + (q-\mu)\omega_K^n)$$

$$\times \text{Tr}\{(L_\mu^{n'} K_{q-\mu}^{n'})^\dagger, [L_\mu^n K_{q-\mu}^n, Q_m]]\} \tag{11}$$

Compared to Eq. 3, the double sum on μ and μ' has been replaced by a sum on q and a sum on μ, with the relation $\mu' = q - \mu$. A_n and $A_{n'}$ index the strength of each relaxation mechanism. For autocorrelation (n=n'), the product A_n^2 is the mean square amplitude of the local fluctuating field for that particular interaction. For cross-correlation interaction (n ≠ n), $A_n \cdot A_{n'}$ represent possible interference between the two mechanisms. L and K are associated with the spin angular momentum operators in the case of dipole-dipole, or L with the external field, in the case of CSA. Under the secular approximation (Redfield, 1965), cross relaxation does not occur between longitudinal and transverse operators.

The calculation of the rate constant is developed here for the CSA contribution to the transverse relaxation of the inphase ^{15}N magnetization. Under the influence of the CSA mechanism the rate expression takes on the very simple form shown below:

$$R_Q(Q_m) = \frac{5}{3h^2} \sum_{q=1}^{-1} \begin{pmatrix} 1 & 1 & 2 \\ 0 & q & -q \end{pmatrix}^2 A_{CSA}^2 \, J(Iq\omega_K I) \, \text{Tr}\{[K_q^{2\dagger},[K_q^2,(Q_m)]]\} \tag{12}$$

This simplification stems from restricting the index μ to 0, for the CSA mechanism, and thus, q to -1, 0, 1. K represents the ^{15}N spin angular momentum operators expressed in the spherical tensor basis. Thus the rates that govern the time evolution of each quantum mechanical observable are readily computed. Such a calculation is demonstrated for the rate of decay of the inphase nitrogen ^{15}N magnetization:

$$Q_m = N_x = \frac{1}{\sqrt{2}} (N_{-1} - N_{+1}) = \frac{1}{2} (N_+ + N_-) \tag{13}$$

N_x is the Cartesian component of the ^{15}N angular momentum operator expressed as a function of its spherical tensor components, N_{-1} and N_{+1}. Also given is the relation with the raising and lowering operators, N_+ and N_- respectively. The expression of the relaxation rate constant becomes:

$$R_N(N_x) = \frac{5}{3h^2} \sum_{q=1}^{-1} \begin{pmatrix} 1 & 1 & 2 \\ 0 & q & -q \end{pmatrix}^2 A_{CSA}^2 \, J(Iq\omega_K I) \, \text{Tr}\{[K_q^{2\dagger},[K_q^2, N_{-1} - N_{+1})]]\} \tag{14}$$

Use of the commutation rules for the angular momentum:

$$[N_{\pm 1}, N_r] = \pm \left(-\frac{1}{\sqrt{2}} \right) [(2 - r(r \pm 1)]^{1/2} N_{r\pm 1}, \; [N_0, N_r] = rN_r \tag{15}$$

Table 1. Expressions of the most common relaxation rate constant for a H-X spin system. The table contains the coefficient of the values of the spectral density function at the relevant frequencies. A and B are the dipolar and the CSA constants respectively and are given in Eq. 26. C is the mixed dipolar-CSA constant (Goldman, 1984):

$$C = \frac{\gamma_X \cdot \gamma_H \cdot \omega_X \cdot \Delta}{r^3} \quad (26')$$

$\rho_{HH'}$ is the dipolar contribution to the relaxation rate of the proton (H-X) due to the neighbouring protons (H'). Upper part: auto-relaxation. Lower part: cross-relaxation. $X_+ = X_x + i.X_y$.

Rate Constants	J(0)	$J(\omega_X)$	$J(\omega_H - \omega_X)$	$J(\omega_H)$	$J(\omega_H + \omega_X)$	$\rho_{HH'}$
$R_X(X_z)$		3A + B	A		6.A	
$R_X(X_x)$	2A + 2B/3	3A/2 + B/2	A/2	3A	3A	
$R_X(2H_zX_z)$		3A + B		3A		1
$R_X(2H_zX_x)$	2A + 2B/3	3A/2 + B/2	A/2		3A	1
$R_H(H_z)$		3A	A		6A	1
$R_X(H_z{-}{>}X_z)$			$-A$		6A	
$R_X(X_z{-}{>}2H_zX_z)$		C				
$R_X(X_+{-}{>}2H_zX_+)$	C/3	C/2				

with r = -1, 0 or +1, and insertion into Eq. 14 of the values for the Wigner 3J symbols for μ=0 and q=1,0,-1 which are $-1/\sqrt{10}$, $2/\sqrt{30}$, $-1/\sqrt{10}$ respectively (Zare, 1988) leads finally to:

$$R_N(N_x) = \frac{2}{3} A_{CSA}^2 J(0) + \frac{1}{2} A_{CSA}^2 J(\omega_N) \tag{16}$$

The dipolar contribution should be added to compute the total relaxation rate constant. Expression for the most common relaxation rate constants are given in Table 1. They contain a term related to the type of interaction responsible for the relaxation, and the spectral density function.

3. MEASUREMENT OF RELAXATION RATES

Measurement of relaxation rates depends on perturbing the spin-systems away from their equilibrium values and devising sequences to monitor their return to equilibrium. Current strategies use the following building blocks that are briefly commented upon below: (a) Double INEPT/DEPT transfer of polarization from proton to heteroatoms and back to protons for signal enhancement (Burum & Ernst, 1980; Bax et al., 1990; Davis, 1990); (b) Pulsed Field gradients (Maudsley et al., 1978; Bax et al., 1980; Hurd, 1990; Kay et al. 1992a) or spinlocked pulses (Otting & Wüthrich, 1988; Messerle et al., 1989) or presaturation of solvent resonance position in absence of gradients, to suppress the huge water peak; (c) Minimization of radiation damping; (d) Muting of interference effects due to cross-correlation between dipolar and chemical-shielding anisotropy relaxation mechanisms, and evolution of antiphase to inphase magnetization under scalar-coupling (Boyd et al., 1990; Palmer et al., 1991; Kay et al., 1992b); and (e) Preservation of two-orthogonal paths of magnetization to enhance the sensitivity of experiments (Cavanaugh & Rance, 1993).

Double INEPT/DEPT is used to excite protons and refocus the magnetization to the desired spin order, then the desired spin-order is relaxation-encoded during variable delays, chemical shift labelled with the Larmor precession frequency of the heteroatom, and finally transferred back to protons for detection. This leads to gains in signal-to-noise (S/N) of 300 for proton-proton excitation-detection compared to 30 and 10 for either proton-nitrogen or nitrogen-proton excitation-detection respectively; in the case of carbon, the S/N gain of proton-proton excitation-detection over proton-carbon, carbon-proton, carbon-carbon excitation-detection is 32, 8, 4, and 1 respectively.

Until recently, most relaxation experiments carried out in water have used direct presaturation of the water resonance or radiofrequency field pulses to reduce the intense water resonance. This initial presaturation of the solvent line can be used under conditions of low pH and or slow exchange of amides with solvent. Otherwise, long scrambling pulses can be used (Otting & Wüthrich, 1988; Messerle et al. 1989). Here the magnetizations not associated with the X-H spin system prior to t_2 detection are randomized with long millisecond ^{1}H pulses at full power and of the same phase as the desired coherences but orthogonal to the undesired coherences such as the solvent. These pulses are preferably applied when the desired magnetization is of the form of proton antiphase coherence (e.g. $2H_xN_z$) during the reverse INEPT, or just prior to detection.

Introduction of shielded gradients into high resolution NMR enabled the application of pulsed-field gradients to select desired NMR signals almost free of artifacts, with excellent water suppression and reduced t_1 noise. Two variants are currently employed in relaxation measurements: gradients combined with the Cavanaugh-Palmer-Rance (CPR) preservation of coherence pathways method (e.g. Cavanaugh & Rance, 1993) as demonstrated by Kay et al. (1992b) or gradients combined with selective pulses that ensure that water is efficiently suppressed leaving the desired signals intact (Sklenar et al., 1993). The major disadvantage of the former is the signal losses due to longer pulse-sequence whereas the latter suffers from signal losses for resonances on the edges of the spectrum for protein samples with large chemical shift dispersion as found in non-helical proteins, for example.

All gradient-enhanced versions also use the water "flip-back" strategy (Grzesiek & Bax, 1993; Stonehouse et al., 1994) to ensure the water magnetization is little perturbed and oriented along the +z axis during most of the experiment, especially just prior to acquisition to minimize the saturation of water. This is especially useful for samples at high pH with rapidly exchanging amide protons (Li & Montelione, 1994).

Pulse sequences for the following relaxation parameters have been published: the relaxation rate for ^{15}N longitudinal magnetization, $R_N(N_z)$; the relaxation rates of in-phase ^{15}N coherence, $R_N(N_x)$; the heteronuclear NOE, XNOE, in order to measure the cross relaxation rate, $R_N(H_z->N_z)$; the auto-relaxation rate of the heteronuclear longitudinal two-spin order, $R_{HN}(2H_zN_z)$; the auto-relaxation rate of antiphase two-spin order, $R_{HN}(2H_zN_x)$; the transverse relaxation rate of the amide proton ^{1}H in an ^{15}N enriched amide group, $R_H(H_x)$; and the spin-lattice relaxation rate of the amide proton ^{1}H, $R_H(H_z)$. Amplitude modulated versions of the first five (Peng & Wagner, 1992) sequences as well as sensitivity-enhanced versions for the first three (Palmer et al., 1991), all of which use presaturation or purge pulses, are well established in the literature; gradient-enhanced versions of the first three (Farrow et al., 1994) and all seven (Dayie & Wagner, 1994) are more recent. The proton relaxation measurements have been used to show that deuteration might extend the size of proteins accessible to NMR spectroscopy by a factor of two (Markus et al., 1994). The first four relaxation rates ($R_N(N_z)$, $R_N(N_x)$, $R_N(H_z->N_z)$ and $R_{HN}(2H_zN_z)$) are the most useful for mapping the frequencies of motion along the protein backbone.

Fig. 2 to 5 shows the pulse schemes for sensitivity-enhanced measurements of relaxation rates of the X nucleus in the H-X group, avoiding solvent presaturation. The sequence devised for measuring the relaxation of the transverse magnetizations in fact yields the spin-lattice relaxation rate in the rotating frame, $R_X(X_{\rho z'})$. In the case of intermediate exchange, $R_X(X_{\rho z'})$ is equal to $R_X(X_{xy})$ only when strong spin-lock field is applied. Unless

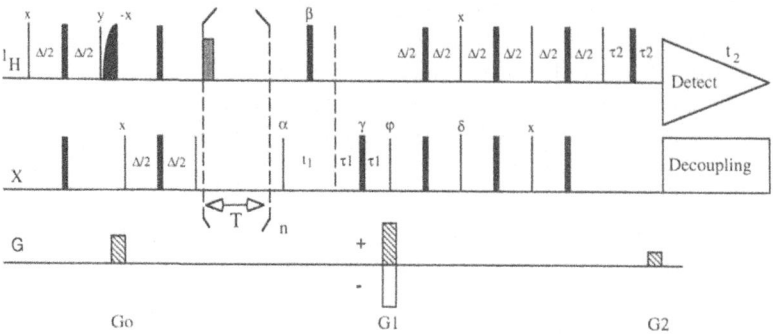

Figure 2. Longitudinal relaxation measurements: $R_X(X_z)$: The X spin (^{15}N, ^{13}C) magnetization is inverted for the variable delay T after a refocused INEPT. Cross-relaxation effects throughout the relaxation period are suppressed by saturating the protons during the T delay by a train of hard 90° (or 120°) proton pulses. This saturation reduces the longitudinal magnetization kinetics to a one-spin system and suppresses the dipolar-CSA cross-correlation effects as discussed by Boyd et. al. (1990). Thin and thick vertical bars ars 90° and 180° pulses respectively. The shaped pulse is a half gaussian pulse of 2 ms duration applied at the water carrier frequency. The gradient Go (typically 1 ms, 0.5 T m^{-1}) is used to dephase any residual transverse magnetization. The ratio G2/G1 is equal to the ratio of the proton to X nucleus gyromagnetic ratio. Typically G2 is set to 2 T m^{-1} for a 4 ms duration. The sign of G1 is alternated every FID, together with the phase φ. The complete phase cycling is given in Dayie and Wagner (1994). $\Delta = 1/2J_{HX}$. τ_1 and τ_2 are delays allowing for the application of gradient and for the recovery.

the spin-lock is on resonance for all X spins, $R_X(X_{\rho z'})$ will not be identical to $R_X(X_{xy})$ (Peng et al. 1991). For a spectral width that results in tip angles of < 80°, the relaxation rates are better measured with at least 2 different spin-lock carrier positions such that the desired spectral range is covered with minimal off-resonance errors. While the use of the CPMG pulse train to measure $R_X(X_{xy})$ avoids this problem altogether provided the spacing between consecutive 180° refocusing pulses is much smaller than $1/2\,J_{HX}$, it does introduce J-coupling modulation of the spin-echo envelope when used in ^{13}C-labelled samples. Additionally, measurements of $R_X(X_{\rho z'})$ as a function of spin-lock field strength under favorable conditions

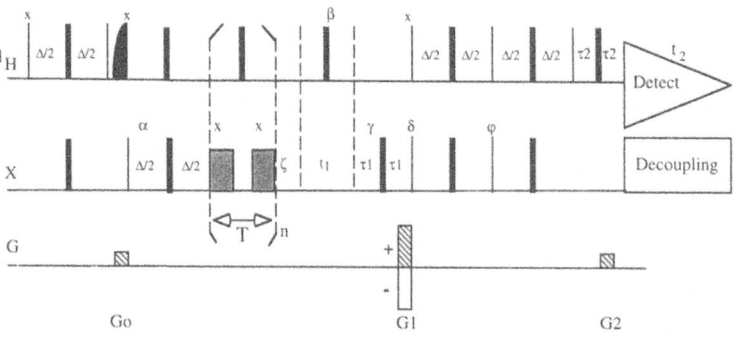

Figure 3. Transverse relaxation measurements, $R_X(X_{xy})$: The sequences uses a low power spin-lock consisting of same phase 180° pulses to maintain transverse X_x magnetization for the relaxation delay T. Attached protons are periodically pulsed with hard 180° pulses inserted between fixed lengths ($\approx$ 3ms) of spin-locking to diminish contributions from inhomogeneities in the static field, to prevent the oscillation between in-phase coherence and antiphase (e.g. $2H_z\,N_y$) coherence, and suppress effects of the dipolar-CSA cross-correlation. Gradients and delays are similar to those in Fig. 2.

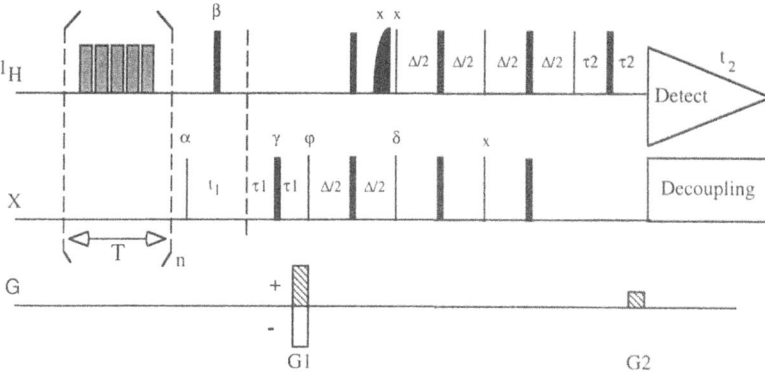

Figure 4. Measurement of the cross relaxation rate, $R_X(H^X_z \to X_z)$: Gradients and delays are similar to those in Fig. 2. Typically, two spectra are recorded. In one, the protons are not saturated and the X magnetization is assumed to be at the equilibrium Zeeman value X_{eq}. In the other, the protons are saturated to force the $<X_z>$ magnetization to a steady state intensity, X_{sat}. If the NOE values are small, then a series of spectra should be acquired for various values of τ to get an estimate for the error in the recorded NOE values. The maximum intensity of the nuclear Overhauser, η, effect is obtained using the relation: $\eta = (X_{sat} - X_{eq})/X_{eq}$ and is related to the cross relaxation rate by $\eta = (\gamma_H / \gamma_X).(R_X(X_z \to H_z) / R_X(X_z))$.

enables the identification of regions involved in slow conformational exchange processes (e.g. Lane & Lefèvre, 1994).

Pulse sequences with presaturation of water resonance have been reported for ^{15}N relaxation rate measurements that incorporate constant time and accordion spectroscopy; reduction in measuring time of 1/2 to 1/3 have been reported (Mandel & Palmer, 1994). Recently pulse sequences for $^{13}C_\alpha$-^1H (Yamazaki et al., 1994) and carbonyl ^{13}C (Dayie & Wagner, 1995a) measurements have been reported; but they will not be discussed here.

4. MODELS FOR CORRELATION FUNCTIONS AND SPECTRAL DENSITY FUNCTIONS

The overall tumbling of biomolecules in a non-viscous solvent is usually assumed to be a small-amplitude diffusive process (Debye, 1929; Favro, 1965; Huntress, 1975). As shown above (Eq. 3 and 5), the stochastic Hamiltonian can be partitioned into spin and lattice components, where the lattice components are characterized by the correlation function described in Eq. 8. Under the assumption of a rigid spherical top, the autocorrelation function has the widely used monoexponential decaying expression:

$$G_n(\tau) = <F_n^2(\tau)> \exp\left(-\frac{\tau}{\tau_c}\right) = \frac{1}{5} \exp\left(-\frac{\tau}{\tau_c}\right) = G(\tau) \tag{17}$$

In this case, the spectral density function, which is the cosine Fourier transform of the correlation function is a simple Lorentzian:

$$J(\omega) = \frac{2}{5} \int_0^{+\infty} \exp\left(-\frac{\tau}{\tau_c}\right) \cos(\omega t) \, d\tau = \frac{2}{5} \frac{\tau_c}{1 + \omega^2 \tau_c^2} \tag{18}$$

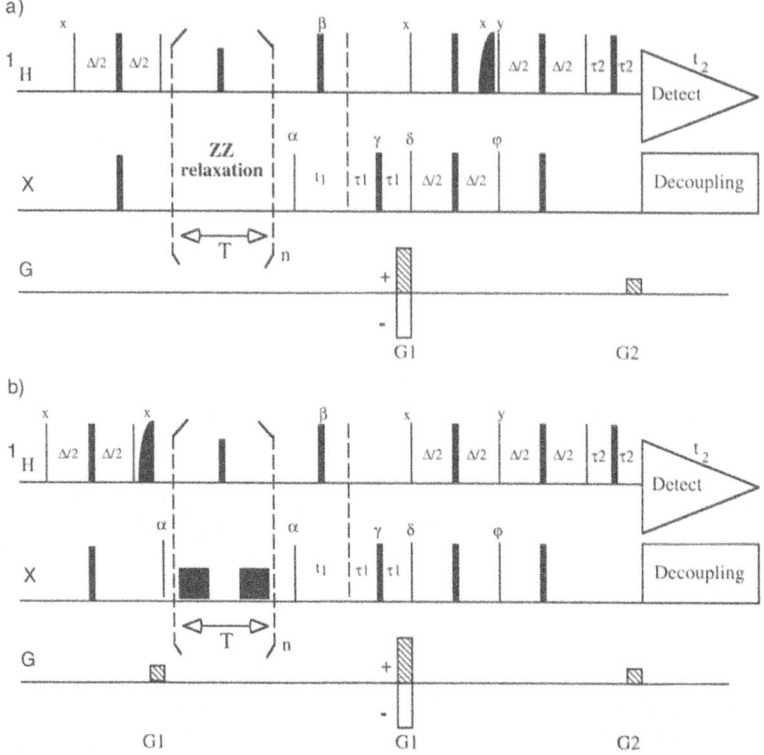

Figure 5. Measurements of two-spin orders: Gradients and delays are similar to those in Fig. 2. a) $R_X(2H_zX_z)$ longitudinal two spin order is created immediately after the second 90° proton pulse, b) $R_X(2H_zX_{xy})$: it is then converted into antiphase coherence after the relaxation period for subsequent t_1 labeling. In the case of antiphase measurements, this coherence is maintained with low power spin-lock prior to the t_1 labeling, similar to the one used for $R_X(X_{xy})$ measurements. In this case, only one 180° pulse is applied to attached protons during the relaxation delay. An identical 180° pulse is placed in the middle of the relaxation delay for the $R_X(2H_zX_z)$ sequence.

G (τ) can be calculated using the inverse Fourier transform of $J(\omega)$. The area under $J(\omega)$ is equal to the correlation function for $\tau = 0$ and therefore is constant:

$$G(0) = \int_0^{+\infty} J(\omega)\, d\omega = \frac{1}{5}$$

$$(19)$$

Inversely the spectral density function at zero frequency has a value proportional to the integral of the correlation function, which, in the case of an isotropic tumbling, is equal to the correlation time:

$$J(0) = \int_0^{+\infty} \exp\left(-\frac{\tau}{\tau_c}\right) d\tau = \frac{2}{5}\tau_c$$

$$(20)$$

The result of these two features is that, when the correlation time is long (slow motion), the spectral density gathers its intensity in the low frequency range, while it becomes flat and spreads over a wide frequency range when τ_c is short (fast motion).

The motions of N-H, C-H vectors, or tensors in macromolecules are segmental and made up of the overall tumbling of the particle, combined with internal motion which can be slower, of the same order or faster than the correlation time τ_c ; thus the correlation function has a complicated shape.

Internal motions in the slow regime do not contribute to the relaxation through transitions between the energy levels, as described above. However, they may affect the shape of the resonance line by acting continuously on the Larmor frequency of the nucleus. This occurs for a molecule exchanging between several states (at least two) in the interme- diate regime (e.g. Lane & Lefèvre , 1994). The change in the orientation of the molecule due to Brownian tumbling has the same effect through the modulation of the local magnetic field. Because this process does not imply energy exchange it is said to be an adiabatic process. It therefore affects the value of the spectral density function at zero frequency and only the relaxation of transverse magnetization.

When the correlation time for the internal dynamics is similar to the time for the reorientation of the whole macromolecule, the two motions become indistinguishable. Fast internal motion affects the effective correlation time of the vectors and the relaxation processes. One generally admits that internal and overall motions are uncorrelated, so that the correlation function can be decomposed into a weighted sum of contributions:

$$G(\tau) = \sum_i a_i \, G_i(\tau) \qquad \text{with} \quad \sum_i a_i = 1$$

$$(21)$$

which results in a similar expression in the frequency domain for the spectral density function:

$$J(\omega) = \sum_i a_i \, J_i(\omega)$$

$$(22)$$

This general formalism is used in many studies attempting to identify the various components of motions in macromolecules. Some approaches start from a specific model of motion: a vector freely rotating about an axis attached to a tumbling sphere leads to a three components spectral density function (Woessner, 1962). This model is generalized to a multiple bond chain freely diffusing by independent rotations around the bonds (Wallach, 1967; Levine et al., 1973; 1974; London & Avitabile, 1976). This latter model is further refined by restricting the conformational space available to the chain (Wittebort & Szabo, 1978). This excluded volume is introduced to realistically take in account the steric constraints experienced by a chain imbedded within the matrix of other side chains. An alternative to the restricted diffusion model is the jump model (Wittebort & Szabo, 1978; London, 1980) where the atoms jump from one node to another on a lattice representing sterically allowed conformations with the constraint, equivalent to the excluded volume, that two atoms cannot occupy the same node.

These models have been mainly applied to side chains. Modelling the backbone motions in proteins is a more difficult task because the restrictions of motions depend on the structure and the environment of each residue. Molecular dynamic simulation may be used to account for some of the relaxation data (e.g. Kordel et al., 1992b). However trajectories are still restricted to the very short time scale. Thus large amplitude movements, which are likely to be the effective movements for relaxation, may be too rare events to be correctly sampled by these simulations. The use of lattice modelling of proteins (Dill et al, 1995) may be a way to efficiently explore large amplitude motions but has not been used yet for this purpose.

Other approaches rely more on the mathematical analysis of the spectral density function trying to identify the weights (a_i) and the correlation times (τ_c) contributing to Eq. 22. Jardetzky and co-workers (King & Jardetzky, 1978; King et al., 1978) using a Markovian formalism, investigated the possibility of determining the various components from a set of relaxation experiments at various fields. The minimal number of components allowing a good fit with the relaxation data is found by progressively increasing the number of contributions. A simplified version of this approach was introduced by Lipari and Szabo (1982) who use only two contributions, one for the overall tumbling and the other for internal dynamics. While the two approaches are similar, but with different motivations and reach different conclusions, Lipari and Szabo argued that rapid motion can be described by no more than two parameters: The coefficient for the spectral density function corresponding to the tumbling and interpreted as an order parameter S^2, and the correlation time for internal motion, τ_i, is defined as the integral of the internal correlation function. The spectral density function has the simple expression:

$$J(\omega) = S^2 \cdot \frac{2}{5} \cdot \frac{\tau_c}{1 + (\omega \cdot \tau_c)^2} + (1 - S^2) \cdot \frac{2}{5} \cdot \frac{\tau_e}{1 + (\omega \cdot \tau_e)^2}; \quad \tau_e^{-1} = \tau_c^{-1} + \tau_i^{-1} \tag{23}$$

The correlation times and the order parameters are calculated by fitting three relaxation rate constants, $R_X(X_z)$, $R_X(X_x)$ and $R_X(H_z\text{->}X_z)$. An exchange rate term, R_{ex}, is added to the transverse relaxation rate constant and used as a variable parameter during the fitting procedure, in order to take into account possible slow motions lying in the intermediate regime (e.g. Kay et. al, 1989; Clore et. al. 1990). As the number of parameters is larger than the number of independent data, the fully extended approach (comprising the third component and the exchange term) cannot be used by brute force, and the fitting procedure is performed stepwise, by determining first the overall correlation time, then adding the very fast internal motion and finally the exchange terms if the performance of the fitting requires it (e.g. Mandel et al., 1994).

The use of the Lipari-Szabo formalism and its extension to fit data by adding progressively new contributions until data are correctly fitted clearly becomes an underdetermined problem when too many contributions are added with respect to the number of available data. Once the various contributions to the spectral density function are found, it remains the problem of determining the kind of motions responsible for them. We are therefore back to the use of specific motional models.

5. DIRECT DETERMINATION OF FULL AND REDUCED SPECTRAL DENSITY FUNCTION

The conventional approach to relaxation data analysis that involves fitting the set of three relaxation parameters $R_X(X_z)$, $R_X(X_x)$ and $R_X(H_z\text{->}X_z)$ to a model of spectral density function as discussed above is clearly not truly model-free. Recently Peng and Wagner (1992) proposed a direct determination of the values of the spectral density function independent of any mechanistic models at the five frequencies of 0, ω_X, ω_{H-X}, ω_H and ω_{H+X}. The five relaxation rates needed to determine the five unknown spectral density functions are chosen to be: $R_X(X_z)$, $R_X(X_x)$, $R_X(H_z\text{->}X_z)$, $R_X(2H_zX_z)$, and $R_X(2H_zX_x)$. However, the relaxation rates for the 2 spin order magnetizations are contaminated by spin-lattice relaxation rate of the directly attached amide proton to other non-amide protons; both measurements are, therefore, more error prone. The additional measurement of the longitudinal relaxation rate of the amide proton $R_H(H_z)$ enables a simple analytic solution to the system of linear equations relating the six relaxation parameters and the spectral density function values complete with the dipolar term for the relaxation of the proton by other protons ($\rho_{HH'}$):

$$
\begin{bmatrix}
R_X(X_z) \\
R_X(X_x) \\
R_X(H_z{\rightarrow}X_z) \\
R_X(2X_zH_z) \\
R_X(2X_xH_z) \\
R_H(H_z)
\end{bmatrix}
=
\begin{bmatrix}
0 & E & A & 0 & 6A & 0 \\
\dfrac{2E}{3} & \dfrac{E}{2} & \dfrac{A}{2} & 3A & 3A & 0 \\
0 & 0 & -A & 0 & 6A & 0 \\
0 & E & 0 & 3A & 0 & 1 \\
\dfrac{2E}{3} & \dfrac{E}{2} & \dfrac{A}{2} & 0 & 3A & 1 \\
0 & 0 & A & 3A & 6A & 1
\end{bmatrix}
\times
\begin{bmatrix}
J(0) \\
J(\omega_X) \\
J(\omega_H-\omega_X) \\
J(\omega_H) \\
J(\omega_H+\omega_X) \\
\rho_{HH'}
\end{bmatrix}
\tag{24}
$$

which is equivalent to:

$$
R = C \cdot J \tag{25}
$$

with

$$
A = \left(\frac{\mu_o}{4\pi}\right)^2 \cdot \frac{\gamma_H^2\,\gamma_X^2\,\hbar^2}{4r_{XH}^6}, \quad B = \frac{\Delta^2\omega_x^2}{3}, \quad E = 3A + B \tag{26}
$$

μ_o is the permeability of the vacuum ($4.\pi.10^{-7}$), and Δ the chemical shift anisotropy (-160 ppm for nitrogen (Hiyama et al.) and 25-35 ppm for $^{13}C^\alpha$–carbon (Ye et al., 1993)). The values of $J(\omega)$ can be obtained from the system described in Eq. 24 by inverting the matrix:

$$
J = C^{-1} \cdot R \tag{27}
$$

The calculation of the spectral density function using the complete approach described above has been performed on three ^{15}N enriched proteins: the protease inhibitor eglin c, at three different magnetic fields (Peng & Wagner, 1995), on the DNA binding domain of the transcription factor GAL4(1-65) (Lefèvre et al., 1995) at 11.74 Tesla, and on the N-terminal 14kD domain of the actin binding protein Villin at 11.74 Tesla (Markus et al., 1995). The very thorough analysis on eglin c shows that the segmental dynamics of the protein in the C terminus and the inhibitory loop is well reflected in the low frequency values of the spectral density function ($J(0)$ and $J(\omega_N)$), while the behavior of the function is less correlated with the structural elements at high frequencies. The same observation is made on GAL4(1-65) and Villin 14T where even negative values of $J(\omega)$ around the proton frequency are obtained. These anomalies are attributable to the high sensitivity of the high frequency values of the spectral density function to small errors in the rate constants. These values are small and calculated in Eq. 27, from a linear combination of large relaxation rate constants. Nevertheless, the multiple frequency study on eglin c shows that in the range 400 to 600 MHz, the spectral density function is quite flat. This suggests that the values of $J(\omega)$ for the proton, the double quantum and the zero quantum frequencies can be equalized in Eq. 24. Three values can be left in the vector J: $J(0)$, $J(\omega_X)$ and a unique average value at the proton frequency, $J_{avg}(\omega_H)$. The three first relaxation rate constants of vector R are now sufficient to determine the reduced spectral density function set:

$$R_{red} = C_{red} \cdot J_{red}, \qquad C_{red} = \begin{bmatrix} 0 & E & 7A \\ \dfrac{2E}{3} & \dfrac{E}{2} & \dfrac{13A}{2} \\ 0 & 0 & 5A \end{bmatrix}$$

(28)

It is noteworthy that if a fourth relaxation rate constant is added, using either the relaxation of the J-ordered ($2H_z\,X_z$) or the antiphase ($2H_z\,X_x$) states, the longitudinal relaxation rate constant of the proton itself can be calculated. This is of significant value as the measurement of this constant is very difficult due to the multi-exponential character of the relaxation curve. In practice it is preferable to use the measurements of J-ordered state ($2H_z\,X_z$) because they are more accurate than $2H_z\,X_x$.

As shown by simulations (Lefèvre et al. 1995), this simplification does not alter the values of J(0) and J(ω_X), and $J_{avg}(\omega_H)$ is very similar to J(ω_H -ω_X). The values $J_{avg}(\omega_H)$ appear now to be less erratic because they are determined by the cross relaxation rate constants.

6. ANALYSIS OF THE SPECTRAL DENSITY FUNCTION IN PROTEINS

The analysis of variation of the spectral density functions with the polypeptide sequence, secondary, and tertiary structure of five proteins are presented here. First those of two ^{15}N enriched proteins which do not seem to experience slow motion are discussed: the relaxation data for Eglin c as reported by Peng and Wagner (1992; 1995), and those for Calcium loaded Calbindin from the work of Chazin and co-workers (Kördel et al., 1992; Akke et al., 1993). This is followed by three which show some degree of millisecond types of motion, Villin 14T (Markus et al. 1995; Dayie & Wagner, 1995b), GAL4(1-65) (Lefèvre et al. 1995), and PMPD2 (Mer et al., 1995). The spectral density functions are calculated from the three relaxation rates: $R_X(X_z)$, $R_X(X_x)$ and $R_X(H_z->X_z)$.

The value of the spectral density function at zero frequency is the most diagnostic of nano-second to millisecond motion (Fig. 6–8). Regions of well-defined secondary structure have uniformly high J(0) values. The loops connecting the helices and beta-sheets show reduced J(0) values indicative of enhanced mobility relative to the core residues forming helices and beta-sheets. For example in Villin 14T (Markus et al., 1995; Dayie & Wagner, 1995b), the loops are differentially constricted; the most mobile appears to connect β2 and β4. The last six c-terminal residues show the least restriction to motion, progressively getting floppier and floppier until the last residue Lys126 appears completely unrestricted. Picoseconds motions are better reflected in the values of the spectral density function at the proton frequencies. These higher frequency components display behavior complimentary to those observed at lower frequency. Regions of least restriction have high values at the proton frequencies whereas regions of high restriction exhibit low values. This compensatory behavior is revealed by the anticorrelation of J(0) and J(ω_H) as discussed below. A similar behavior is seen in the other proteins: the values of J(0) are lower where the backbone is more mobile in the N terminus and the inhibitory loop for Eglin C, in both terminal ends and the loop between the two EF hands for Calbindin (Fig. 6).

The general pattern is therefore the following: J(ω) varies smoothly as a function of sequence. For instance, J(0) decreases slowly when the sequence enters a loop, reaches a minimum in the middle of the loop and then increases progressively back to the value of the less mobile part of the backbone (Fig. 6). This is also true for mobile terminal ends where J(0) drops from high to low value in several step residues. This behavior suggests an

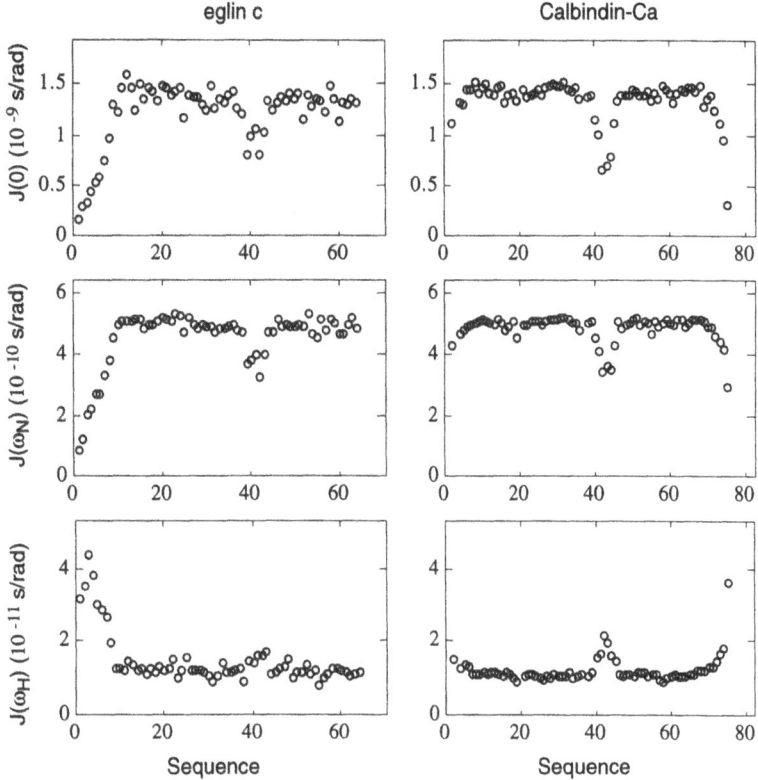

Figure 6. Spectral density functions at the three frequencies: 0 (top), ω_N (middle) and ω_H (bottom) plotted as a function of the protein sequence for eglin c (left) and for Calcium loaded Calbindin (right), both at 35°C. For eglin c, helices span from F10 to V14 and V18 to Y29, and the inhibitory loop between P38 and V49. In Calbindin, four helices are found E4-A14, K25-E35, L46-D54 and F63i73. The flexible loop between the two EF hands is F36-T45.

amplification along the sequence of the movement initiated at flexible points. The further the residue is located from these points, the larger is the amplitude of its motion. The apparent mobile sequence does not need to be all flexible. This has been demonstrated for the inhibitory loop of eglin C which has a well define structure and moves with respect of the rest of the protein because of two hinges on each side of the loop (Hybert et al., 1992).

6.1. Linear Correlation of $J(\omega_x)$ with $J(0)$: Determination of the Correlation Times for the Main Motions

The problem is now to disentangle the effects of overall tumbling from internal motion contributions to the spectral density function. Here, we make use of the striking empirical observation of the linear relationship between $J(0)$ and $J(\omega_x)$, as shown in Fig. 9 first reported by Lefèvre et al. (1995). Assuming that in flexible molecules the correlation function describing the reorientation of an internuclear vector is a composite function (Brooks et al, 1988) such that the overall tumbling and internal motions are uncorrelated, one may express the spectral density function as a weighted sum as described earlier (Eq. 22). Combined with our empirical observations, we can write the following equations valid for all residues and various combinations of a_i subject to the normalization condition $\Sigma_i\ a_i = 1$:

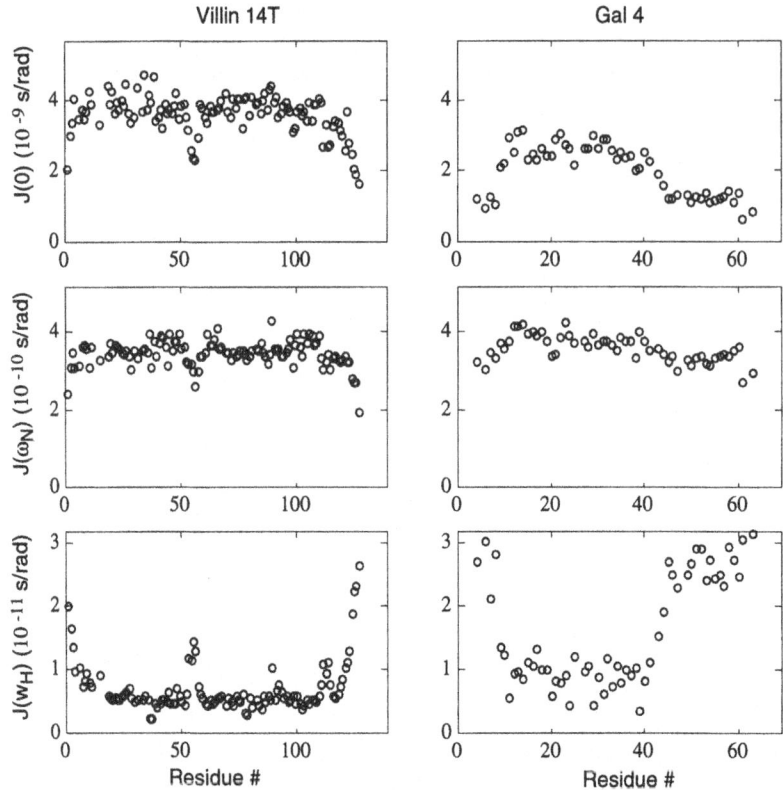

Figure 7. Spectral density functions at three frequencies, as in figure 6, plotted as a function of the protein sequence for the N terminal fragment of Villin 14T at 25°C and for Gal4(1-65) at 10°C. For Villin 14T, helices span L3-G9, Q71-L88 and E104-F110. In Gal 4, the sequences in helix are A10-K18, and L27-N34. Cysteins chelating the two zincs are C11, C14, C21, C28, C31 and C38.

$$\sum_i a_i J_i(\omega_X) = \alpha \cdot \sum_i a_i \cdot J_i(0) + \beta \qquad (29)$$

where α and β are the slope and the intercept of the linear fit to the experimental data. Without loss of generality, all the $J_i(\omega)$ components should obey the same linear relationship:

$$J_i(\omega_X) = \alpha \cdot J_i(0) + \beta \qquad (30)$$

Using the simplest form of the correlation function, a decaying mono-exponential such that the corresponding spectral density is Lorentzian, Eq. 30 leads to a third degree equation in τ:

$$\tau^3 \left[0.4 \cdot \alpha \cdot \omega_X^2\right] + \tau^2 \left\{\beta \cdot \omega_X^2\right\} + \tau \left[0.4 \cdot (\alpha - 1)\right] + \beta = 0 \qquad (31)$$

where the positive roots give the correlation time of the various motions contributing to the low frequency part of spectral density functions. In particular the overall correlation time is easily extracted from the plot of $J(\omega_N)$ versus $J(0)$. Table 2 shows the parameters from the linear fit for six proteins together with the published values for the overall correlation. In all cases the first root has no physical meaning, while the second and the third, on the basis of their magnitudes are assigned to the overall tumbling of the molecule and the internal motion respectively.

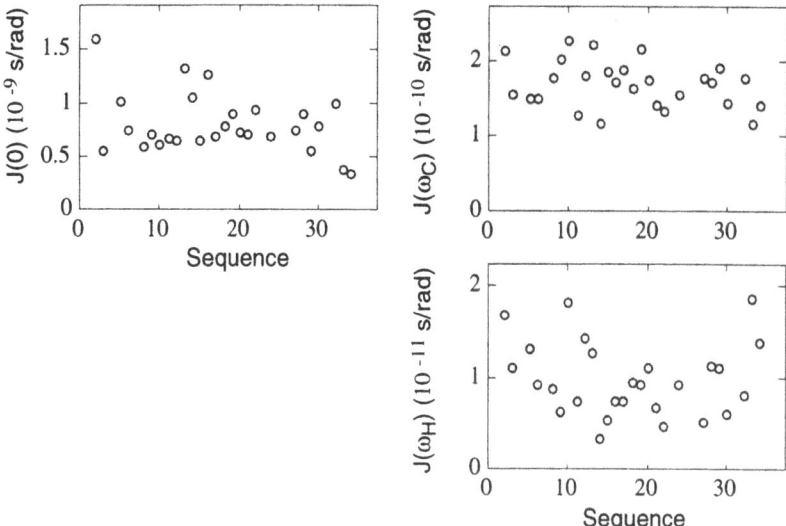

Figure 8. Spectral density function of PMP-D2 at 35 °C. This protease inhibitor folds in three antiparallel beta strands (Q8-K11, N15-C19 and W25-R29), maintained by three disulfide bridges (C4-C19, C14-C32 and C17-C27).

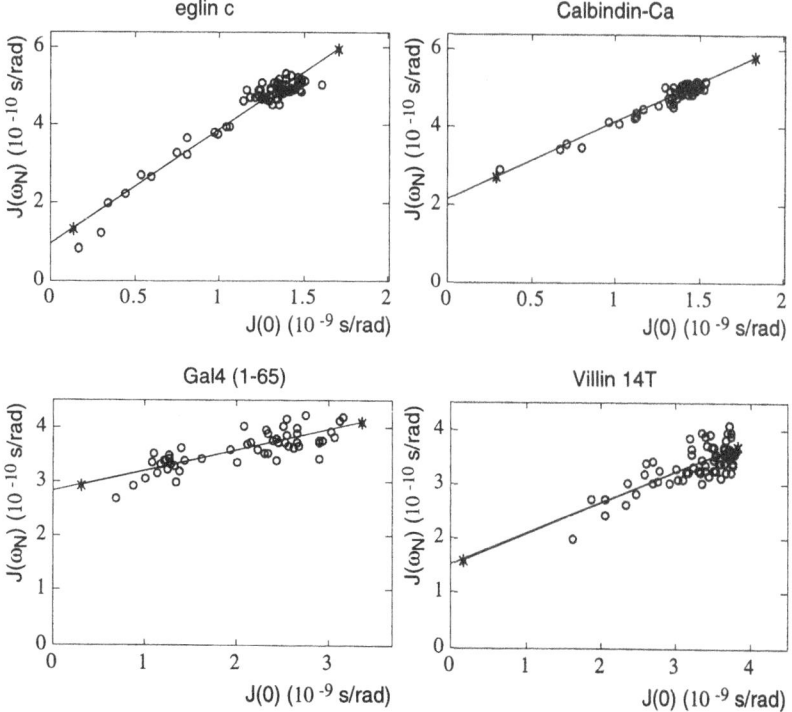

Figure 9. Plots of $J(\omega_N)$ as a function of $J(0)$ for four different proteins. All experimental data points are plotted except for Villin 14T which exhibit a lot of exchange process in the intermediate range. In this case, the data set has been purged by removing those for which $J(0)$ is greater than $3.8 \ 10^{-9}$ s and which lie after the highest positive root in the above representation. -: linear least square fit. *: positive roots of equation (31).

Table 2. Dynamic parameters deduced from heteronuclear relaxation data for six proteins. τ_c and τ_i are the correlation times deduced from Eq. 31 for the overall tumbling and the general internal motion respectively. The relaxation data and the correlation time τ_m are taken from the references indicated in the text. Naa is the number of aminoacids, T the temperature at which the relaxation data where collected and X the nature of the heteronucleus. (a): the relaxation data set was purged by removing the data for which J(0) is higher than a threshold value (3.8 ns/rad and 0.9 ns/rad for Villin 14T and PMP-D2 respectively), prior to the correlation time calculation. (b) deduced from the translational diffusion constant measured with a gradient enhanced spin echo method (Kieffer & Lefèvre, 1996)

Protein	Naa	T(°C)	X	τ_c (ns)	τ_i (ps)	τ_m (ns)
Eglin c	70	36	^{15}N	4.26	347	3.9-4.1
Gal4 (1-65)	65	10	^{15}N	8.37	779	8.6
Villin 14T	126	25	^{15}N	9.53[a]	405	10.2
Calbindin-Ca	75	27	^{15}N	4.56	721	4.25
Apo-neocarzinostatin	113	35	^{13}C	6.67	177	5.7
PMP-D2	35	35	^{13}C	2.69[a]	357	2.6[b]

Now, the respective contribution to the spectral density function of overall (hereafter indexed with o) and internal motions (indexed with k) can be calculated by modelling J(ω) according to Eq. 22:

$$J(\omega) = a_o \cdot J_o(\omega) + \sum_k a_k \cdot J_k(\omega) \tag{32}$$

$J_o(\omega)$ is given a pure Lorentzian shape, using the correlation time found previously (with Eq. 31), or a composite shape if the molecule is anisotropic and the orientation of each H-X vector in respect with the major axis is known (vide infra). With the three experimental J(ω) values, only three unknowns can be determined from the above equation. It is expected that the rapid movements inside the macromolecule are characterized by a rather flat spectral density function in the low frequency domain, such that one can make the following approximation:

$$\sum_k a_k \cdot J_k(\omega_X) = \sum_k a_k \cdot J_k(0) \tag{33}$$

This assumption reduces the number of unknowns to three:

$$a_o, \quad J_{int}(0) = \sum_k a_k \cdot J_k(0), \quad J_{int}(\omega_H) = \sum_k a_k \cdot J_k(\omega_H) \tag{34}$$

which can be determined by solving the system of Eq. 32.

The results are shown for eglin c and calcium loaded Calbindin in Fig. 10. It should be noted that a_o is analogous to the order parameter S^2 defined above. However, there is no correlation between the internal correlation times defined in the model free approach (Lipari & Szabo, 1982a) and the values of the internal spectral density function, $J_{int}(0)$. The observed linear relationship between J(ω_x) and J(0) is recovered when $J_{int}(0)$ is plotted against $1-a_o$ (Fig. 11). When extrapolated to $a_o = 0$, where the contribution of the overall tumbling is null, $J_{int}(0)$ has the value of the Lorentzian function, calculated at zero frequency, corresponding to the internal correlation time found with Eq. 31:

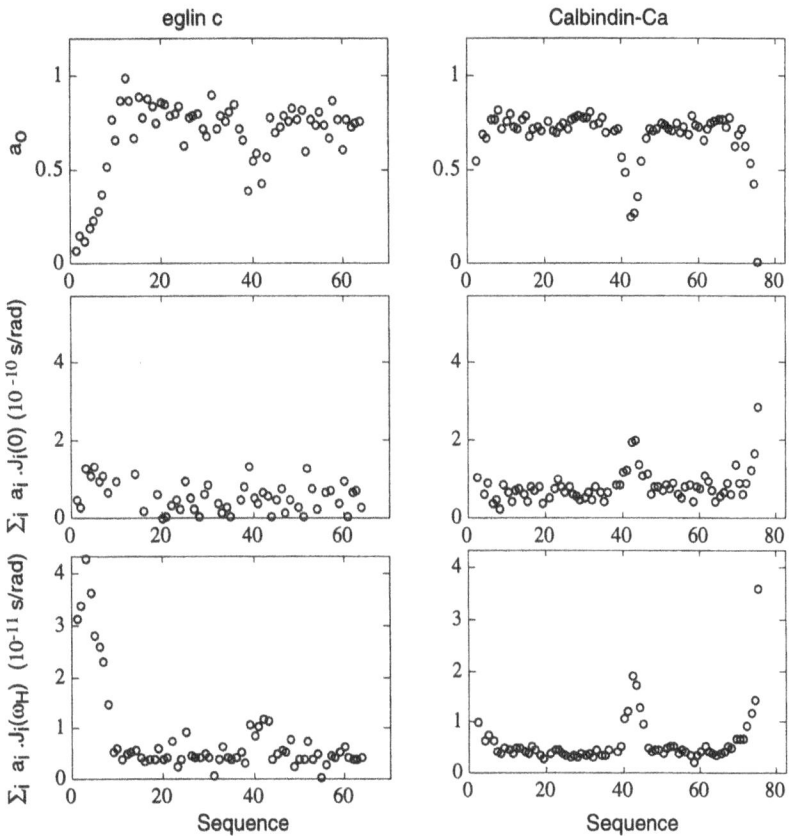

Figure 10. Decomposition of the spectral density function for eglin c and calcium loaded Calbindin: a_o is the weight of the overall tumbling contribution, $\Sigma a_i J_i(0)$ and $\Sigma a_i J_i(\omega_H)$ are the values of the internal spectral density function at low and high frequency (see Eqs. 32-34).

$$J_{int}(0) = \frac{2}{5} \cdot \tau_{int} \qquad (35)$$

τ_{int}, which is actually the third root of Eq. 31, appears as a general correlation time for the internal motions. Interestingly, although more noisy, the same linear relationship is found between $J_{int}(\omega_H)$ and $1-a_o$ (Fig. 11). The values of $J_{int}(0)$ and $J_{int}(\omega_H)$ extrapolated at $a_o = 0$ do not fall on a unique Lorentzian, suggesting that the spectral density function for the internal motion is itself a composite function, as noted by the sum over contributions k in Eq. 34, or that it does not have a Lorentzian shape.

The observations made above may reflect the fact that the movements which are significant for relaxation, that is with an amplitude which significantly affects the dipolar and the CSA interaction, all lie in the same frequency range. The spectral density functions contributing from the overall and internal motions would be almost constant from one residue to the other, and the resulting spectral density function would be modulated by the amplitude of the motion, reflected in the weights a_o and a_k.

6.2. Extracting the Exchange Contribution from the Spectral Density Function

The value obtained for the spectral density function at zero frequency may be contaminated by an adiabatic contribution from motions which are slower than the overall

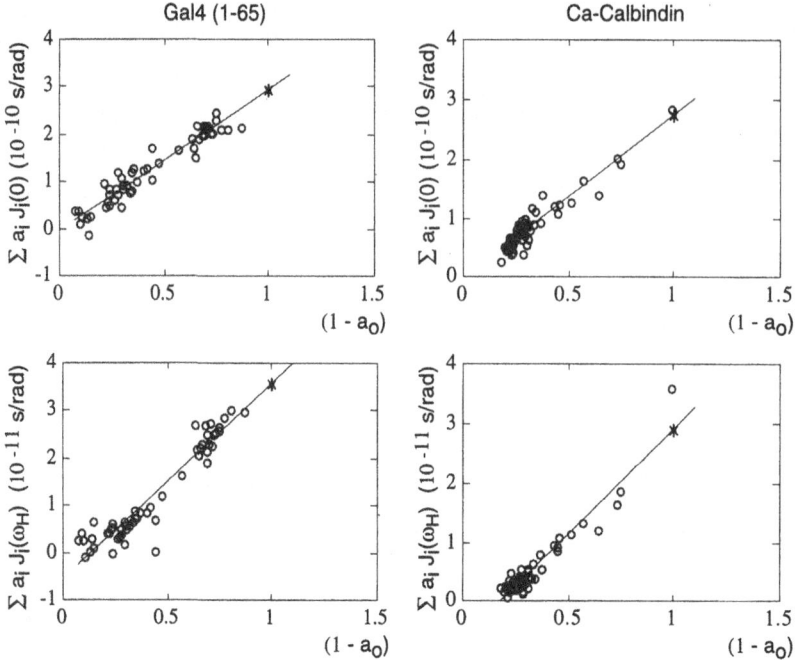

Figure 11. Plots of the contribution of internal motions to the spectral density function at low (top) and high (bottom) frequencies as a function of $1-a_o$. Data are shown for Gal4 (1-65) and calcium loaded Calbindin.

tumbling of the molecule and lie in the microsecond to millisecond range. Such a contribution increases the transverse relaxation rate and hence $J(0)$ thus:

$$R_{ex} = \frac{3E}{2} \left| J^{exp}(0) - J^{cal}(0) \right| \tag{36}$$

$J^{exp}(0)$ and $J^{cal}(0)$ are respectively the experimentally observed and theoretically calculated spectral density function at zero frequency, and E is the weighted sum of the dipolar and CSA interaction constants (see Eq. 26). $J^{cal}(0)$ can be obtained from a fit of the experimental $J(0)$ as a function of sequence, with a smooth function (like a third degree polynomial or an average taken on a sliding window). The experimental points which diverge from the fitted curve by more than a threshold value are removed from the data set. The threshold can be calculated from a root mean square deviation, and the anisotropy of the molecule may be taken in account in the calculation (vide infra). The fitting and sorting procedure is carried out twice or more, as needed, until the experimental data are close to the fitted curve. An example of this method applied to the GAL4(1-65) relaxation data is shown in Fig. 12.

Assuming an exchange between two states A and B, characterized by the on and off rates k_1 and k_{-1} respectively, a resonance frequency difference Δv, and populations P_A and P_B in the A and B states, the expression of R_{ex} is (Deverell, 1970; Sandström, 1982):

$$R_{ex} = \frac{4 \cdot \pi^2 \cdot \Delta v^2 \cdot P_A \cdot P_B \, \tau_{ex}}{1 + (\tau_{ex} \cdot \omega_1)^2} \tag{37}$$

where $(k_1 + k_{-1})^{-1} = \tau_{ex}$; ω_1 is the strength, expressed in angular frequency, of the B_1 field utilized to spin lock the magnetization state in the xy plane. The exchange time τ_{ex} can be extracted from a set of data acquired with different spin lock field strengths (Deverell, 1970).

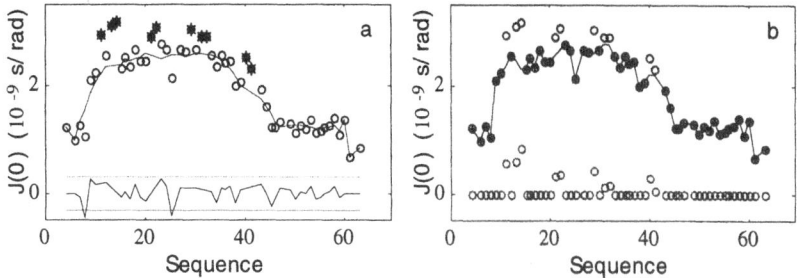

Figure 12. Analysis of exchange contribution to J(0) for Gal4 (1-65) at 10°C. a) Top: the data were smoothed twice with an averaging sliding window. J(0) values greater than one standard deviation (*) were removed after the first smoothing. Bottom: difference between the average and the experimental values and threshold values placed at ± 2 standard deviation. b) Top: the missing points are reconstructed from the conserved data (x) using here a spline function. Bottom: the exchange term is calculated from the difference between the experimental and the recalculated J(0) values.

6.3. Effect of the Anisotropy of the Molecule

When the shape of the protein departs from a simple spherical model, the contribution of the overall tumbling to the motion of the X-H vectors depends on their orientation in the molecule-fixed frame. The simplest models, other than spheres, are ellipsoids defined by a major (a) and a minor (b) axis. Oblate ellipsoids are generated by revolution of a 2D ellipse around the minor axis (this gives a pumpkin shape). Revolution around the major axis results in prolate ellipsoid (like a rugby ball). In the following, we will concentrate on the latter shape which is the most commonly observed for macromolecules. Two correlation times now define the tumbling of the prolate ellipsoid: one for rotational diffusion of the major axis (τ_a, by combined rotations around the minor axes) and another common one for the diffusion of each minor axis (τ_b, by combined rotations around the major and the other minor axis). These correlation times can be calculated from the hydrodynamic expressions worked out by Perrin (1934; see also Cantor & Shimmel, 1980):

$$\tau_a = \tau_s \cdot F_b; \quad \tau_b = \tau_s \cdot 2\left[\frac{1}{F_a} + \frac{1}{F_b}\right]^{-1} \tag{38}$$

where τ_s is the correlation time for a sphere with a volume v equivalent to the volume of the ellipsoid (v = (4/3) $\pi.a.b^2$). In a medium of viscosity η, at a temperature T (°K):

$$\tau_s = \frac{\eta \cdot v}{k_B \cdot T} \tag{39}$$

F_a and F_b are the ratio of the frictional coefficients for rotation about the major and the minor axis respectively to the frictional coefficient of the equivalent sphere. These ratios depends only on the axial ratio p = a/b:

$$F_a = \frac{4(1 - p^{-2})}{3(2 - S \cdot p^{-2})}, \quad F_b = \frac{4(1 - p^{-4})}{3 \cdot p^{-2}[S(2 - p^{-2}) - 2]}, \quad S = \frac{\ln\left(\frac{1 + \sqrt{1 - p^{-2}}}{p}\right)}{2\sqrt{1 - p^{-2}}} \tag{40}$$

The corresponding equations for an oblate ellipsoid can be found in Cantor and Shimmel (1980). The contribution of the overall tumbling to the spectral density function for a particular vector making an angle Θ with the major axis can be calculated from the expressions given by Woessner (1962):

$$J(\omega) = a_1 \cdot J(\omega,\tau_1) + a_2 \cdot J(\omega,\tau_2) + a_3 \cdot J(\omega,\tau_3) \tag{41}$$

where $J(\omega,\tau_i)$ has the classical expression of a Lorentzian corresponding to a correlation time τ_i (see Eq. 24):

$$\tau_1 = \tau_a, \quad \tau_2 = \frac{6 \cdot \tau_a \cdot \tau_b}{\tau_a + 5 \cdot \tau_b}, \quad \tau_3 = \frac{3 \cdot \tau_a \cdot \tau_b}{2 \cdot \tau_a + \tau_b} \tag{42}$$

and:

$$a_1 = \left(\frac{3\cos^2\Theta - 1}{2}\right)^2, \quad a_2 = 3\sin^2\Theta \cos^2\Theta, \quad a_3 = \frac{3\sin^4\Theta}{4} \tag{43}$$

An example of such a calculation is given in Fig. 13 where the theoretical values of $J(\omega)$, assuming a rigid body, are plotted as a function of the sequence for a small protease inhibitor, PMP-D2. The three dimensional structure of the protein is known (Mer et al., 1995). The tensor of inertia was calculated, giving a weight of one to all atoms (in fact, as all atoms are more or less evenly spread through the molecule, it does not make much difference whether the true or unity atomic weights are taken for the calculation). The tensor is then diagonalized in order to find the axes. The largest projections of the atom positions on each axis give the magnitude of the ellipsoid axes. In the present case, the dimensions of the axes were 40Å x 20Å x 20Å. The protein appears as a prolate ellipsoid with an axial ratio p=2. On the other hand, the correlation times τ_a and τ_b were deduced from the diffusion constant of the molecule, measured with a gradient stimulated spin echo experiment (Kieffer & Lefèvre, 1996). Knowing the angles made by the C_α-H_α vectors with the long axis of the molecule (angle β), and assuming a rigid molecule, the theoretical values of $J(\omega)$ were calculated according to (41).

Experimental $J(0)$ values are in general smaller than these theoretical values which were calculated without accounting for internal motion. However, in few places where they are much higher, clearly pointing to a participation of an exchange process. It should be noticed that unless the axial ratio is very large (p>2), the effect of the anisotropy is much smaller than the effect of internal motion, as can be seen on Fig. 13.

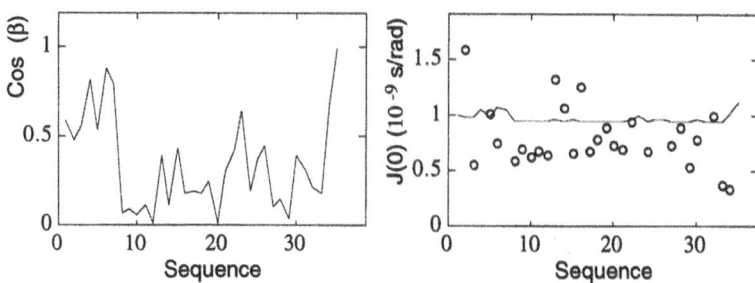

Figure 13. Effect of anisotropy on the spectral density function of PMP-D2 which has a prolate ellipsoid shape with an axial ratio of 2. Left: orientation of the vectors in respect with the major axis (angle β). Right: plot of the experimental (o) and the theoretical (-) value of $J(0)$, taking in account the anisotropy of the protein and assuming no internal motions.

7. CONCLUSION

After the initial excitement that NMR can solve protein structures at atomic resolution, attention is shifting to aspects of mobility, ligand binding, and mapping of interaction sites. The rapid progress in experimental techniques (Nirmala & Wagner, 1988; Kay et al., 1989; Palmer et al., 1992; Peng & Wagner, 1992; Dayie & Wagner, 1994; Farrow et al. 1995) and important conceptual advances (Lipari & Szabo, 1982a,b; Peng & Wagner, 1992; Wagner, 1994; Ishima & Nagayama, 1995; Farrow *et al.* 1995; Lefèvre *et al.* 1995) have made heteronuclear ^{1}H-^{15}N relaxation studies of proteins almost routine. The task that faces scientists now is identifying the various types of motions and establish their relevance to structure and function.

Firstly, it is not yet understood whether the motion observed by heteronuclear relaxation is an activated process or simply results from vibrations and libration of valence bonds and dihedral angles with very low energy barrier. A tentative answer may be sought from considering an Arrhenius plot with the internal correlation time found for various proteins at various temperature (see table 2). The resulting activation energy is about 7 kcal.mol^{-1}. This of course does not refer to the same molecule and should be considered with care. However, should the result be confirmed, we may be able to understand the determinants of such activated motion.

Secondly, while the past decade has taught us that we can probe millisecond to picoseconds motions with relaxation studies with ease, it is not clear yet how those motions couple ultimately to modulate the functioning of such macromolecules. In a sense this third dimension is still an open-ended question, and the coming years should see more increase in comparisons of experimental rates and values of spectral density functions with molecular dynamics simulations. This it is hoped might provide better insight internal motions in proteins than we can currently glean from order parameters.

REFERENCES

Abragam, A.(1961) The Principles of Nuclear Magnetism, Clarendon Press, Oxford, Chap. VIII.

Akke, M, Skelton, N.J., Kördel, J., Palmer, A.G., Chazin, W.J. (1993) Biochemistry, *32*, 9832-9843.

Bax, A., De Jong, P.G., Mehlkopf, A.F., Smidt, A. (1980) Chem. Phys. Lett. *69*, 567-570.

Bax, A., Ikura, M, Kay, L.E., Torchia, D.A., Tschudin, R. (1990) J. Magn. Reson., *86*, 304-318.

Boyd, J, Hommel, U., Campbell, I.D. (1990) Chem. Phys. Lett., *175*, 477-482.

Brooks, C.L. III, Karplus, M., Pettitt, B. M. (1988) Proteins: A Theoerectical Perspective of Dynamics, Structure, and Thermodynamics, John Wiley and Sons: New York, pp 1-259.

Burum, D.P., Ernst, R.R. (1980) J. Magn. Reson., *39*, 163-168.

Cantor, C.R., Schimmel, P.R. (1980) Biophysical Chemistry: The Techniques for Study of Biological Structure and Function, W.H. Freeman and Co., New York, pp 562-565.

Cavanaugh, J.,Rance, M. (1993) Ann. Rep. NMR Spectrosc., *27*, 1-58.

Clore, M.G., Szabo, A., Bax, A., Kay, L.E., Driscoll, P.C., Gronenborn, A.M. (1990) J. Am. Chem. Soc., *112*, 4989-4991.

Davis, D.G. (1990) J. Magn. Reson., *90*, 589-596.

Dayie, K.T., Wagner, G. (1994) J. Magn. Reson. A, *111*, 121-126.

Dayie, K.T., Wagner, G. (1995a) J. Magn. Reson. B, *109*,105-108.

Dayie, K.T., Wagner, G. (1995b) In preparation.

Debye, P. (1929) Polar Molecules, The Chemical Catalogue: New York, pp. 1-172.

Deverell, C., Morgan, R.E., Strange, J.H. (1970) Mol. Phys., *18*, 553-559.

Dill, K.A., Bromberg, S., Yue, K., Fiebig, K.M., Yee, D.P., Thomas, P.D., Chan, H.S. (1995) Protein Science, *4*, 561-602.

Edmonds, A.R. (1957) Angular Momentum in Quantum Mechanics, Princeton Univ. Press: Princeton.

Farrow, N.A., Zhang, O.W., Forman-Kay, J.D., Kay, L.E. (1995) Biochemistry, *34*, 868-878.

Favro, L.D. (1960) Phys. Rev., *119*, 53-62.

Goldman, M. (1984) J.Magn.Reson., *60*, 437-452.

Goldman, M. (1988) Quantum Description of High Resolution NMR in Liquids, Clarendon Press: Oxford.

Grzesiek, S., Bax, A. (1993) J. Am. Chem. Soc., *115*, 12593-12594.

Hiyama, Y., Niu, C., Silverton, J.V., Bavoso, A., Torchia, D.A. (1988) J. Am. Chem. Soc., *110*, 2378-2383.

Huntress, W.T. (1970) Adv. Magn. Reson., *4*, 1-37.

Hurd, R.E. (1990) J. Magn. Reson., *87*, 422-428.

Hybert, S.G., Golberg, M.S., Havel, T.F., Wagner, G. (1992) Protein Science, *1*, 736-751.

Ishima, R., Nagayama, K. (1995) J. Magn. Reson. B, *108*, 73-76.

Ishima, R, Nagayama, K. (1995) Biochemistry, *34*, 3162-3171.

Kay, L.E., Torchia, D.A., Bax, A. (1989) Biochemistry, *28*, 8972-8979.

Kay, L.E., Keifer, P., Saarinen,T. (1992) J. Am. Chem. Soc., *114*, 10663-10665.

Kay, L.E., Nicholson, L.K., Delaglio, F., Bax, A., Torhcia, D.A. (1992) J. Magn. Reson., *97*, 359 375.

Kieffer, B., Lefèvre, J.F. (1996) manuscript in preparation

King, R., Jardetzky, O. (1978) Chem. Phys. Lett., *55*, 15-18.

King, R., Mass, R., Gassner, M., Nanda, R.K., Conover, W.W., Jardetzky, O. (1978) Biophys. J., *6*, 103

Kordel, J. Skelton, N.J., Akke, M, Palmer, A.G., Chazin, W.J. (1992) Biochemistry, *31*, 4856-4866.

Lane, A.N., Lefèvre, J-F. (1994) In Methods in Enzymology: Nuclear Magnetic Resonance, eds. T.L. James and N.J. Oppenheimer, San Diego, pp. 596-619.

Lefèvre, J-F., Dayie, K.T., Peng, J.W., Wagner, G. (1995) Biochemistry, submitted.

Li, Y-C., Montelione, G.T. (1994) J. Magn. Reson.B, *105*, 45-51.

Lipari, G., Szabo, A. (1982a) J. Am. Chem. Soc.,*104*, 4546-4559.

Lipari, G., Szabo, A. (1982b) J. Am. Chem. Soc.,*104*, 4559-4570.

Levine Y.K., Partington, P., Roberts, G.C.K. (1973) Mol. Phys., *25*, 497-514.

Levine Y.K., Birsdall, M.J.M., Lee, A.G., Metcalfe, J.C. Partington, P., Roberts, G.C.K. (1974) J. Chem. Phys., *60*, 2890-2899.

London, R.E., Avitable, J. (1978) J. Am. Chem. Soc., *100*, 7159-7165.

London, R.E. (1980) In Magetic Resonance in Biology, Cohen J.S., Ed., Wiley: New York, p1.

Mandel, A. M., Palmer III, A.G. (1994) J. Magn. Reson. A, *110*, 62-72.

Mandel, A. M., Akke, M., Palmer III, A.G. (1995) J. Mol. Biol., *246*, 144-163.

Markus, M. A., Dayie, K.T., Matsudaira, P., Wagner, G. (1994) J. Magn. Reson. B, In press.

Markus, M. A., Dayie, K.T., Matsudaira, P., Wagner, G. (1995) Biochemistry, In press.

Maudsley, A.A., Wokaun, A., Ernst, R.R. (1978) Chem. Phys. Lett., *55*, 9-14.

McCammon, J.A., Harvey, S.C. (1987) Dynamics of proteins and nucleic acids, Cambridge University Press, Cambridge, pp 1-234.

Mer, G., Dejaegere, A., Stote, R., Kieffer, B., Lefèvre, J.F. (1995) J. Phys. Chem, in press

Meserle, B.A., Wider, G., Otting, G., Weber, C., Wüthrich, K. (1989) J. Magn. Reson., *85*, 608 613.

Nirmala, Wagner, G. (1988) J. Am. Chem. Soc., *110*, 7557-

Otting, G., Wüthrich, K. (1988) J. Magn. Reson., *76*, 569-574.

Palmer, A.G. III, Skelton, N.J., Chazin, W.J., Wright, P.E., Rance, M. (1992) Mol. Phys., *75*, 699-711.

Peng, J.W., Wagner, G. (1992) J.Magn.Reson., *98*, 308-332.

Peng, J.W., Wagner, G. (1994) In Understanding Chemical Reactivity: Nuclear Magnetic Resonance Probes of Molecular Dynamics, (R. Tycko, Ed), Kluwer Academic Publishers:Boston, pp. 373-454.

Peng, J.W., Wagner, G. (1995) Biochemistry, In press.

Perrin, F. (1934) J. Phys. Rad. Ser. VII, *5*, 497-511.

Redfield, A. G. (1965) Adv. Magn. Reson., *1*, 1-32.

Richarz, R., Nagayama, K., Wüthrich, K. (1980) Biochemistry, *19*, 5189-5196.

Rose, M.E. (1957) Elementary Theory of Angular Momentum, J. Wiley: New York.

Sandström, J. (1982) Dynamic NMR Spectroscopy. Academic Press, New York.

Sklenar, V., Piotto, M., Leppik, R., Saudek, V. (1993) J. Magn. Reson., *102*, 241-245.

Stonehouse, J., Shaw, G.L., Keeler, J., Laue, E.D. (1994) J. Magn. Reson. A, *107*, 178-184.

Wallach, D.J. (1967) J. Chem. Phys., *47*, 5258-5268.

Wittebort, R. J., Szabo A. (1978) J. Chem. Phys., *69*, 1723-1736.

Woessner, D.E. (1962) J. Chem. Phys., *36*, 1-4.

Yamazaki, T., Muhandiram, R., Kay, L.E. (1994 J. Am. Chem. Soc., *116*, 8266.

Ye, C., Fu, R., Hu, J., Hou, L., Ding, S. (1993) Mag. Reson. Chem., *31*, 699-704.

Zare, R. (1988) Angular Momentum, John Wiley & Sons: New York.

12

HOW CONVENTIONAL ANTIGENS AND SUPERANTIGENS INTERACT WITH THE HUMAN MHC CLASS II MOLECULE HLA-DR1

Ted Jardetzky

Department of Biochemistry, Molecular Biology and Cell Biology
Northwestern University
2153 Sheridan Road, Evanston, Illinois 60208-3500

1. INTRODUCTION

Reliable molecular recognition is an important aspect of an effective immune response. In order to ensure a specific response against the appropriate pathogens, the immune system has evolved a network of regulatory mechanisms that distinguish between self and non-self at the molecular level. Three proteins of this network play a key role in this process, by interacting directly with antigens. These proteins are the antibody, the T cell receptor (TcR) and the Major Histocompatibility Complex (MHC) glycoproteins. Antibodies and the TcRs share a similar mechanism for generating a huge diversity of antigen combining sites by recombination of germ-line encoded gene segments. The subsequent diversity of these two proteins is found associated with individual cells, which clonally express unique antigen specificities at their cell surface. Antibodies are expressed by the B cell and the TcR is expressed by the T cell, where each cell is poised to recognize a distinct antigenic epitope. The TcR is further restricted to the recognition of peptide antigens bound to self MHC glycoproteins, after undergoing a developmental selection process in the thymus.

In contrast to these other two molecules, the MHC molecules do not undergo recombination events and they are not expressed as receptors of clonally distinct cells. MHC molecules, although they also must interact with foreign antigens, do so in a manner which does not require the generation of binding site diversity associated with antibodies and TcRs. MHC molecules are cell-surface glycoproteins and they are extremely polymorphic between individuals (Germain and Margulies, 1993), thereby providing cellular markers that enable the immune system to distinguish between self and non-self. The molecular mechanism by which this occurs has been elucidated in the last 5-10 years and many aspects of MHC function have been clarified by the study of the crystallographic structures of these molecules (Bjorkman et al., 1987; Bjorkman et al., 1987; Brown et al., 1993). Two classes of MHC molecules have been identified which play distinct roles in the generation of the immune response. Here we will focus on the MHC class II molecule HLA-DR1 and it interaction with both conventional antigen and bacterial superantigens.

Dynamics and the Problem of Recognition in Biological Macromolecules, edited by Jardetzky and Lefèvre
Plenum Press, New York, 1996

The function of MHC class II molecules is to present peptide fragments of foreign proteins to T cells of the immune system (Germain and Margulies, 1993). Peptide antigen presentation can be carried out by many specialized cells of the immune system, such as B cells and macrophages. B cells expressing antibodies that are specific for an intact protein antigen can be especially effective at internalizing foreign proteins, proteolyzing these proteins into short 13-25 amino acid fragments and assembling a complex with a newly synthesized MHC class II molecule. The foreign peptide complexed to the self MHC molecule is then displayed on the presenting cell surface where T cells can form specific interactions and respond by releasing lymphokines that induce B cell maturation and proliferation. Thus MHC class II molecules play a critical regulatory role in the antibody response. In the absence of a peptide:MHC complex to stimulate T cell responses, the majority of B cells will not produce soluble antibody in response to foreign protein antigen.

This places an important requirement on the functional properties of the MHC molecule and its interaction with peptides. The MHC molecule must be able to recognize many different peptide sequences, in order to insure that the panel of self MHC molecules that one inherits is sufficient to guarantee an immune response to just about any foreign protein. How is the peptide binding site of the MHC molecule constructed and how is it able to accommodate so many different peptides in its binding site? Although the structure determination of the human MHC class II molecule HLA-DR1 (Brown et al., 1993) has allowed us to begin to address these questions, many questions regarding the recognition of peptides remain to be answered. One important functional observation is that peptide binding and dissociation from MHC class II molecules is slow, with half lives that last on the order of days. Peptide loading of MHC class II molecules may be accelerated by a recently discovered protein, HLA-DM (Sloan et al., 1995), which may interact with other MHC class II molecules directly to increase the rate of peptide binding and dissociation in specific peptide loading compartments. However, the dissociation rates of peptides at the cell surface remain slow and are a distinguishing feature of these proteins. We have analyzed the density in the HLA-DR1 peptide binding site that corresponds to a mixture of endogenous peptides that copurify with the class II molecule as isolated from a lymphoblastoid cell line (Jardetzky et al., 1995). The density for this mixture is continuous and well defined, indicating that there is substantial similarity in the conformation and interactions of different peptides within the HLA-DR1 peptide binding site.

In addition to conventional peptide antigens, MHC class II molecules form interactions with a class of T cell stimulating proteins known as superantigens (Thibodeau and Sékaly, 1995). In the case of superantigen stimulation of T cells, the specificity of the TcR interaction with the MHC molecule is overridden by the superantigen. Both bacterial and viral proteins have been identified which can act as superantigens. These proteins have been shown to bind to both MHC molecules and TcR outside of the conventional antigen binding sites of each of these molecules. Different bacterial superantigens may have at least three distinct binding sites to MHC molecules, while the viral superantigen that has been most thoroughly studied, the mouse mammary tumor virus superantigen, may define yet another mode for the recognition of the MHC molecule by these proteins (Jardetzky, 1995). In contrast, all of the superantigens seem to share the same binding site on the T cell receptor and this has been mapped to the outer face of the variable domain of the TcR beta chain (Vβ). The solution of the crystal structure of the beta chain of the TcR has verified that this domain assumes an immunoglobulin fold and the predicted interaction site for superantigens lies adjacent to the hypervariable loops of the conventional antigen combining site of the protein (Bentley et al., 1995). The structure determination of the HLA-DR1 molecule in a complex with two bacterial superantigens, S. aureus enterotoxin B (SEB) (Jardetzky et al., 1994) and Toxic Shock Syndrome Toxin (TSST-1) (Kim et al., 1994), has elucidated two overlapping binding sites for these toxins on the same MHC molecules and raised the possibility that these toxins may bind to distinct peptide-loaded populations of MHC class II molecules at the cell surface. The analysis and comparison of these structures has helped clarify the functional characteristics of this family of bacterial toxins and has provided some insight in to the formation of ternary complexes with T cell receptors.

2. COMMON INTERACTIONS IN PEPTIDE BINDING TO HLA-DR1

The MHC class II molecule is composed of two chains (alpha and beta), both of which are anchored to the cell membrane with transmembrane domains. The crystallization of the HLA-DR1 from cells involves proteolysis of these transmembrane domains in order to release the soluble extracellular domains. This material crystallizes alone and in complexes with bacterial superantigens (SEB and TSST-1). HLA-DR1 purified from cell lines has 85-90% of its peptide-binding site filled with a mixture of peptides derived from proteins produced by the cell. Some of these endogenous peptides have been sequenced and identified for the protein used in determining the crystal structure of the DR1:SEB complex (Chicz et al., 1992; Chicz et al., 1993).

The elucidation and refinement of the DR1:SEB structure has allowed the visualization of the electron density which is due to bound endogenous-peptides (figure 1). This density is of high quality and allows the identification of the conformation of bound peptide(s) that is 13 amino acids long. By comparison, endogenous peptide sequences have demonstrated that peptides from 14-25 amino acids in length are typically found bound to HLA-DR1. The observed peptide electron density does not have side chain characteristics that correspond to one of the predominant endogenous-peptides, and at positions that point out of the DR1 peptide-binding site the quality of the electron density is lower. The observed features of this electron density may therefore represent the superposition of different peptide sequences. The ability to easily interpret this electron density suggests that peptides may

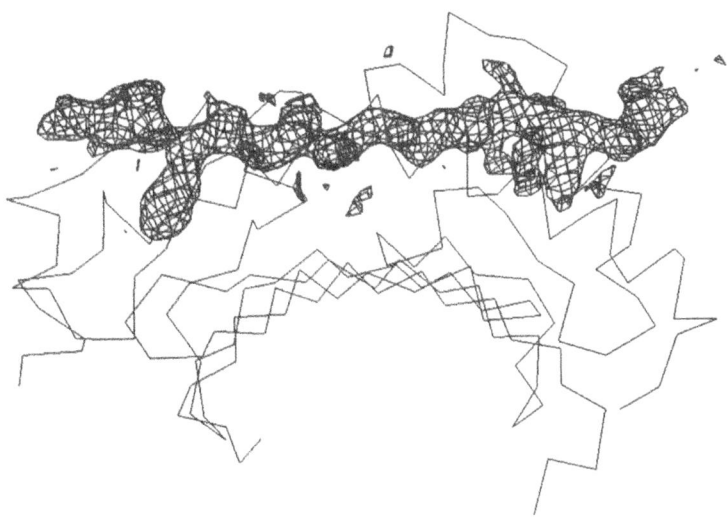

Figure 1. Electron density observed for endogenous peptides bound to HLA-DR1. A side view of the peptide-binding site is shown. The beta1 helix is in front and the alpha1 helix is behind. Note the large extension of electron density near the left end of the binding site that points downwards towards the beta sheet of the binding site.

share a common structure upon binding to the DR1 peptide-binding site. This idea is further supported by the similarity in conformation of peptides bound to two other HLA-DR molecules, as observed by X-ray crystallography (Stern et al., 1994; Ghosh and Wiley, 1995).

The endogenous-peptide electron density has been modeled as a 13 amino acids peptide and an analysis of the interactions with HLA-DR1 residues carried out. Given the inability to assign the side chain electron density to a unique endogenous-peptide sequence, we have concentrated on the peptide main-chain interactions with HLA-DR1. These interactions may be especially important, since residues which are conserved in all MHC class II molecule sequences could potentially form a subset of interactions with conserved atoms of all peptides. If a common set of MHC-peptide interactions is found in different MHC peptide structures, then these may provide direct insights into the common functional characteristics of the peptide:MHC interaction.

Figure 2 shows a representation of potential hydrogen bond interactions between the DR1 molecule and the endogenous-peptide model. Given the resolution limit of the structure (2.7Å) and the current refinement status, the identification of these hydrogen bonds remains somewhat speculative. However, the quality of the electron density, prior to the building of any peptide models, including the observation of peptide carbonyl bumps at this resolution, and the proximity DR1 residues to this density argues for the formation of a common and specific set of interactions with the endogenous-peptides. There are 14 potential hydrogen bonds found in this complex and many of these are also observed in the crystal structure of HLA-DR1 with a single influenza peptide (Stern et al., 1994) and in the crystal structure of a mixture of peptides bound to the HLA-DR3 molecule (P. Ghosh and D. Wiley, personal communication).

The interactions between peptide main-chain atoms and DR1 side chains are evenly distributed throughout the peptide binding site, leaving the N- and C- termini of bound peptides free to extend out either end. Three of the residues that interact with the bound peptides are asparagines and these form bidentate hydrogen bonds with the peptide main-chain. Similar interactions of the amide side chains of asparagine and glutamine have been identified in refined protein structures and may be involved in establishing some of the conformational constraints of the bound peptides (Le Questel et al., 1993; Jardetzky et al., 1995).

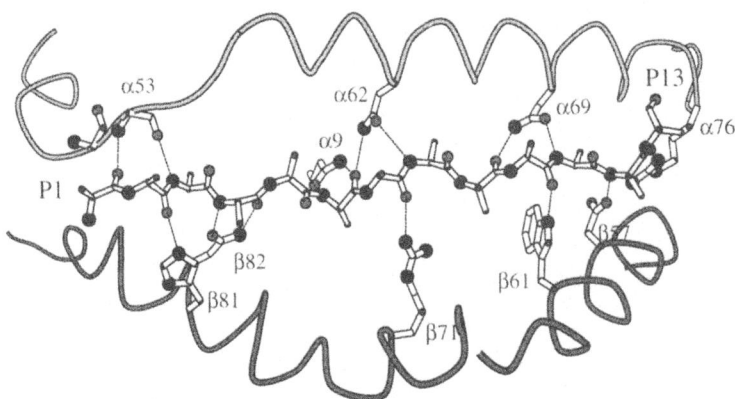

Figure 2. Potentail hydrogen bond interactions between HLA-DR1 residues and the endogenous peptides. Hydrogen bonds are formed between the peptide main-chain and the MHC class II molecule throughout the peptide-binding site. Bidentate hydrogen bonds are formed between asparagine residues α62, α69 and β82 with the peptide main chain. Figures 2 through 6 were generated by the program MOLSCRIPT (Kraulis, 1991).

The conformation of the endogenous-peptides bound to HLA-DR1 approximates a left-handed poly-proline II helix, especially in the central residues 3-11 of the 13mer. This conformation has been observed in other globular protein structures, as well as in proline-containing peptides that bind to Src homology 3 domains (SH3). This helical conformation is demonstrated in figure 3, viewed from above the peptide-binding site. In this representation the helical pitch is clarified by connecting the C-beta atoms of adjacent peptide residues, rather than a more traditional C-alpha representation. The nearly three-fold repeat of the helix leads to the projection of three residues out of the peptide binding (peptide positions P2, P5, P8). On either side of these amino acid positions, side chains contact the DR1 peptide binding site in rather shallow (P3, P4, P6, P7) or deep pockets (P1, P9). The helical repeat of the peptide insures a large surface area of interaction with the DR1 molecule, while providing access to the peptide main-chain atoms for the formation of hydrogen bonds described above.

Our previous studies have shown that it is possible to substitute the majority of peptide side chains to alanine, and retain high affinity and slow dissociation rate constants in binding to HLA-DR1, for a 13 amino acid peptide (Jardetzky et al., 1990). This was demonstrated with a peptide with the sequence NH2-AAYAAAAAAKAAA-COOH, which contains an important hydrophobic anchor residue (tyrosine) near the N-terminus of the peptide. While other peptide side chain interactions with the DR1 can be important in increasing affinity, this minimal peptide has equilibrium and kinetic binding constants that are typical if not better than antigenic peptides. While the relative importance of side chain and main chain interactions in determining peptide binding affinity will clearly vary with peptide sequence, some subset of interactions between the DR1 molecule and this polyalanine peptide may provide further insight into general aspects of the MHC-peptide interaction. Perhaps the endogenous-peptides provide the first step in furthering our understanding of these mechanisms, by suggesting a common conformation and interaction scheme for many peptides.

3. SEB AND TSST-1 BINDING TO HLA-DR1

While peptide binding to HLA-DR1 apparently has some conserved characteristics, the interactions of superantigens with DR1 provides a more puzzling problem to understand.

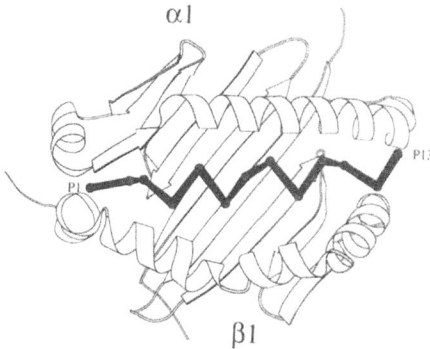

α1

β1

Figure 3. The polyproline type II helical conformation of peptides bound to HLA-DR1. A top view of the HLA-DR1 peptide binding site, showing the left-handed helical twist of bound peptide that is similar to polyproline type II helix (ppII). The peptide C-beta atoms are shown connected by rods. The conformation of bound peptides may be more constrained due to the formation of hydrogen bonds between the MHC molecule and the peptide main-chain throughout the peptide-binding site. In addition, bidentate hydrogen bonds between asparagine residues and the peptide main chain may induce a preference for this conformation.

Superantigens bind to MHC molecules outside of their antigen-binding sites and these toxic proteins are able to interact with a variety of T cell receptors, leading to massive T cell activation. Biochemical and crystallographic studies of a family of bacterial superantigens indicates that these proteins bind to distinct, though perhaps overlapping sites on the MHC molecule. Two of these have been elucidated in the structure solutions of the DR1:SEB DR1:TSST1 complexes.

The result of crystallographic studies is summarized in Figures 4 and 5, which shows the complexes formed between HLA-DR1 and SEB (top) or TSST1 (bottom). The two toxins

A

B

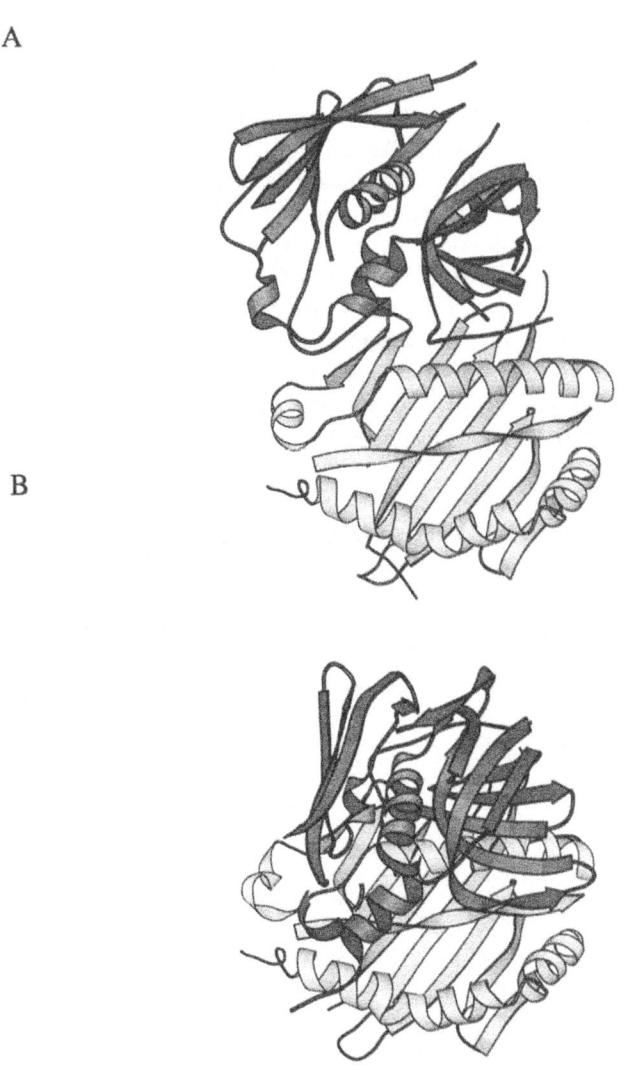

Figure 4. (a) Top view of the DR1:SEB complex. Only the HLA-DR1 alpha1 and beta1 domains are shown binding to SEB. The view is with the alpha1 domain of HLA-DR1 above and the beta1 domain below. SEB binds only to the alpha1 domain of the HLA-DR1 molecule. (b) Top view of the DR1:TSST-1 complex. Only the HLA-DR1 alpha1 and beta1 domains are shown with TSST-1. The view is as in (a). TSST-1 binds to both the alpha1 domain and beta1 domain of the HLA-DR1 molecule, with potential interactions to the C-terminal region bound peptides.

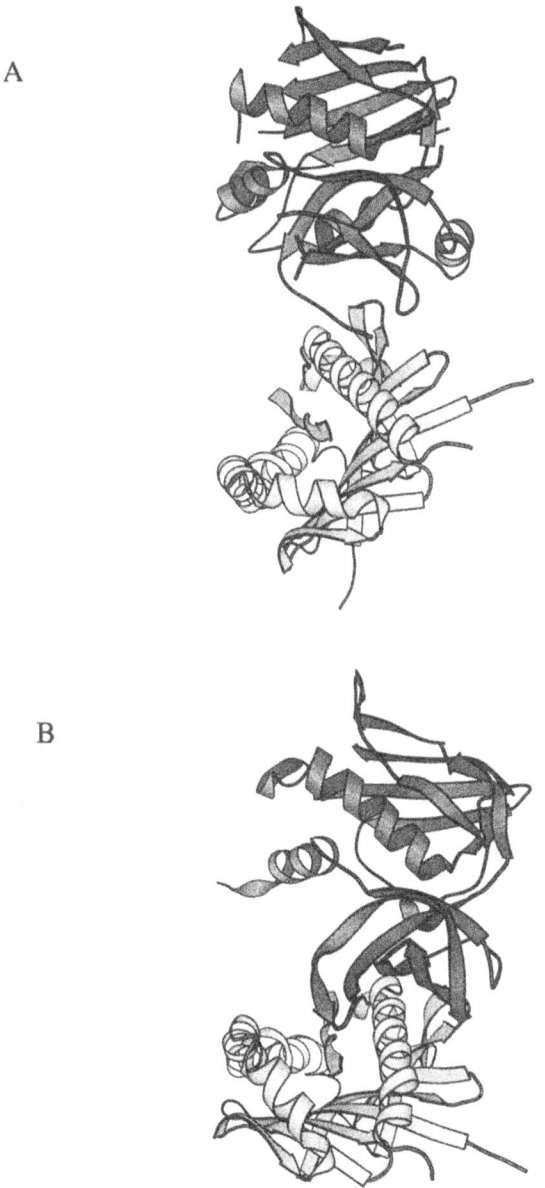

Figure 5. (a) End view of the DR1:SEB complex. Only the HLA-DR1 alpha1 and beta1 domains are shown interacting with SEB. DR1 alpha2 and beta2 domains are not shown, but lie below the peptide binding site as shown. The view is looking down the MHC peptide binding site, with the alpha1 domain of HLA-DR1 to the right and the beta1 domain to the left. The C-terminal domain of SEB extends away from the DR1 peptide-binding site. Residues implicated in direct TCR interactions line a wall formed by SEB above and to one side of the antigen binding site. (b) End view of the DR1:TSST-1 complex. Only the HLA-DR1 alpha1 and beta1 domains are shown with TSST-1. DR1 alpha2 and beta2 domains are not shown, but lie below the peptide binding site. The view is looking down the MHC peptide binding site, with the alpha1 domain of HLA-DR1 to the right and the beta1 domain to the left. In contrast to the DR1:SEB structure, the C-terminal domain of TSST-1 extends above the DR1 peptide-binding site. Residues from TSST-1 implicated in direct TCR interactions lie along the long C-terminal alpha helix, which is above the MHC antigen binding site.

bind to the DR1 domains which form the antigen-binding site. In the case of SEB (Jardetzky et al., 1994), interactions are exclusively formed with residues from one chain of the DR1 molecule (the alpha chain). For TSST-1 (Kim et al., 1994), a similar set of residues of the DR1 molecule are used in binding, but this toxin has additional interactions with both the beta1 domain of DR1 as well as potential interactions with peptides that are bound to DR1. Both toxins share an interaction with an open groove on the MHC surface that is formed by two loops of the DR1 alpha1 domain and the long alpha helix above these loops (figure 6). The relative orientations of the two toxins in these complexes is characterized by a significant rotation from the SEB position to one side of the MHC peptide binding site to the TSST1 position above and across the alpha1/beta1 superdomain.

Previous mutational studies have shown that SEB and TSST-1 binding is sensitive to a common set of DR residues (Thibodeau et al., 1994; Thibodeau and Sékaly, 1995). These include the DR alpha chain residues methionine 36 and lysine 39, which are found in the

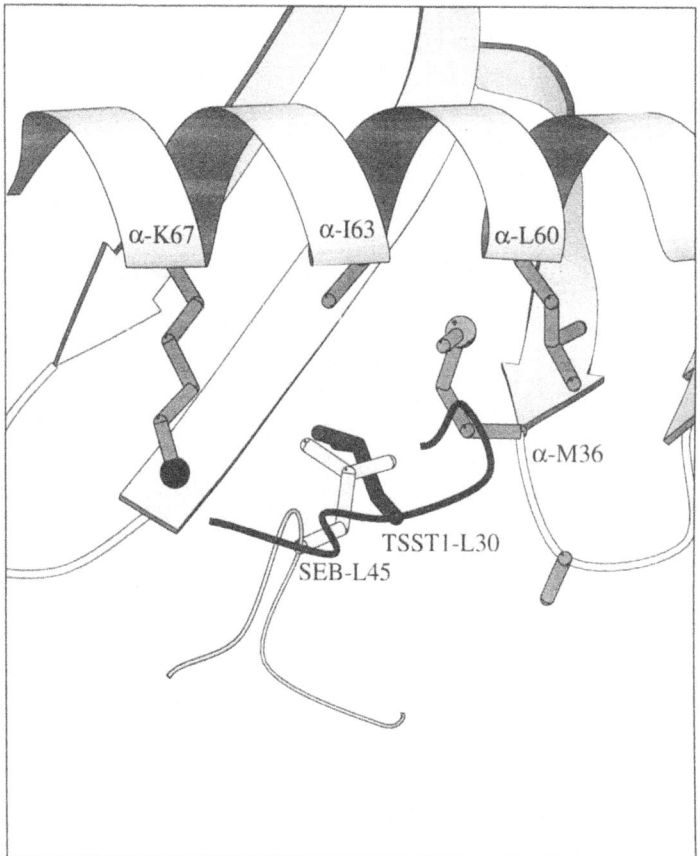

Figure 6. Conserved interaction of a leucine residue in SEB and TSST-1 with the HLA-DR1 alpha chain. Leucine 45 of SEB (light gray) and leucine 30 of TSST-1 (dark gray) are shown superimposed in their binding site on the HLA-DR1 molecule. The main-chain paths of the neighboring superantigen residues are shown as coils, demonstrating that these residues of the superantigens do not superimpose due to the different geometry of complex formation shown in figures 4 and 5. In spite of these very different binding modes, the two superantigens utilize a leucine in the same position to interact with HLA-DR1. A region of the MHC alpha chain is shown that forms the hydrophobic pocket for these leucines, bounded by the loop between strands 1 and 2 (left side in the figure), the loop between strands 3 and 4 (right side in the figure), and the alpha1 helix.

DR1 loop between strands 3 and 4 in the alpha1 domain (figure 6). Although these two residues are important for the binding of both toxins to DR molecules, the interactions that are formed differ. Methionine 36 contributes to a hydrophobic patch between the DR1 loops that interacts with hydrophibic residues of the two toxins. For SEB this region is defined by a ridge of hydrophobic residues that interact with the open groove on the MHC surface that includes Phenylalanine 44, leucine 45, phenylalanine 46 and 47. In TSST1, leucine 30, which corresponds structurally to SEB leucine 45, forms an analogous set of interactions with DR1 (figure 6). However, the interaction of lysine 39 with the two toxins is different. For SEB, this lysine is involved in a buried salt bridge between the two proteins, with glutamic acid 67 and two additional tyrosines from the SEB molecule. These interactions form along the central face of the N-terminal beta barrel domain of the toxin. In contrast, the TSST1 molecule is rotated significantly away from the SEB position, leaving lysine 39 to form hydrogen bonds to a threonine residue in a different region of the superantigen fold.

The two crystallographic complexes neatly explain how these toxins are sensitive to common set of DR1 mutations, but other observations indicate that additional toxin binding sites on the MHC surface remain to be mapped, as well as implicate MHC-bound peptides in influencing the affinity of toxins for MHC molecules. The evidence for additional toxin binding sites is particularly strong for another relative of the S aureus toxins, SEA. Mutations of the MHC beta chain residue (His 81), disrupt the binding of this toxin to MHC molecules. His81 is clearly not involved in the crystallographically observed interactions between SEB and TSST1 with HLA-DR1. For SEA, this interaction may be mediated by a Zn atom bound to residues of the C-terminal domain of the toxin, again suggesting a completely different mode of binding to HLA-DR molecules (Fraser and Hudson, 1993). Comparison of the sequences of the SEA and SEB toxins led to the suggestion that SEA may also be able to form the same complex with MHC molecules as SEB and recent functional studies support this possibility. It remains to be established if this is an important functional characteristic that could be common to many of these toxins. The ability of SEA to bind to two distinct sites of MHC molecules may lead to crosslinking of these proteins at the cell surface and could be important in signal transduction events that could influence T cell stimulation.

In addition, the peptides that are bound to the MHC molecules may influence the affinity of different toxins. Early studies had established that SEB and TSST1 binding to MHC molecules at the cell surface is not competitive (Thibodeau et al., 1994). This evidence was used to argue that the toxins bind to distinct regions of the MHC surface. These studies have been reproduced using HLA-DR1 transfected cell lines, where the lack of competition between the toxins and the sensitivity to the same MHC mutations can be established without the surface expression of other MHC proteins. In fact, the expression of HLA-DR1 in different cell lines leads to changes in the relative binding of SEB and TSST1, as if cell-specific peptides created different subsets of MHC molecules that each toxin can then bind. For TSST1, this can be reasonably understood in the context of the crystal structure, in which the toxin clearly can interact with peptides bound to the MHC molecule. However, for SEB there is no direct interaction with bound peptides, nor does the toxin cover any region of bound peptides (Jardetzky et al., 1994). This suggests that indirect conformational effects of peptides may influence the SEB interaction. It will be important to establish how this may occur as it could have significant ramifications for our understanding of peptide:MHC interactions. This interplay of peptide and superantigen interactions with MHC class II molecules may be an important component of the pathology of bacterial infections.

REFERENCES

Bentley, G. A., G. Boulot, K. Karjalainen and R. A. Mariuzza (1995). Science 267, 1984-7.

Bjorkman, P. J., M. A. Saper, B. Samraoui, W. S. Bennett, J. L. Strominger and D. C. Wiley (1987). Nature *329*, 512-18.

Bjorkman, P. J., M. A. Saper, B. Samraoui, W. S. Bennett, J. L. Strominger and D. C. Wiley (1987). Nature *329*, 506-12.

Brown, J. H., T. S. Jardetzky, J. C. Gorga, L. J. Stern, R. G. Urban, J. L. Strominger and D. C. Wiley (1993). Nature *364*, 33-9.

Chicz, R. M., R. G. Urban, J. C. Gorga, D. A. Vignali, W. S. Lane and J. L. Strominger (1993). J. Exp. Med. *178*, 27-47.

Chicz, R. M., R. G. Urban, W. S. Laue, J. C. Gorga, L. J. Stern, D. A. A. Vignali and J. L. Strominger (1992). Nature *358*, 764-8.

Fraser, J. D. and K. R. Hudson (1993). Res. Immunol. *144*, 188-93.

Germain, R. M. and D. H. Margulies (1993). Annu. Rev. Immunol. *11*, 403-450.

Ghosh, P. and D. C. Wiley (1995).

Jardetzky, T. S. (1995). Structural studies of the interaction of superantigens with class II Major Histocompatibility Complex molecules. *Bacterial Superantigens: Structure, Function and Therapeutic Potential* Eds. J. Thibodeau and R.-P. Sékaly. Austin, R.G. Landes. 67-82.

Jardetzky, T. S., J. H. Brown, J. C. Gorga, L. J. Stern, R. G. Urban, Y.-I. Chi, C. Stauffacher, J. L. Strominger and D. C. Wiley (1994). Nature *368*, 711-718.

Jardetzky, T. S., J. H. Brown, J. C. Gorga, L. J. Stern, R. G. Urban, J. L. Strominger and D. C. Wiley (1995). Proc. Natl. Acad. Sci., Submitted

Jardetzky, T. S., J. C. Gorga, R. Busch, J. Rothbard, J. L. Strominger and D. C. Wiley (1990). Embo J *9*, 1797-803.

Kim, J., R. G. Urban, J. L. Strominger and D. C. Wiley (1994). Science *266*, 1870-1874.

Kraulis, P. J. (1991). J. Appl. Cryst. *24*, 945-9.

Le Questel, J. Y., D. G. Morris, P. H. Maccallum, R. Poet and E. J. Milner-White (1993). J. Mol. Biol. *231*, 888-96.

Sloan, V. S., P. Cameron, G. Porter, M. Gammon, M. Amaya, E. Mellins and D. Zaller (1995). Nature *375*, 802-806.

Stern, L. J., J. H. Brown, T. S. Jardetzky, J. C. Gorga, R. G. Urban, J. L. Strominger and D. C. Wiley (1994). Nature *368*, 215-221.

Thibodeau, J., I. Cloutier, N. Labreque, W. Mourad, T. Jardetzky and R.-P. Sékaly (1994). Science *266*, 1874-1878.

Thibodeau, J. and R.-P. Sékaly, Eds.(1995). *Bacterial Superantigens: Structure, Function and Therapeutic Potential*. Austin, R.G. Landes.

13

SIMULATING THE DYNAMICS OF THE DNA DOUBLE HELIX IN SOLUTION

Miriam Hirshberg[1] and Michael Levitt[2]

[1] Protein Structure Group
National Institute for Medical Research
The Ridgeway, Mill Hill, London NW7 1AA, United Kingdom
[2] Beckman Laboratory of Structural Biology
Department of Structural Biology, Stanford Medical School
Stanford, California 94305

1. INTRODUCTION

Deoxyribonucleic acid (DNA) is the most important molecular structure in biology: its aesthetic double-helical structure has come to symbolize the dramatic progress made in molecular biology over the past fifty years. In spite of its central role in all living systems, remarkably little is known about the detailed structure, dynamics and energetics of DNA. Although Watson and Crick's model building study predicted the double-helix structure over 40 years ago (Watson and Crick, 1953) and the structure was soon confirmed by fibre diffraction (Langridge et al., 1960), it was not until 1981 that a three-dimensional structure of DNA was obtained from single-crystal diffraction (Drew et al., 1981). Since then, there have been over two hundred A, B and Z form DNA double-helical structures deposited in the Brookhaven Protein Data Bank. All the 51 different crystal structures for B-form DNA, the biologically important state, are of dodecamer or decamer double-helical sequences. 29 of the 31 dodecamers pack in exactly the same way found originally by Drew et al. (1981) and show more disorder than seen in most protein crystals (resolution ranges from 1.9 Å to 3 Å). In fact, even at very low temperatures (16 K) the X-ray resolution is moderate at 2.7 Å (Drew and Dickerson, 1982). Nuclear magnetic resonance (NMR) studies done in solution at room temperature also show DNA to be more mobile than many globular proteins (Patel et al., 1987). NMR determination of DNA structural details in solution is more difficult than it is for proteins: in DNA there are very few closely packed hydrogen atoms to give the needed inter-hydrogen NOE distances. The situation is more even more difficult as the effect of sequence on DNA double helix structure is very subtle. All double helices have complementary base-pairs and look alike. Changing the sequence causes very small changes to the relative orientation of the bases, the disposition of the sugar-phosphate back-bone, etc. (Dickerson et al., 1982). In addition, DNA is a "soft" structure easily deformed by proteins that bind to it (Kim et al., 1990; Kim et al., 1993a; Kim et al., 1993b).

Dynamics and the Problem of Recognition in Biological Macromolecules, edited by Jardetzky and Lefèvre
Plenum Press, New York, 1996

Faced with these difficulties with determining structure experimentally, there is a clear need to turn to theoretical methods. For the DNA double helix, such methods have probably been more important than for globular protein structures. Watson and Crick's (1953) famous modeling of the double helix was theoretical in nature. The first computer studies of the double helix (Levitt 1978; Trifonov and Sussmann, 1979) also made interesting predictions. Both indicated that DNA could be smoothly bent for modest changes in energy. Levitt (1978) also predicted that the bases would have a propeller twist not seen in the refined fibre-diffraction structures (Arnott and Hukins, 1972) but subsequently confirmed by the first single-crystal structure (Drew et al., 1981).

These early studies employed energy minimization to find strain-free, low energy structures. The first molecular dynamics simulations of DNA (Levitt, 1983; Tidor et al., 1983) were done in vacuo. They showed that DNA was even more flexible than predicted by the minimization studies. At that time simulations of protein dynamics were also done in vacuo but for DNA with its highly-charged phosphate back-bone and easily deformable structure, lack of solvent was more serious.

The inclusion of explicit solvent and counterions is more important in DNA simulations than in protein simulations due to differences in the overall structure and in the charge density of these two molecules. DNA, with its extended structure and large surface/volume ratio is much more susceptible to distortion due to environment than a more compact globular protein. Because DNA is a polyelectrolyte, electrostatic forces between the phosphate groups can be very disruptive if solvent and counterions are not taken into account properly. In the past, it was possible to improve the structural stability of DNA during in vacuo simulations by (a) omitting of all the net charges (Levitt, 1983); (b) scaling the charges using a distance-dependent dielectric function (Tidor et al., 1983); (c) introduction of "hydrated" counterions (Singh et al., 1985) and (d) imposing boundary conditions on the DNA to treat it as part of infinitely long molecule (Prabhakaran and Harvey, 1985).

Several simulations in solution with counterions have been carried out on small DNA oligomers. Seibel (Seibel et al., 1985), followed the dynamics of the pentamer fragment d(CGCGA)-d(TCGCG) for 106 ps, using TIPS3P water molecules and a 10Å cutoff distance for the nonbonded interactions. In an 80 ps study of the molecular dynamics of the octamer d(CGCAACGC)-d(GCGTTGCG) (van Gunsteren et al., 1986) used the united atom approximation, SPC water molecules and a twin-range cutoff for the evaluation of the nonbonded interactions. Zielinski and Shibata (1990) computed a 60 ps trajectory of $d(G_6)$-$d(C_6)$ in a droplet of water, using the united atom approximation and SPC water molecules. These pioneering simulations were too short and involved pieces of DNA too small to indicate if DNA is indeed stable in molecular dynamics calculations. After our work had been completed (Hirshberg, 1990), there were reports of longer simulations on a dodecamer of DNA in solution. In a 140 ps simulations, Swaminathan et al. (1991) showed that it was necessary to use additional hydrogen bonding restraints to preserve the Watson-Crick base pairs in solution. Miaskiewicz et al. (1993) did not include such restraints and found that after 100 ps the double helix had become very distorted.

In 1988, after showing that protein molecular dynamics is stable in solution (Levitt and Sharon, 1988), we turned our efforts to the accurate simulation of the DNA double helix in solution. Getting protocols designed for the simulation of proteins in solution to work for DNA was not an easy task. We first attempted to use exactly the same form of force field for both proteins and nucleic acids. In our potential energy function (Levitt, 1983; Levitt et al., 1995a), nonbonded forces are represented by a conventional 6-12 Lennard-Jones term for van der Waals interactions and a Coulombic term for electrostatic interactions. There are no special terms for hydrogen bonds, which arise solely from the balance between attractive, directionally dependent electrostatics and van der Waal's repulsion. Ionized groups carry their full charges and these interactions are not attenuated by any dielectric constant. Instead, the explicit water molecules reduce effective interactions by orienting their dipole moments with the electrostatic field. As a further complication, our desire for sufficiently long simulations to ensure proper sampling means we must truncate the nonbonded forces at short range (6 Å). To ensure numerical stability, we also insist that the total energy be conserved.

To meet these stringent requirements for proteins (Levitt and Sharon, 1988), it was assumed that electrostatic interactions are smoothly truncated between neutral groups rather than between individual atoms. This leads to smaller discontinuities at the cutoff distance and energy was well-conserved (Levitt, 1989). Net charges can never form neutral groups, but, as proteins have few such groups, the method was successful at maintaining the native structure of a small protein in solution. The DNA double helix is more of a problem. The atomic partial charges on the aromatic bases and the ester oxygens in the backbone cannot easily be arranged into neutral groups. There is also a net negative charge on every phosphate group. Initial attempts to apply the methods used successfully for proteins to DNA, caused the double-helix to break apart in less than 50 ps. Reproducing the known stability of a macromolecule in solution is a basic requirement for any simulation and we re-examined the assumptions behind neutral group truncation. By using a different truncation scheme (Levitt et al., 1995a), which truncates van der Waals and electrostatic interactions between atoms not groups, we were able to use arbitrary atomic partial charges on the base atoms. Because this scheme employs a very smooth truncation function involving force shifting, energy was conserved as well as before. In fact, this new method also worked as well as the earlier method for simulations proteins (Daggett and Levitt, 1993) and for simulations of pure water (Levitt et al., 1995b).

Here we report the results of preliminary analysis of 500 ps molecular dynamics simulations of two short base-paired segments of DNA in solution. The overall properties of the systems are analyzed first. Then we discuss the characteristics of the base-pair hydrogen bonds, followed by structural analysis of several helical parameters. Finally, we examine properties of the surrounding ions and water molecules.

With this improved treatment of nonbonded interaction, the double helix is completely stable for the time simulated without needing any restraints, special hydrogen bond functions or reduction of net charges. The dynamics does show interesting structural deformations that include broken G-C base-pairs at the ends of the helix, breakage and reformation of an internal A-T base-pair and a narrowing of the minor groove in the A-tract of one of the sequences studied.

The work reported here was done from 1987 to 1990 and is taken from Hirshberg's Ph.D. thesis (Hirshberg, 1990).

2. METHODS

The molecular dynamics simulations of DNA in solution were carried out for a three-component system consisting of a segment of double-helical DNA, sufficient sodium counter-ions to neutralize the net negative charge on the DNA, and a large number of water molecules. The initial system was constructed in three steps. First, the DNA structure was taken from a single crystal structure or generated as a uniform structure from fiber diffraction data. In all cases presented here, the uniform structures were of the B-form, with a helical rise of 3.38Å and a helical twist of 36.0° (Arnott and Hukins, 1972). All polar and non-polar hydrogen atoms were included. Hydroxyl groups were added at the 5'- and 3'-ends. In a second step, a rectangular box was constructed around the DNA and filled with F3C water molecules (Levitt et al., 1995b). The sides of this box were set so that in the initial structure no water molecule is closer than 8 Å to a wall of the box. Finally, the whole system was electrically neutralized by the addition of sodium ions counterions (one Na^+ for each of the

22 phosphate groups) placed at random positions 5Å from one of the phosphate group oxygen atoms.

The potential energy of the entire system was evaluated using the function and energy parameters given in Levitt et al., (1995a). Discontinuities in the potential function were eliminated using an atom-based force shifting formalism mentioned in Allen and Tildesley (1987) and detailed in Levitt et al., (1995a). The simulations were performed using periodic boundary conditions, integrating the equations of motion by means of the Beeman method (Beeman, 1976) with an integration time step of 0.002 ps. The nonbonded interactions were evaluated using a list of nonbonded pairs within 8Å. The list was updated every ten time-steps and the energy calculated to a cutoff distance of 6Å (Levitt et al., 1995a); this ensures continuous energies and derivatives while maintaining computational efficiency. While this range for van der Waals and electrostatic interactions may seem unreasonably short, it is important to stress that it has worked well for a large number of simulations of pure water (Levitt et al., 1995b), small solutes in water, and macromolecules in solution (Daggett and Levitt, 1994). In fact, the success of the DNA simulations reported here offer additional support to the usefulness of short range cutoffs for molecular dynamics simulations in solution.

Each simulation was preceded by equilibration consisting of two alternating cycles of 2000 steps of conjugate energy minimization followed by 2000 steps of dynamics. During the first cycle, the DNA was kept fixed, while during the second cycle the water molecules and counterions were kept fixed. After the equilibrating stage, the system was heated to the desired temperature by generating mass-scaled random impulses between pairs of randomly selected atoms (this trick increases the kinetic energy but conserves momentum). After the system reached the desired temperature (the first 50 ps), it was left un-perturbed unless the mean temperature differed from the desired temperature by more than 5%, in which case the velocities were rescaled to give the correct temperature. Because energy conservation is so good here, such rescaling of velocities needed to be done only once during the four 500 ps trajectories (for the DK run). Data were collected every 0.2 ps, corresponding to a total of 2500 coordinate sets for a trajectory of 500 ps.

3. RESULTS AND DISCUSSION

3.1. Overview

The results presented here involve 500 ps molecular dynamics simulations of two dodecamers: d(CGCA$_6$GCG)-d(CGCT$_6$GCG) (called HL, see Table 1) and the self-complementary dodecamer d(CGCGAATTGCGC)$_2$ (called DK). This choice was dictated by (a) the amount of experimental data available for both these sequences and (b) the similarity of the sequences excepts for the run of adenines (A-tracts) in HL. Such A-tracts are considered the main determinant of sequence directed DNA bending (Nelson et al., 1987; Nadeau and Crothers, 1989; Diekmann and von Kitzing, 1992; Diekmann et al. 1992).

Do the results of the simulations depend on the initial geometry of the macromolecule? We have performed two separated dynamics simulations for each of the sequences starting from two different starting structures, (a) the standard uniform B-DNA and (b) the structure derived from X-ray studies, which typically deviate from each other by less than 1.5 Å RMS. The calculated structural and dynamical properties indicate that the results are very similar for the two starting structures. Here we present results of only one member of each pair, the DK trajectory for d(CGCGAATTGCGC)$_6$ and the HL trajectory d(CGCA$_6$GCG) (HL). Both DK and HL start from the uniform B-DNA geometry derived by fibre diffraction.

Table 1. Structures used for simulations

Sequence	Initial structure	Temp(K)	Period(ps)	Name
(i) Simulation in solution with counterions				
d(CGCGAATTGCGC)$_2$	Uniform[1]	300	500	DK
d(CGCGAATTGCGC)$_2$	X-ray[2]	300	500	DKX
d(CGCA$_6$GCG)-d(CGCT$_6$GCG)	Uniform[1]	300	500	HL
d(CGCA$_6$GCG)-d(CGCT$_6$GCG)	X-ray[3]	300	500	HLX
(ii) Simulation in vacuo				
d(CGCA$_6$GCG)-d(CGCT$_6$GCG)	Uniform[1]	300	200	HLV

[1]The coordinates of uniform B-DNA are based on the fiber-diffraction cylindrical coordinates tabulated in Arnott and Hukins (1972).
[2]The X-ray coordinates were given to us by H. Nelson (Nelson and Klug, 1985) and are now available from the Brookhaven Protein Data Bank as set 1D98.
[3]The X-ray coordinates (Drew et al., 1981) were taken from the Brookhaven Protein Data Bank (set 1BNA)

During the simulations volume and temperature were kept constant and structural and dynamical properties were calculated over the last 400ps (100 ps to 500 ps). Various parameters of each simulation are summarized in Table 2.

In what follows we present the results of the preliminary analysis of these two trajectories. The overall properties of the systems are analyzed first. Then we discuss the characteristics of the base-pair hydrogen bonds, followed by structural analysis of several helical parameters. Finally, we examine properties of the surrounding ions and water molecules.

3.2. Overall Properties of the System

3.2.1. Energy Conservation. The data in Table 3 and Figure 1 indicate that our simulation do conserve energy very well. The temperature of the whole system is stable,

Table 2. Parameters of solution simulations

Property	DK	HL
Box size		
a (Å)	35.48	35.47
b (Å)	36.36	36.34
c (Å)	60.15	60.19
Volume (Å^3)	77597	77584
Number of atoms in		
DNA	758	760
Na$^+$ Ions	22	22
Waters	6462	6456
Total atoms	7242	7238
Water molecules	2154)	2152
Molecular dynamics		
Temperature (K)	300	300
Period (ps)	500	500
Time step (ps)	0.002	0.002
Cutoff distance (Å)	6	6
No. nonbonded pairs	99,000	95,000
Cpu (min/ps)[1]	74	68

[1]Computer times for one MIPS R3000 (25 Mhz) processor of a slow 1989-vintage Silicon Graphics IRIS-4D 240.

Table 3. Conservation of energy and temperature in solution

	HL	DK
Total Energy		
Average energy (kcal-mol^{-1})	-14702 ± 4	-14785 ± 5
Drift (kcal-mol^{-1}-ps^{-1})	0.01	0.04
Temperature		
Average System (K)	299.7 ± 1.2	300.3 ± 1.3
Average DNA (K)	298.3 ± 1	299.1 ± 1
Drift over 450 ps (K)	0.3	1.2

Time-averages are calculated over a period of 400 ps (100–500 ps).
The drift along the trajectory is calculated as the slope of the best fit line to a
plot of the total energy against time after the equilibration temperature has ended
at 50 ps.

there are no temperature imbalances between the DNA and the water and the total energy is
well-conserved. This is a very strong validation of our smooth truncation scheme as the DNA
system involves very strong electrostatic interactions.

3.3.2. Structural Deviations from Initial Structure. The time dependence of the root
mean square (RMS) deviation of the simulated structure from the initial geometry, offers a
simple means to detect whether the double helix is stable during the simulation. In Figure 2
and in Table 4 we present the time dependence of the RMS deviation from uniform A-form
and uniform B-form DNA as well as from the corresponding X-ray structure. Also shown
are RMS deviation during in-vacuo simulation of HL.

The overall picture is of a stable system. For each trajectory, the RMS reaches a
plateau after about 100 ps and both sequences remain closer to the B-form starting structure
than to the corresponding uniform A-form structure. For both simulations, the time-variation

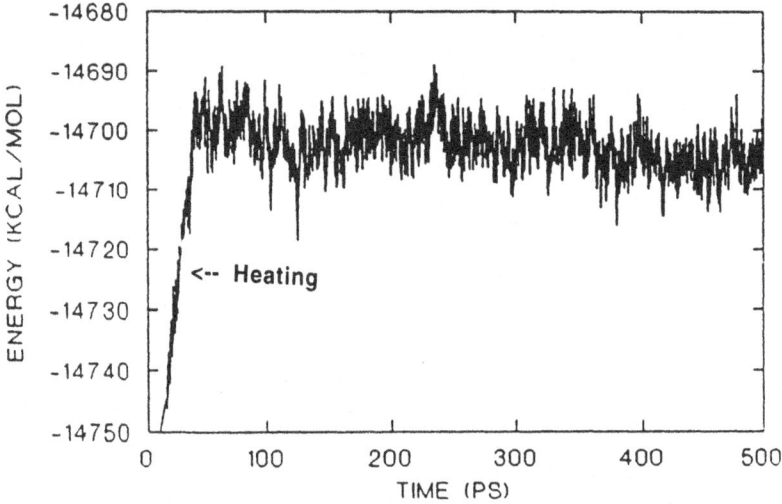

Figure 1. The variation of the total energy with time along the 500 ps trajectory of the HL simulation shows
that the total energy is very well conserved. The increase of the energy for the initial 50 ps is due to equilibration.
The subsequent downward drift of energy is very slow changing by 5 kcal-mol^{-1} in 450 ps (0.01 kcal-mol^{-1}
-ps^{-1}). Expressed as a fraction of the total kinetic energy of 6514 kcal-mol^{-1} (3kT/2 for every atom) the drift
corresponds to 0.1% or 0.3 K in temperature over 450 ps.

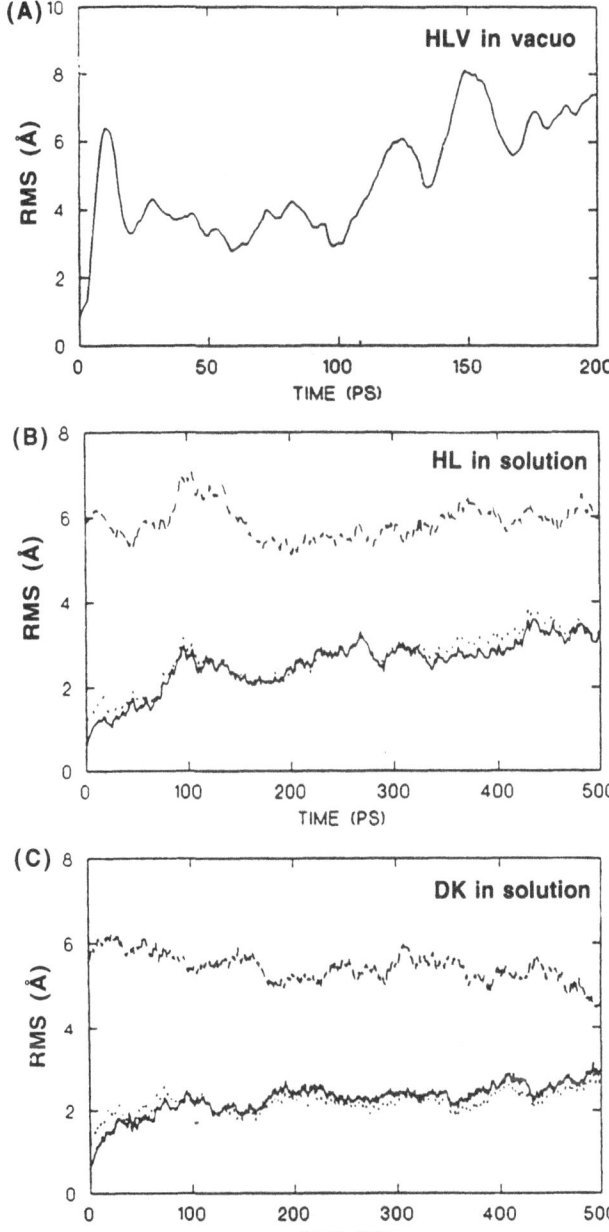

Figure 2. Time dependence of the RMS deviation of the simulated structure from the uniform B-DNA (solid line), the uniform A-DNA (dashed line) and the corresponding X-ray structure (dotted line) for (A) the HLV simulation in vacuo, (B) the HL simulation in solution and (C) the DK simulation in solution. The deviation from the starting structure is much larger for the in vacuo simulation than for that in solution. Note that both structures simulated in solution remain close to B-form; the structure does not become significantly A-form-like.

of RMS from the uniform B-form (the initial model) coincides almost exactly with the time-variation of RMS from the corresponding X-ray structure. Similar results were observed for the simulations which started from the DNA X-ray structure (DKX and HLX). The RMS for the in vacuo simulation (see Fig. 4A) shows large oscillations, corresponding to major displacements along the helical axis. For comparison, in vacuo simulations of a small protein (Levitt and Sharon, 1988), show much smaller deformations. This confirms that DNA with is extended helical structure is much more sensitive to environment than a small globular protein.

Table 4. Time average RMS deviations

RMS for	From		
	B-DNA[1]	A-DNA[2]	X-ray
HL			
All atom	2.776	5.867	2.894[3]
Base	2.342	5.094	2.632
Backbone	3.100	6.467	3.100
DK			
All atoms	2.409	5.334	2.252[4]
Base	1.983	4.468	1.989
Backbone	2.722	5.008	2.545

Time average RMS(Å) is calculated relative to [1]uniform B-DNA, [2]uniform A-DNA, [3]X-ray for HL (PDB set 1D98, Nelson et al., 1987),and [4]X-ray for DK (PDB set 1BNA, Drew et al., 1981).
The RMS is calculated for all bases excepts those at the ends of the double helix. It is averaged over the last 400 ps of the trajectory.

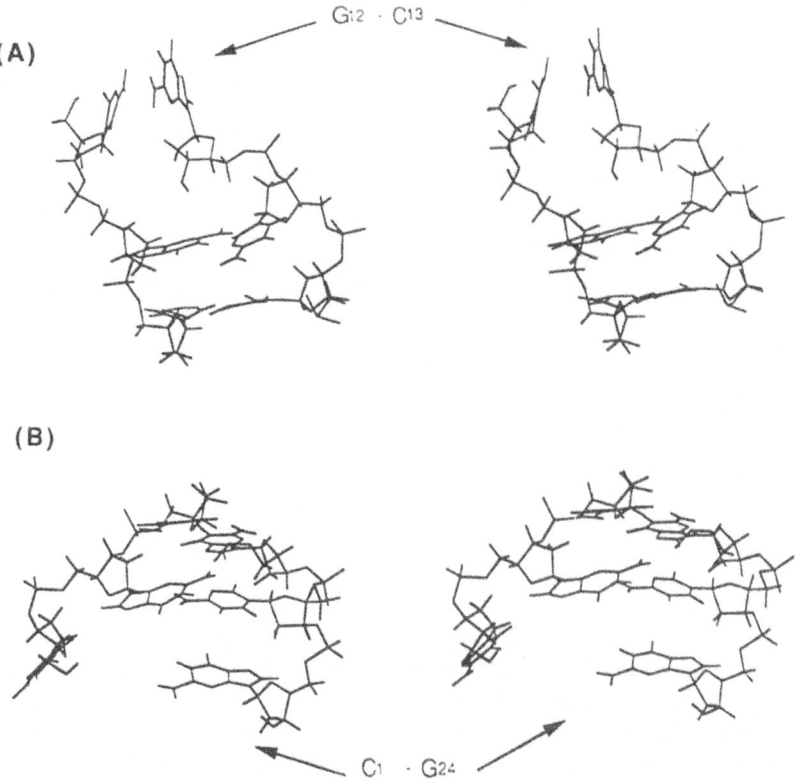

Figure 3. (A) The percentage of time a hydrogen bond exists between the imino proton and the imino nitrogen calculated over the last 400ps for the DK simulation (clear circles) and the HL simulation (filled circles). (B) the numbering of the bases in the HL and DK sequences.

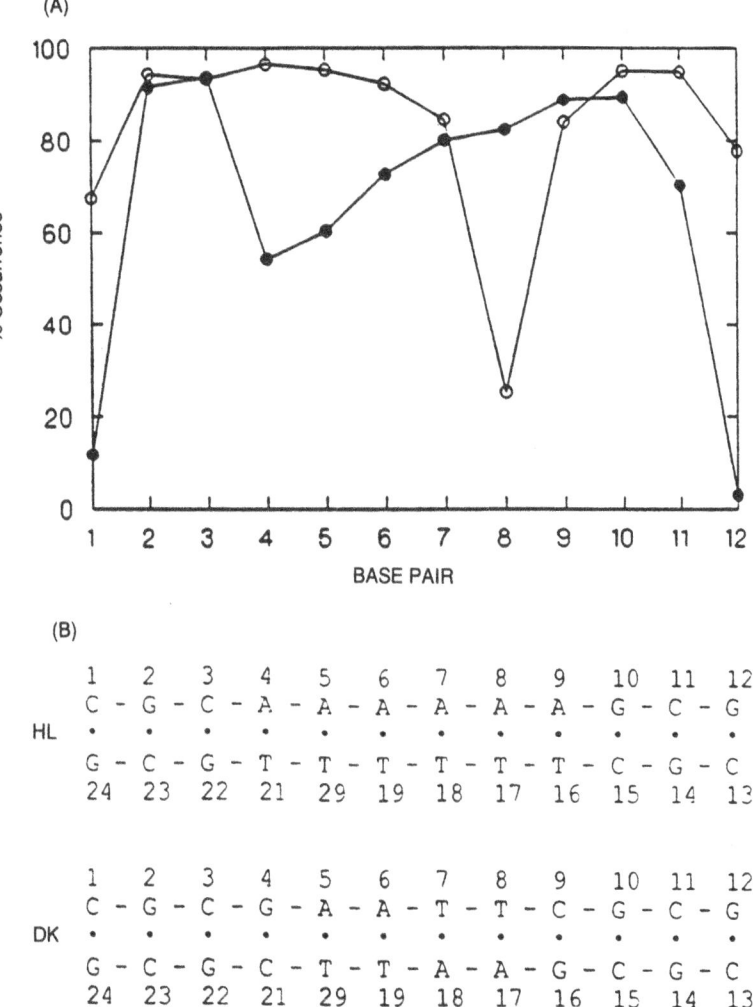

(A)

(B)

```
        1    2    3    4    5    6    7    8    9   10   11   12
        C  - G  - C  - A  - A  - A  - A  - A  - A  - G  - C  - G
   HL   •    •    •    •    •    •    •    •    •    •    •    •
        G  - C  - G  - T  - T  - T  - T  - T  - T  - C  - G  - C
       24   23   22   21   29   19   18   17   16   15   14   13


        1    2    3    4    5    6    7    8    9   10   11   12
        C  - G  - C  - G  - A  - A  - T  - T  - C  - G  - C  - G
   DK   •    •    •    •    •    •    •    •    •    •    •    •
        G  - C  - G  - C  - T  - T  - A  - A  - G  - C  - G  - C
       24   23   22   21   29   19   18   17   16   15   14   13
```

Figure 4. A stereoscopic views of the HL structure at 100 ps showing the two ends of the double-helix. In both cases hydrogen bonds in the terminal G-C base-pairs break. In (A) the two unpaired bases stack onto one another. In (B), the G_{24} base stacks onto the preceding base pair.

3.3. Hydrogen Bond Dynamics

During the molecular dynamics simulations the base pairs remain normally intact with stable Watson-Crick hydrogen bonds. The dynamic stability of a hydrogen bond is measured by the percentage of time it exists (P) and by the average duration of its existence, τ. Tables 5 and 6 show P, τ, the average bond length and the bond angle of all the Watson-Crick hydrogen bonds in the HL and the DK simulations.

Throughout the simulations the central hydrogen bond between the imino proton and the imino nitrogen is more stable than the side bonds between carbonyl oxygen and amino hydrogen, exhibiting shorter length, smaller angles and higher P values. The stabilities of the hydrogen bonds in the terminal base pairs are smaller than those for within the sequence, reflecting the fraying of the ends of the double helix. In the HL simulation, all three hydrogen bonds of the G_{12}-C_{13} terminal base pair broke and did not form again. Such fraying, which

Table 5. Properties of Watson-Crick hydrogen bonds in the HL simulation

Base-pair	Hydrogen bond			% Occurrence[1] P	Life-time[2] τ(ps)	Angle (°)	Length (Å)
C_1-G_{24}	H42	⋯	O6	2.5	1.1	24.5	2.18
	N3	⋯	H1	11.7	5.9	16.5	2.01
	O2	⋯	H21	13.7	3.2	20.5	2.12
G_2-C_{23}	O6	⋯	H42	79.1	5.0	18.2	2.01
	H1	⋯	N3	91.5	11.5	16.6	1.92
	H21	⋯	O2	85.5	7.8	18.2	2.00
C_3-G_{22}	H42	⋯	O6	81.3	6.1	16.6	1.99
	N3	⋯	H1	93.7	16.4	16.3	1.93
	O2	⋯	H21	82.0	6.0	17.7	2.04
A_4-T_{21}	H61	⋯	O4	43.9	2.7	18.4	2.15
	H1	⋯	N3	54.3	3.5	19.0	2.00
A_5-T_{20}	H61	⋯	O4	48.4	3.0	18.0	2.03
	N1	⋯	H3	60.4	4.0	18.1	2.00
A_6-T_{19}	H61	⋯	O4	60.1	3.1	18.3	2.07
	N1	⋯	H3	72.7	4.1	17.8	2.00
A_7-T_{18}	H61	⋯	O4	59.1	3.3	18.5	2.13
	H1	⋯	N3	80.1	5.2	17.6	2.00
A_8-T_{17}	H61	⋯	O4	66.3	3.3	17.7	2.07
	N1	⋯	H3	82.4	7.1	17.3	1.99
A_9-T_{16}	H61	⋯	O4	72.6	4.2	18.0	2.08
	N1	⋯	H3	88.8	8.3	17.2	1.98
G_{10}-C_{15}	O6	⋯	H42	80.6	5.1	16.8	2.01
	H1	⋯	N3	90.5	13.4	16.3	1.94
	H21	⋯	O2	89.3	9.2	18.0	2.05
C_{11}-G_{14}	H42	⋯	O6	57.1	4.6	18.5	2.09
	N3	⋯	H1	70.3	23.5	16.4	1.93
	O2	⋯	H21	61.6	4.0	18.8	2.00
G_{12}-C_{13}	O6	⋯	H42	2.7	3.7	19.6	2.12
	H1	⋯	N3	3.0	3.0	21.6	2.05
	H21	⋯	O2	3.0	4.1	25.4	2.10

The percent occurrence, P, of a hydrogen bond is the percentage of the total time simulated for which the bond exists. The upper limit for hydrogen bonds lengths (O..H or N..H distance) is 2.6Å and that for hydrogen bond angles is 35° (O..N-H or N..N-H angle).
The life-time, τ, is the mean time between formation and breakage of the hydrogen bond.
The time averages of the hydrogen bond length (O..H) and hydrogen bond angle (were calculated over a period of 400ps. Their standard deviations are 0.2Å and 8°, respectively.

has been observed in solution by NMR (Patel et al., 1987) occurs in all of our simulations. It is interesting to note that the stability of hydrogen bonds in the base pairs next to the terminal ones is hardly affected by the fraying process. Figure 4 shows typical conformations adopted by terminal base pairs.

Bifurcated hydrogen bonds, as suggested by the X-ray structure of HL (Nelson et al., 1987), occurred only rarely in the HL trajectory. The highest percentage occurrence of 6.8% was found for the bifurcated hydrogen bond between the amino hydrogen of A_6 and the carbonyl O4 of T_{20}. The typical percentage occurrence (P) of a bifurcated hydrogen bond was found to be less than 1%. The criterion used for detecting a bifurcated hydrogen bond was more generous than for a normal hydrogen bond, with upper limits on the length of 3.5 Å and on the angle of 45°.

The low probability of the central hydrogen bond of the A_8-T_{17} base pair in the DK system (Fig. 3) is a consequence of the spontaneous breaking of this pair in an otherwise

Table 6. Properties of Watson-Crick hydrogen bonds in the DK simulation

Base-pair	Hydrogen bond			% Occurrence[1] P	Life-time[2] τ(ps)	Angle (°)	Length (Å)
C_1-G_{24}	H42	⋯	O6	51.3	3.7	18.9	2.09
	N3	⋯	H1	67.5	10.4	16.3	1.95
	O2	⋯	H21	44.3	2.9	19.0	2.08
G_2-C_{23}	O6	⋯	H42	79.3	5.0	8.0	2.03
	H1	⋯	N3	94.3	15.7	16.6	1.96
	H21	⋯	O2	86.3	7.5	17.3	2.04
C_3-G_{22}	H42	⋯	O6	83.5	6.2	18.9	2.05
	N3	⋯	H1	93.3	15.5	16.9	1.94
	O2	⋯	H21	89.8	10.3	16.3	2.00
G_4-C_{21}	O6	⋯	H42	86.8	7.7	16.9	2.00
	H1	⋯	N3	96.5	24.3	16.4	1.93
	H21	⋯	O2	86.5	7.2	17.6	2.06
A_5-T_{20}	H61	⋯	O4	64.0	3.2	20.3	2.14
	N1	⋯	H3	95.3	23.8	16.3	1.92
A_6-T_{19}	H61	⋯	O4	41.5	2.1	19.5	2.20
	N1	⋯	H3	92.3	13.3	16.7	1.95
T_7-A_{18}	O4	⋯	H61	71.3	3.6	18.0	2.06
	H3	⋯	N1	84.5	6.6	17.3	2.02
T_8-A_{17}	O4	⋯	61	24.3	5.4	18.2	2.07
	H3	⋯	N1	25.3	6.3	17.7	1.99
C_9-G_{16}	H42	⋯	O6	47.5	2.8	18.6	2.09
	N3	⋯	H1	84.0	7.5	17.5	1.99
	O2	⋯	H21	82.8	6.8	17.8	2.02
G_{10}-C_{15}	O6	⋯	H42	76.0	4.8	17.7	1.98
	H1	⋯	N3	95.0	20.0	15.9	1.94
	H21	⋯	O2	86.3	7.5	17.8	2.06
C_{11}-G_{14}	H42	⋯	O6	83.5	6.7	16.7	2.06
	N3	⋯	H1	94.8	18.1	16.1	1.92
	O2	⋯	H21	88.0	9.3	17.4	1.99
G_{12}-C_{13}	O6	⋯	H42	58.3	3.2	19.1	2.05
	H1	⋯	N3	77.8	6.5	17.1	1.95
	H21	⋯	O2	72.0	4.1	19.0	2.02

See legend to Table 5.

intact double helix. The base pair remained in an open configuration for about 100 ps, after which time the two Watson-Crick hydrogen bonds formed again remaining intact throughout the rest of the simulation. Such a spontaneous breaking was also observed for the HL sequence at the beginning of the A-tract, but this event lasted only 50 ps (Fig. 5). The fact that the breaking of one base pair was not followed by the melting of adjacent base pairs was also observed in NMR studies by Leroy (Leroy et al., 1988b), who suggested that the opening of a base pair is a localized event. This is confirmed by the simulation.

The change of the probability for hydrogen bond formation along the A-tract in the HL system (Fig. 3) suggests an increase of the stability of AT base pairs in the direction 5' to 3' along the A-tract. NMR studies on this sequence and on other sequences known to generate curved DNA, feature anomalously long life times for the AT base pairs in the A-tracts (Leroy et al., 1988a). Such measurements obtained by catalyzed exchange of imino protons cannot directly be related to the calculated stabilities. Nevertheless, the anomaly showing in our simulation is striking both by its magnitude and directionality. It suggests that long-range effects are correctly modeled in our simulations.

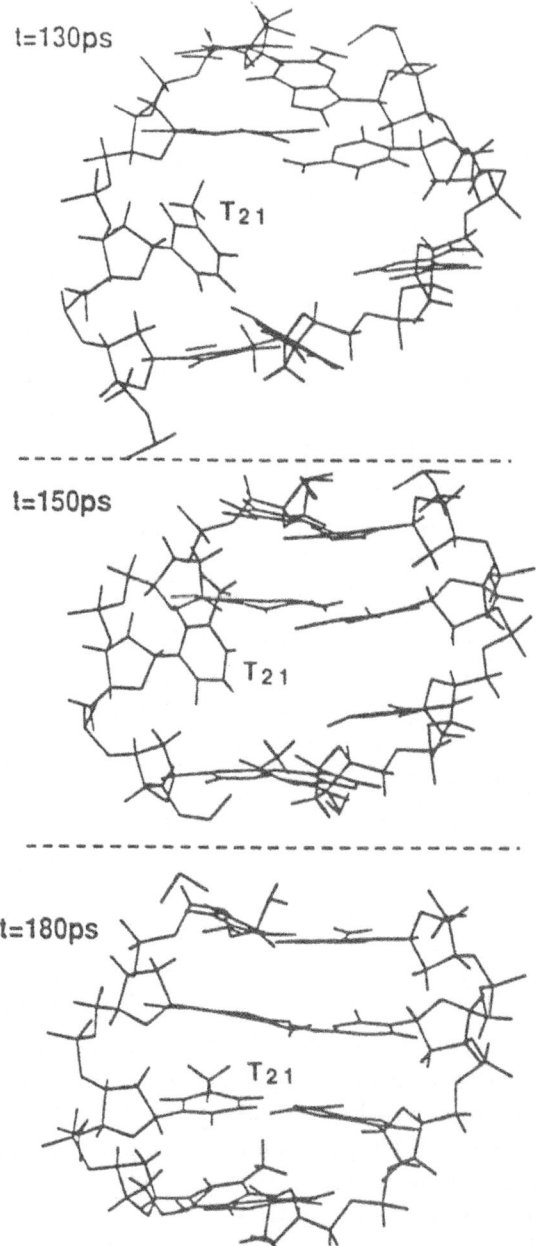

Figure 5. Snapshots from HL trajectory showing the opening and closing of the internal T_{21}-A_4 base pair. The pair is formed in the starting structure, breaks after 130 ps, remains open for 50 ps and then reforms.

3.4. Structural Properties

3.4.1. Torsion Angles and the Sugar Conformation. The backbone torsion angles of DNA are defined in Fig. 6 and Fig. 7 shows the histograms for the six backbone torsion angles for both simulations. The two DNA sequences (DK and HL) show roughly similar distributions of torsion angles along the trajectories, suggesting that the sequence-dependence of these angles is not strong. Each angle generally assumes values from one or two different ranges. The width of the angular distributions reflect the inherent flexibility of the backbone as the presented data are collected over most of the trajectory (100 to 500 ps) and over all nucleotides in the sequence.

The α and γ torsion angles have both two preferred ranges, (t, g-) and (g+,t), respectively (g+ is 0° to 120°, t is 120° to 240°, g- is 240° to 360°). This observation is in accordance with the energy profiles calculated by Pearlman and Kim (1988). These two angles are strongly correlated. The combination (α_t, γ_t) predominates occurring 77% of the time and $(\alpha_{g-}, \gamma_{g+})$ occurs 23% of the time. Harvey (1983) suggested that transitions between the two conformations occur in the B to Z transition. The correlation coefficient for this pair is negative as expected for a crankshaft rotation between a pair of angles separated by the trans torsion angle, β (Singh et al., 1985). The ε and ζ torsion angles are mostly (98% of the time) in the $(\varepsilon_t, \zeta_{g-})$ conformation; the $(\varepsilon_{g-}, \zeta_t)$ conformation, B_{II} (Privé et al., 1987), occurs only 2% of the time. δ is the most flexible torsion angle and is strongly correlated to the sugar pseudorotation angle (Levitt and Warshel, 1978). The χ angle of both purines and pyrimidines remain within the normal range of B-DNA (data not shown).

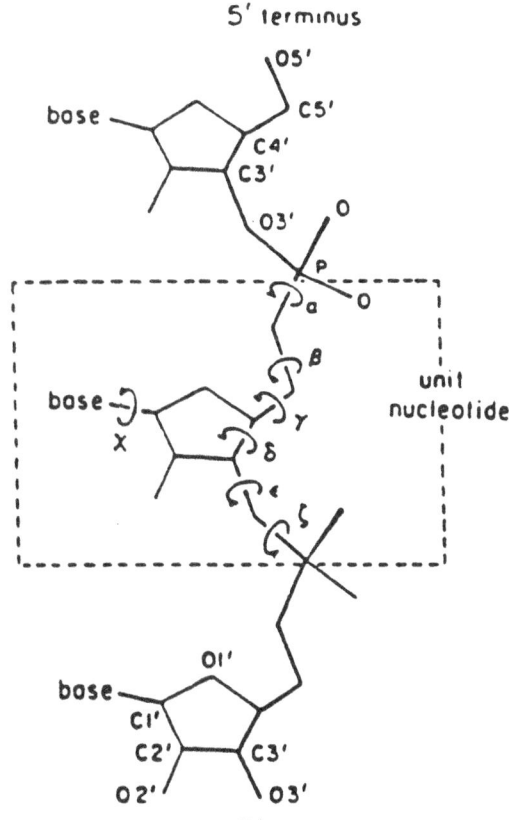

Figure 6. Schematic view of the polynucleotide back-bone. The torsion angles are defined by the four atoms in parenthesis as follows: α (O3'-P-O5'-C5'), β (P-O5'-C5'-C4'), γ (O5'-C5'-C4'-C3'), δ (C5'-C4'-C3'-O3'), ε (C4'-C3'-O3'-P), ζ (C3'-O3'-P-O5'), χ purines (O1'-C1'-N9-C4) and χ pyrimidines (O1'-C1'-N1-C2) (taken from Pearlman and Kim, 1988).

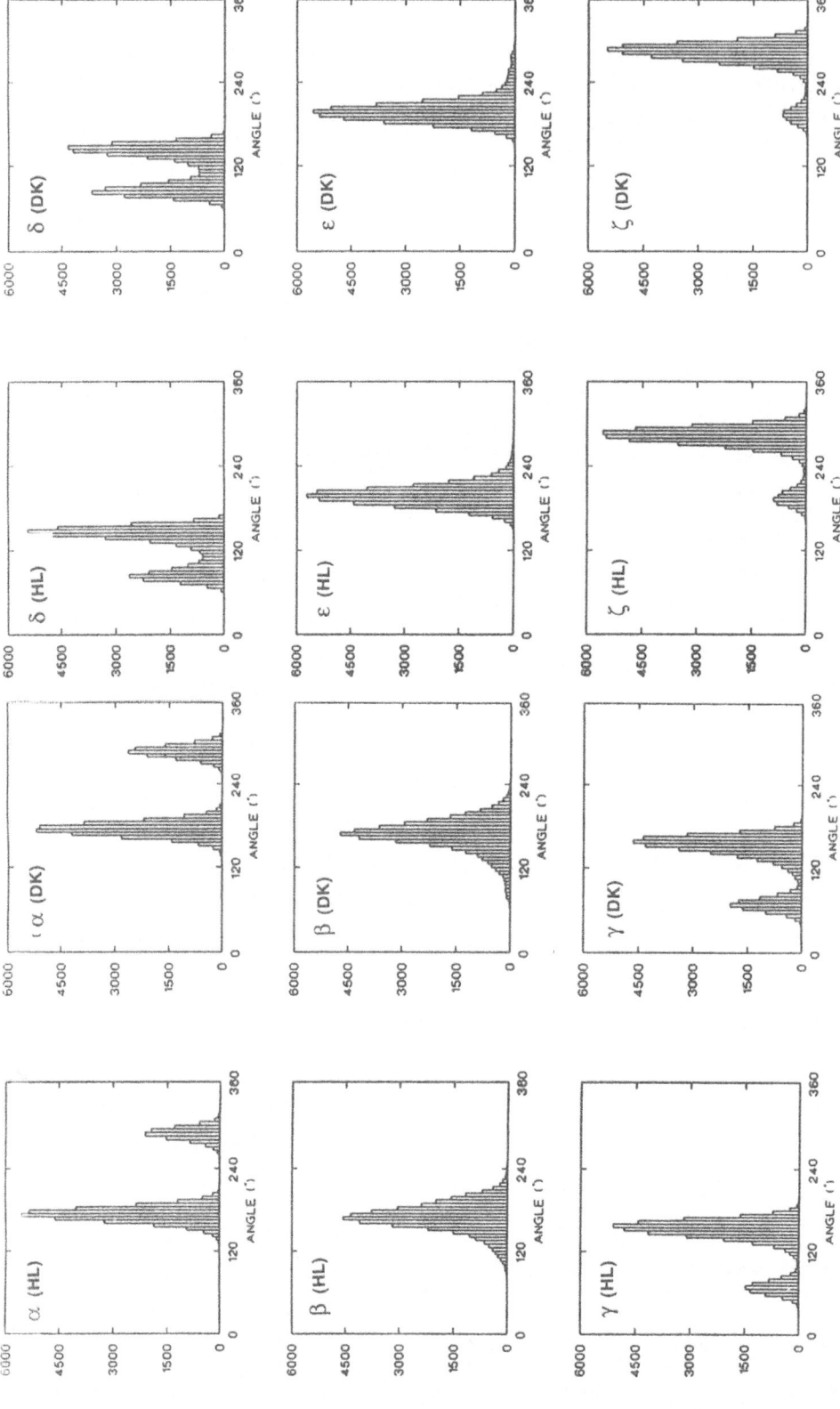

Figure 7. The distributions of the six backbone torsion angles were collected over 400ps (100 ps to 500 ps) of the HL and DK simulations. Data is pooled for all nucleotides except those at the chain termini.

The preferred conformation of the furanose ring was found to be either in the O2'-endo or in the O3'-endo conformation, with equal probabilities. There is, however, a small probability of the O1'-endo conformation, which is on the pseudorotational path between the O2'-endo and the O3'-endo conformation. This distribution reflects the potential energy profile of the furanose ring which has two minima at 3' endo and 2' endo (Levitt and Warshel, 1978).

3.4.2. Helical Parameters. The time-averages of the helical twist angle and the helical rise are given in Table 7. These two parameters conform to the B-form and exhibit relatively small fluctuations around their mean values. In contrast to these two parameters, the tilt, roll and propeller twist angles show large fluctuations during the simulations of both systems. The tilt angle along the A-tract in the HL system is relatively small (around 0°) and the propeller twist angle fluctuates between 20° and 30°.

3.4.3. Minor Groove Width in A-Tracts. The width of the minor groove in sequences containing A-tracts was measured in 2-D NMR experiments and found to vary along the A-tracts (from 5' to 3') (Nadeau and Crothers, 1989; Masato et al., 1988; Masato et al., 1990). Moreover, it was suggested that the gradual compression of the minor groove is the basis of DNA bending. In Table 8 we give the time-average width of the minor groove from our molecular dynamics simulation and compare them with the experimental data from both X-ray and NMR studies. Our results are with good agreement with the NMR data (Masato et al., 1990), showing a gradual compression from the 5' to the 3' ends of the A-tract. Diekmann and co-workers (Diekmann and von Kitzing, 1992; Diekmann et al. 1992) have shown that the narrowing of the minor grove is caused by the lack of a non-hydrogen atom substituent at the 2-position of the adenine ring. Our findings are in agreement with their model.

Table 7. Average helical parameters for the HL and DK simulations

Base-Pair Step	Helix rise (Å)	Helix twist (°)
HL Simulation		
G_2-C_{23}—C_3-C_{22}	3.67 ± 0.49	37.9 ± 2.5
C_3-G_{22}—A_4-T_{21}	4.44 ± 0.33	39.6 ± 3.0
A_4-T_{21}—A_5-T_{20}	4.19 ± 0.72	40.0 ± 2.7
A_5-T_{20}—A_6-T_{19}	3.93 ± 0.55	36.3 ± 3.4
A_6-T_{19}—A_7-T_{18}	3.78 ± 0.47	32.5 ± 3.3
A_7-T_{18}—A_8-T_{17}	3.45 ± 0.37	36.9 ± 4.8
A_8-T_{17}—A_9-T_{16}	3.53 ± 0.41	35.6 ± 3.5
A_9-T_{16}—G_{10}-C_{15}	3.66 ± 0.49	34.4 ± 2.9
G_{10}-C_{15}—C_{11}-G_{14}	4.35 ± 0.52	30.9 ± 5.6
DK Simulation		
G_2-C_{23}—C_3-C_{22}	4.27 ± 0.48	33.8 ± 4.0
C_3-G_{22}—G_4-C_{21}	3.99 ± 0.29	36.4 ± 3.3
G_4-C_{21}—A_5-T_{20}	3.10 ± 0.41	38.4 ± 2.1
A_5-T_{20}—A_6-T_{19}	2.39 ± 0.54	41.3 ± 1.9
A_6-T_{19}—A_7-T_{18}	2.01 ± 0.82	38.7 ± 2.6
T_7-A_{18}—T_8-A_{17}	3.35 ± 0.61	27.4 ± 6.6
T_8-A_{17}—G_9-C_{16}	3.81 ± 0.41	41.7 ± 4.1
G_9-C_{16}—C_{10}-G_{15}	3.44 ± 0.44	39.4 ± 2.3
C_{10}-G_{15}—G_{11}-C_{14}	4.05 ± 0.36	34.6 ± 4.4

Time averages are calculated over the last 400ps of each trajectory. The helical parameters of the first and the last helical steps are not included.

Table 8. Minor groove width along the A tract

Atom pair	Inter-proton Distance (Å)			
	Fiber[1]	X-ray[2]	NMR[3]	MD[4]
$A_4...G_{22}$	5.1	6.2	>4.5	5.8
$A_5...T_{21}$	5.1	4.0	4.2	4.1
$A_6...T_{20}$	5.1	4.0	3.9	4.0
$A_7...T_{19}$	5.1	3.3	3.9	3.6
$A_8...T_{18}$	5.1	4.8	3.8	3.0
$A_9...T_{17}$	5.1	3.8	3.8	3.5

The width of the minor groove is measured as the distance between the H2 atom of an adenine and the H1' atom of a 3' neighboring base on the complementary strand.
[1]Uniform B-DNA (Arnott and Hukins, 1972).
[2]X-ray (Nelson et al., 1987).
[3]NMR (Masato et al., 1988; Masato et al., 1990).
[4]Time average distances were calculated over the last 400 ps of the HL trajectory.

3.5. The Behavior of the Water and the Ions

The water molecules close to the DNA interact extensively with atoms on the DNA backbone and in the two grooves. At a large distance from the DNA the water molecules exhibit bulk properties, similar to the properties calculated from a 500 ps molecular dynamics run of liquid water at room temperature.

The 22 Na^+ ions in both the HL and the DK systems show considerable differences in their diffusional behavior. The motion of each of the ions is monitored by plotting the squared position of each of the ions as a function of time (see Fig. 8). Ions 1 and 2 in exhibit two extreme behaviors; ion 1 moves with approximately linear dependence of the squared shift on time, which indicates a diffusive motion with a diffusion constant close to its bulk value; ion 2 moves very little and stays very close to the DNA. This particular ion, became trapped between the phosphate group and the ribose oxygen of G2 of the HL sequence after the C_1-G_{24} base pair broke down, preventing this base pair from forming again. The behavior of ions 3 and 4 exhibit an intermediate behavior which is common to most of the ions in both systems studied.

4. CONCLUSIONS

Our simulations of DNA are stable in that the energy is well-conserved and the double-helical structure of DNA is preserved without any special hydrogen bond restraints (Swaminathan et al., 1991), attenuated charges (Levitt, 1983), or special dielectric constants (Singh et al., 1985). In fact, the same energy functions we use for DNA (Levitt et al., 1995a) also work well for proteins. Our study of DNA required better energy functions without the restrictions on atom grouping used before (Levitt & Sharon, 1988) and thus led to a major methodological advance.

In both sequences, the structure of the double-helix remains for the entire simulation close to the canonical B-form as judged by the time averaged helical parameters, torsional angles and sugar conformations. The DNA structure is seen to move around this equilibrium B-form double helix by means of torsional angles transitions, twisting about the helical axis,

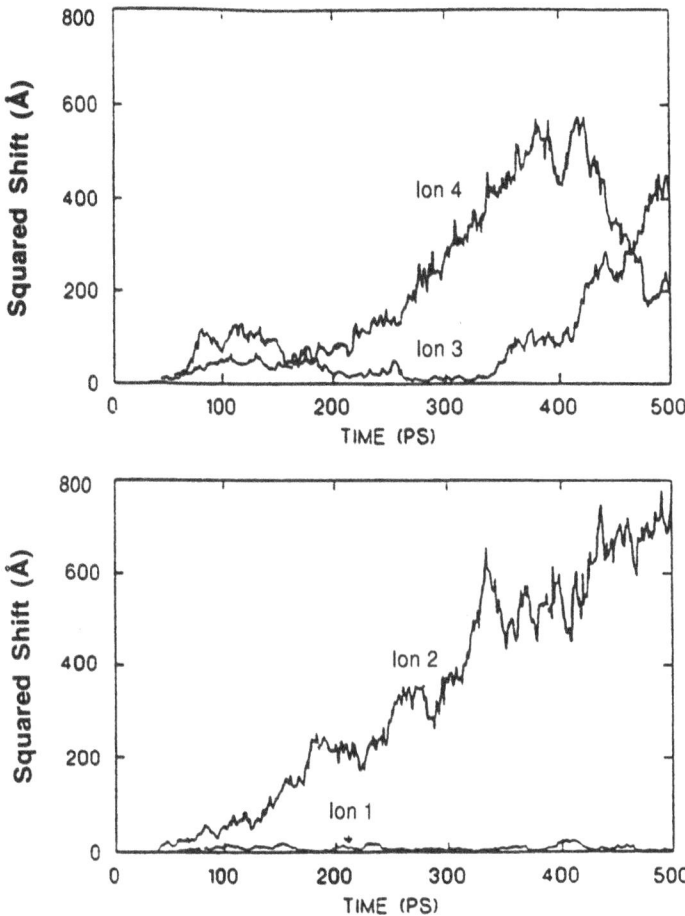

Figure 8. The variation of the squared shift in position of four of the 22 Na$^+$ ions with time over the entire 500 ps of the DK simulation.

reorientation of paired bases, breakage and formation of hydrogen bonds and fraying of terminal base pairs.

The time-average RMS deviation from the crystal structure is found to be larger than for similar simulations of proteins in solution (Levitt & Sharon, 1988; Daggett & Levitt, 1993). This is to be expected if we reflect on the differences in shape of a DNA double-helix and a globular protein structure. The double-helix is a linear structure with a great deal of surface area exposed to solvent and little tertiary packing of nucleotides. Changing a single torsion angle consistently by a few degrees can change the three-dimensional structure significantly. Nevertheless, in the simulations RMS deviation reaches a plateau after about 100 ps and remains stable throughout the rest of the simulation.

The fraying of the terminal base pairs that we find in all our molecular dynamics simulations is observed experimentally (Patel et al., 1987), but was not found in previous simulations. We also find that the un-paired bases continue to interact with one another and with adjacent bases by stacking interactions. Thus, they limit the access of water molecules the base pairs adjacent to the terminal base pairs, which remain intact throughout the simulation.

The most exciting results were obtained from the HL simulation where the six base-pair A-tract showed anomalies in both hydrogen-bond stability and minor-groove width. These properties were not seen in the short, two base-pair, A-tract in the DK simulation, suggesting that they are a consequence of long range affects rather than nearest-neighbors affects. Further analysis of the HL simulation and more simulation on related DNA sequences should provide us with clearer insight into the actual mechanism for DNA bending at atomic level.

5. ACKNOWLEDGMENTS

This work was supported by the Program in Mathematics and Molecular Biology through the National Science Foundation grant number DMB-8720208.

REFERENCES

Allen, M. P. and Tildesley, D. J. (1987) Computer Simulation of Liquids; Clarendon Press: Oxford.

Arnott, S. and Hukins, D. W. L. (1972) *Biochem. Biophys. Res. Comm.* 47, 1504-1509.

Beeman, D. (1976) *J. Computational Physics*, *20*, 130-139.

Daggett, V. and Levitt, M. (1993) Protein Unfolding Pathways Explored Through Molecular Dynamics Simulations. *J. Mol. Biol. 232* 600-618.

Daggett, V. and Levitt, M. (1994) Protein Folding Unfolding Dynamics. *Current Opinions in Structural Biology*, *4* 291-295.

Dickerson, R. E., Drew, H. R., Conner, B. N., Wing, R. M., Fratini, A. V. and Kopka, M. L. (1982) The Anatomy of A-, B-, and Z-DNA. *Science*, *216*, 475-485.

Diekmann, S. and von Kitzing, E. (1992) On the Structural Origin of DNA Curvature. *Structure & Expression. 3*, 57-67.

Diekmann, S., Mazzarelli, J. M., McLaughlin, von Kitzing, E. and Travers, A. A. (1992) DNA Curvature Does Not Require Bifurcated Hydrogen Bonds pr Pyrimidine Methyl Groups. *J. Mol. Biol. 225*, 729-738.

Drew, H. R. and Dickerson, R. E. (1982) Structure of a B-DNA Dodecamer at 16 Kelvin. *Proc. Nat. Acad. Sci. USA*, *79*, 4040-4044.

Drew, H. R., Wing, R. M., Takano, T., Borka, C., Takano, S., Itakura, K. and Dickerson, R. E. (1981) Structure of a B-DNA Dodecamer. Conformation and Dynamics. *Proc. Nat. Acad. Sci. USA, 78*, 2179-2183.

Harvey, S. C. (1983) *Nucl. Acid. Res. 11*, 4867-4878.

Hirshberg, M. (1990) Ph.D. Thesis, Weizmann Institute of Science, Israel.

Kim, Y., Grable, J. C., Love, R., Greene, P. J. and Rosenberg, J. M. (1990) Refinement of Eco RI Endonuclease Crystal Structure. A Revised Protein Chain Tracing. *Science*, *249*, 1307-1310.

Kim, Y., Geiger, J. H., Hahn, S., & Sigler, P. B. (1993a) Crystal structure of a yeast TBP-TATA-box complex. *Nature 365*, 512-520.

Kim, J. L., Nikolov, D. B. and Burley, S. K. (1993b) Co-crystal structure of TBP recognizing the minor groove of a TATA element. *Nature 365*, 520-527.

Langridge, R., Marvin, D. A., Seeds, W. E. Wilson, H. R., Hooper, C. W., Wilkins, M. H. F., and Hamilton, L. D. (1960). *J. Mol. Biol. 2*, 38-64.

Leroy, J. -L., Charretier, E., Kochoyan, M., and Guéron, M. (1988a) *Biochemistry*, *27*, 8894-8898.

Leroy, J. -L., Kochoyan, M., Huynh-Dinh, T and Guéron, M. (1988b) *J. Mol. Biol. 200*, 233-238.

Levitt, M. (1978) How Many Base-Pairs per Turn Does DNA have in Solution and in Chromatin? Some Theoretical Calculations. *Proc. Nat. Acad. Sci. USA 75*, 640-644..

Levitt, M. (1983) Computer Simulation of DNA Double Helix Dynamics. *Cold Spring Harbor Symp. Quant. Biol. 47*, 251-261.

Levitt, M. (1983) Molecular Dynamics of Native Protein: II. Analysis and Nature of Motion. *J.Mol.Biol. 168*, 595-620.

Levitt, M. (1989) Molecular Dynamics of Macromolecules in Water. *Chemica Scripta, 29A*, 197-203.

Levitt, M. and Sharon, R. (1988) Accurate Simulation of Protein Dynamics in Solution. *Proc. Natl. Acad. Sci. USA. 85*, 7557-7561.

Levitt, M. and Warshel, A. (1978) Extreme Conformational Flexibility of the Furanose Ring in DNA and RNA. *J. Am. Chem. Soc. 100*, 2607-2613.

Levitt, M., Hirshberg, M., Sharon, R. and Daggett, V. (1995a) Potential Energy Function and Parameters for Simulations of the Molecular Dynamics of Proteins and Nucleic Acids in Solution. *Computer Physics Communications*, in press.

Levitt, M., Hirshberg, M., Sharon, R. and Daggett, V. (1995b) Calibration and testing of a water model for simulation of the molecular dynamics of proteins and nucleic acids in solution. *J. Chem. Phys.*, submitted.

Masato, K., Sugeta, H. and Kyogoku, Y. (1988) *Nucl. Acid. Res. 18*, 613-618.

Masato, K., Sugeta, H., Kyogoku, Y., Fujii, S., Fujisawa, R. and Ken-ichi, T. (1988) *Nucl. Acid. Res. 16*, 8619-8632.

Miaskiewicz, K., Osman, R. and Weinstein, H. (1993) Molecular Dynamics Simulation of the Hydrated d(CGCGAATTCGCG)$_2$ Dodecamer. *J. Am. Chem. Soc. 115*, 1526-1537. 150 ps MD is unstable after first 60 ps with RMS up to 5 Å. Uses AMBER force field.

Nadeau, J. G. and Crothers, D. M. (1989) *Proc. Natl. Acad. Sci. USA*, *86*, 2622-2626.

Nelson, H. C. M., Finch, J. T., Bonaventura, F. L. and Klug, A. (1987) *Nature*, *333*, 221-226.

Patel, D. J., Shapiro, L. and Hare, D. (1987) *Quart. Rev. Biophys. 20*, 35-112.

Pearlman, D. A. and Kim, S. -H. (1988) *Biopolymers*, *27*, 59-77.

Prabhakaran, M. and Harvey, S. C. (1985) *J. Phys. Chem. 89*, 5767-5771.

Privé, G., Heinemann, U., Chadrasegaran, L. S., Kan, M. L., Kopka, M. and Dickerson, R. E. (1987) *Science*, *238*, 498-504.

Pabhakaran, M. and Harvey, S. C. (1985) *J. Phys. Chem. 89*, 5767-5771.

Seibel, G. L., Singh, U. C. and Kollman, P. (1985) *Proc. Nat. Acad. Sci. USA*, *82*, 6537-6540.

Singh, U. C., Weiner, S. J., and Kollman, P. (1985) *Proc. Nat. Acad. Sci. USA*, *82*, 755-759.

Swaminathan, S., Ravishanker, G. and Beveridge, D. L. (1991) Molecular Dynamics of B-DNA Including Water and Counterions: A 140-ps Trajectory for d(CGCGAATTCGCG) Based on the GROMOS Force Field. *J. Am. Chem. Soc. 113*, 5027-5040. Use restraint to maintain Watson-Crick base pairing. RMS 2.7 Å after 140-ps.

Tidor, B., Irikura, K. K., Brooks, B. R. and Karplus, M. (1983) Dynamics of DNA Oligomers. *J. Biomol. Struct. & Dyn. 1*, 231-252.

Trifonov, E. N. and Sussman, J. L. (1979) Smooth Bending of DNA in Chromatin. In Molecular Mechanisms of Biological Recognition (Balaban, M., ed.) pp. 227-231, Elsevier North-Holland Biomedical Press, New York.

van Gunsteren, W. F., Berendsen, H. J. C., Geurtsen, R. G. and Zwinderman, H. R. J. (1986) *Ann. NY Acad Sci. 482*, 287-301.

Watson, J. D. and Crick, F. H. C. (1953) Molecular Structure of Nucleic Acid: A Structure for Deoxyribonucleic Acid. *Nature*, *171*, 737-740.

Zielinski, T. J. and Shibata, M. (1990) *Biopolymers*, *29*, 1027-1044.

14

DEVELOPMENTS IN NMR STRUCTURE DETERMINATION OF NUCLEIC ACIDS

C. W. Hilbers,[1,2] S. S. Wijmenga,[1] H. Hoppe,[2] and H. A. Heus[2]

[1] SON/NWO National HF-NMR Facility
[2] Laboratory of Biophysical Chemistry
   Nijmegen SON Research Centre for Molecular Design, Structure and
   Synthesis
   Toernooiveld, 6525 ED Nijmegen, The Netherlands

INTRODUCTION

The spectacular developments in high resolution NMR, during the last five to ten years, have established the technique as one of the most important methods in the field of structural biology. With the aid of NMR spectroscopy structures of biomolecules in aqueous solutions can be determined; the other powerful technique playing a key role in structural biology, X-ray diffraction, needs molecules in the crystal form in order to elucidate their structures. In the present contribution some of the NMR methods used to determine nucleic acid structures will be discussed. Compared to protein structure determination, by means of NMR, the elucidation of nucleic acid structures has its peculiar difficulties and as a result has lagged somewhat behind. One reason is that functional nucleic acid molecules are normally very large and not accessible to structure determination by the high resolution techniques X-ray crystallography and NMR spectroscopy. It has been found, however, that in these molecules particular regions may adopt highly intricate folding patterns which are functionally very important. Examples are the ribozyme [Pley et al. (1994); Scott et al., (1995)] and pseudoknot motifs in RNA [Pleij, (1990)] and the H-DNA triple helices [Frank-Kamenetskii and Mirkin, (1995)] and hairpins in DNA [Hilbers et al. (1994)]. DNA and RNA fragments containing these folding motifs are amenable to high resolution structural studies and these investigations form the basis for a reliable description of the structure function relationships in these molecules. Other reasons for the slower progress in NMR structure determination of nucleic acids follow from a further comparison between nucleic acid and protein structural characteristics.

    i. in nucleic acids the nucleotide subunit has six variable backbone angles; in a protein the amino acid subunit has only two, the Φ and the Ψ angles.

    ii. in nucleic acids one encounters four different side chains, the bases A, G, U(T) and C; in proteins one finds twenty different amino acids. This simplification in nucleic acids is a blessing in disguise.

Dynamics and the Problem of Recognition in Biological Macromolecules, edited by Jardetzky and Lefèvre
Plenum Press, New York, 1996

iii. in nucleic acids the bases are attached to the backbone via the furanose sugar moiety; in proteins the amino acid side chain are directly connected to the backbone.

iv. there is one main secondary structural element in nucleic acids, the double helix; in proteins one finds the α-helix (sometimes the 3_{10} - helix) and the extended β-chains forming parallel and/or anti-parallel β-sheets.

v. in nucleic acids one finds an uneven distribution of protons; the sugar-phosphate backbone is proton rich while the bases are proton poor. This is in contrast to proteins where the protons are more or less evenly distributed along the amino acid chain.

One of the main hurdles to be taken in the structure determination of a biomolecule is the interpretation of its NMR spectrum, i.e. the assignment and interpretation of the cross peaks containing the information to derive the structure. Until recently, for nucleic acids the assignment procedure was based on the NOE contacts that can be observed between sugar proton resonances in the backbone and the H6 and H8 resonances of the pyrimidine and purine bases respectively. A second sequential assignment pathway, used in spectrum interpretation, is formed by the NOE contacts between the imino-proton resonances of consecutive base pairs in a double helix. The two pathways can be connected via the NOE connectivities measurable within a GC pair. This approach has been successfully applied and has been reviewed extensively in the literature [Wijmenga et al., (1993)]. Despite its shortcomings, which are discussed in the following, these methods will remain valuable tools in future work, despite newly developed methods involving uniformly isotope enriched material (*vide infra*).

The aforementioned assignment procedure is basically only applicable to nucleotide sequences that are part of helical structures. For instance, it has been shown to break down in the loop regions of hairpin molecules [Blommers et al. (1987); Blommers et al. (1991); Mooren et al. (1994); Heus and Pardi (1991)]. This is an undesirable situation, which may easily lead to erroneous, circular reasoning in a structure determination. In the structure determination of proteins, uniform enrichment with ^{13}C and ^{15}N nuclei has proved to be extremely helpful. For such uniform enriched molecules sequential assignment can then be based on the use of through-bond J-couplings in coherence transfer experiments so that assignments no longer depend on the presence of *à priori* assumed structural characteristics. This not only improves the reliability of assignment procedures but also provides a strongly improved spectral resolution by making hetero-nuclear multiple-dimensional spectroscopy possible. The procedures developed for the uniform enrichment of protein molecules is not applicable to nucleic acids because the nucleic acid fragments that can be studied by NMR do not occur independently in the cell. However, for RNA now procedures are available which make it possible to produce uniformly enriched tailor made molecules on a more or less routine basis.

SEQUENTIAL BACKBONE ASSIGNMENTS IN ^{13}C-LABELED RNA VIA THROUGH-BOND COHERENCE TRANSFER

As has been indicated in the introductory part of this contribution, sequential ^1H-resonance assignments derived from through space contacts (NOE cross peaks) have the disadvantage of conformational dependence. In methods which make use of through-bond coherence transfer this problem can be avoided. Sequential assignment in nucleic acids by means of through-bond coherence transfer can be achieved with the aid of two types of triple resonance experiments [Heus et al., 1994] , namely with the socalled three-dimensional HCP or PCH experiment, which correlates ^1H, ^{13}C and ^{31}P resonances, in combination with a two-dimensional hetero-TOCSY experiment (CCH-TOCSY). Before discussing the effects of and the interplay between the corresponding pulse sequences, we first consider the possible sequential pathways that may be traversed during these experiments. In Fig. 1 the two main pathways are indicated, i.e. one involving $C4'(n) \Rightarrow P(n + 1) \Rightarrow C4'(n + 1)$ governed by the highest coupling constants (~11 Hz) (Fig. 1A, dark grey arrows), and the other encompassing $C3'(n) \Rightarrow P(n + 1) \Rightarrow C5'(n + 1)$ involving coupling constants from 4 to 6 Hz (Fig. 1A, light grey arrows). These pathways can be connected to the spin systems in the sugar rings via the J-couplings between the ^{13}C-spins of ~42 Hz (Fig. 1B). Compared to the situation in proteins, the J-couplings in the nucleic acid backbone are rather small and it is therefore expected that the coherence transfer methods will be of more limited value in nucleic acids. The pulse sequences corresponding to the aforementioned HCP and PCH

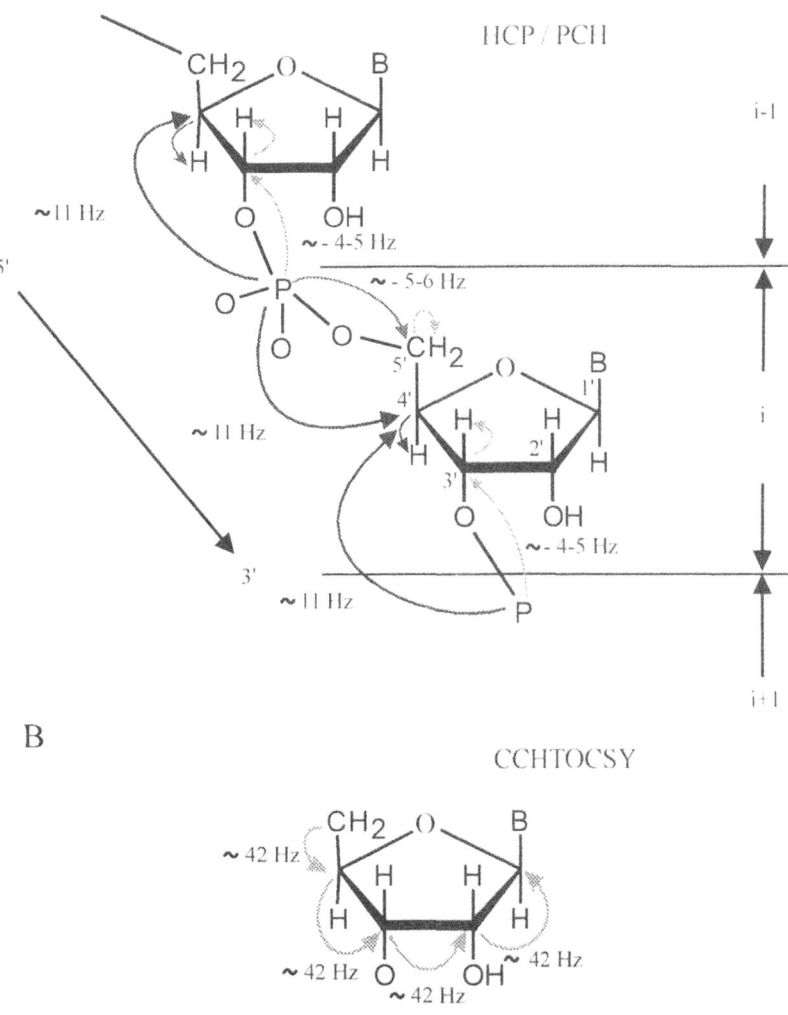

Figure 1. Schematic of the RNA ribose backbone, showing the coherence pathways as indicated by arrows (see text) in the HCP and PCH experiments (A) and in the CCHTOCSY experiment (B).

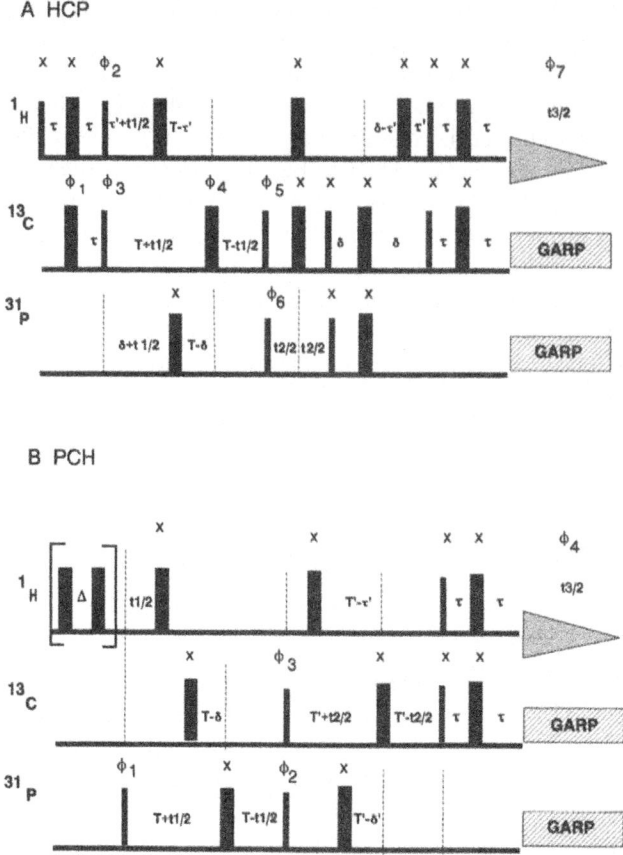

Figure 2. Pulse sequence of the HCP (A) and PCH (B) experiment. The thick and thin bars indicate 90° and 180° pulses, respectively. Phase cycling was as follows: (A HCP) ϕ_1 = x, -x; ϕ_2 = y, -y; ϕ_3 = x (+TPPI(t1)); ϕ_4 = 4x, 4y; ϕ_5 = x, -x; ϕ_6 = x, x, -x, -x (+TPPI(t2)); ϕ_7 = x, x, -x, -x, -x, -x, x, x; (B PCH) ϕ_1 = x (+TPPI(t1)); ϕ_2 = y, -y; ϕ_3 = y, y, -y, -y (+TPPI(t2)); ϕ_4 = x, -x, -x, x.

experiments are presented in Fig. 2A and 2B, respectively. The HCP pulse sequence closely resembles that of the CT-HNCO experiment frequently used in the analysis of protein spectra [Grseziek and Bax, 1992]. Considering the pathways outlined above, the HCP-experiment correlates ^{1}H, ^{13}C and ^{31}P resonances as follows:

			f1		f2		f3
i.	H3'(n)	$\Rightarrow$	C3'(n)	$\Rightarrow$	P(n + 1)	$\Rightarrow$	{C3'(n)}H3'(n)
					P(n + 1)	$\Rightarrow$	{C4'(n)}H4'(n)
ii.	H4'(n)	$\Rightarrow$	C4'(n)	$\Rightarrow$			
					P(n)	$\Rightarrow$	{C4'(n)}H4'(n)
iii.	H5'/H5"(n)	$\Rightarrow$	C5'(n)	$\Rightarrow$	P(n)	$\Rightarrow$	{C5'(n)}H5'/H5"

In the scheme above f1, f2 and f3 denote the ^{13}C, the ^{31}P and the ^{1}H frequency axis respectively; n stands for the nth nucleotide. The cross peaks expected between the ^{1}H, ^{13}C and ^{31}P resonances are indicated in the three-dimensional spectrum in Fig. 3. In the plane, the cross section, perpendicular to the ^{13}C-frequency axis at the C4'(n)

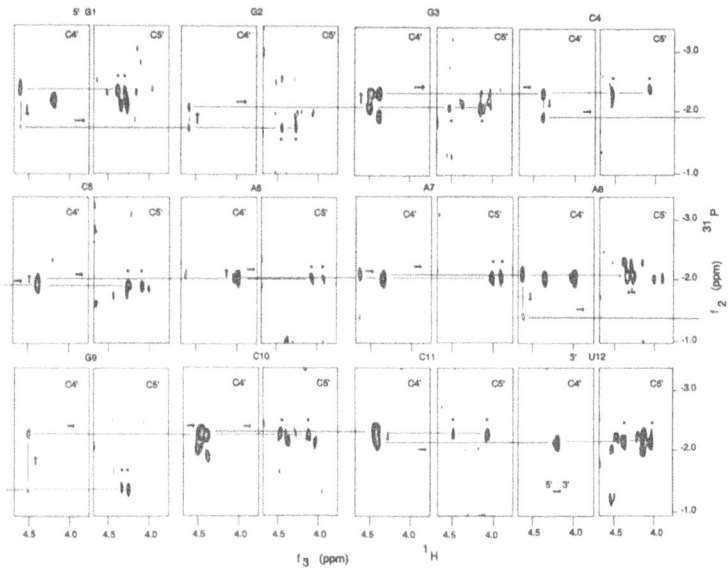

Figure 3. Sequentially arranged ^{31}P, ^{1}H planes from the HCP spectrum (next page), taken at the different C4' and C5' resonance frequencies along the ^{13}C axis. The C4' planes show the sequential walk through the phosphate backbone from residue G1 (5'-end) to the final residue U12 (at the 3'-end) as indicated by the arrows and drawn lines; the two cross peaks in the C5'$_n$ planes indicate the H5'/H5" resonances of residue n; the dotted line connects the H5'/H5" cross peaks of residue n with the P$_n$ cross peak in the C4'$_n$ plane (see text). The HCP was recorded in approximately 20 hours with the following acquisition settings: 16 scans for each FID of 1024 data points (t3), 128 t1 values (t1^{max}/2 = T = δ = 12.5 ms), 28 t2 values, $\tau = \tau' = 1.5$ ms ($\approx 1/4J_{CH}$), T-δ = 5 μs, 1 s relaxation delay with solvent presaturation, low power (3 Watt) GARP decoupling of ^{31}P and ^{13}C, spectral width, 2941 Hz, 400 Hz, and 2941 Hz and carrier position, 73.6 ppm, -2.23 ppm, and 4.75 ppm for ^{13}C, ^{31}P, and ^{1}H, respectively. Typical processing parameters: zero-filling twice in t1, t2 (preceeded by zeropadding to 32 points), and t3; applying a sin^2 window-function shifted by $\pi/5$, $\pi/3$, and $\pi/2$ in t1, t2, and t3, respectively. The chemical shifts are referenced relative to TSP for the ^{1}H and ^{31}P dimensions; for ^{31}P the 0 ppm chemical shift value is obtained by multiplying the ^{1}H TSP frequency by 0.40480793, which corresponds to calibration relative to inorganic phosphate.

resonance position two cross peaks are expected between the H4'(n) resonance and the ^{31}P(n) and the ^{31}P(n + 1) resonances, respectively. In the corresponding cross section at the C3'(n) resonance position one cross peak is expected between the H3'(n) and the P(n + 1) resonance. Similarly, in the cross section at the C5'(n) resonance the H5'(n) and H5"(n) resonance each give rise to a cross peak to P(n). The two cross peaks in the cross section at the C4'(n) resonance belong to the second pathway in the scheme above and can be used for sequential assignment. However, at this point we don't know which of the two cross peaks is to the P(n) or to the P(n + 1) resonance. This can be determined by establishing the resonance position of the C5'-atom that belongs to the same residue as the C4'-atom, i.e. by establishing which of the two C5'-planes, that can, via the two ^{31}P resonances, be connected to the two cross peaks in the C4'-plane, crosses the resonance of the C5'-atom of the same residue as the C4'-atom.

The C4'(n) C5'(n) connection can be decided using either a CCH-TOCSY or a constant-time CCH-TOCSY experiment of which the corresponding pulse sequences are presented in Figs. 4A and 4B, respectively. The pulse sequences start with a series

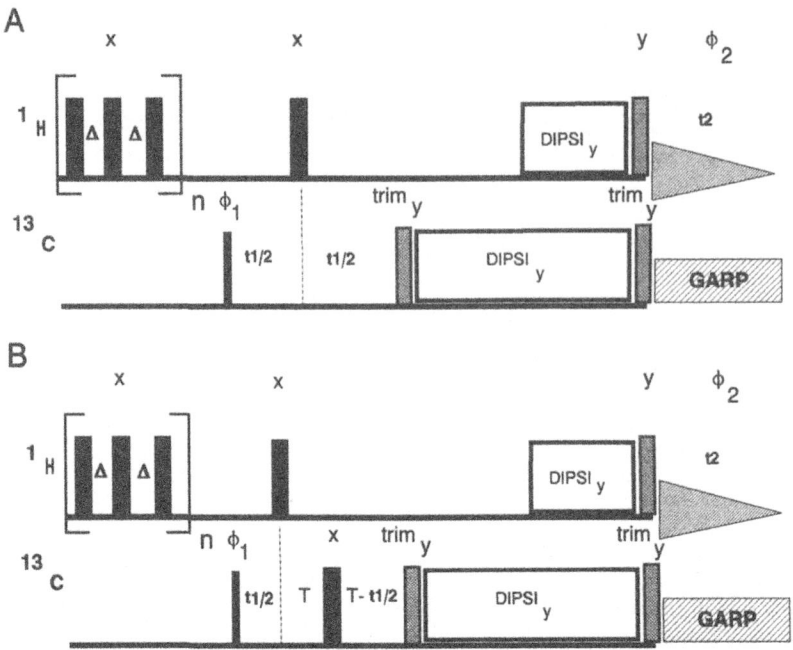

Figure 4. Pulse sequences of the CCH-TOCSY (A) and constant time CCH-TOCSY (B) experiments. Phase cycling is as follows for both A and B: ϕ^1 = x, -x (+TPPI(t1)); ϕ_2 = x, -x. The applied rf-field in the DIPSI correspond typically to a 90° pulse of 28 sec; the cross polarization period corresponds to about 6.2 msec ($\approx 1/J_{CH}$); the homonuclear (C-C) mixing to about 18 msec in order obtain sufficient transfer through the J-coupled ^{13}C sugar ring spin system; the constant time period T is typically set ≈ 12 msec ($\approx 1/2J_{CC}$).

of 180° ^{1}H pulses to saturate the ^{1}H magnetization and thereby create NOE transfer to the carbon spin system. Subsequently, after labeling the ^{13}C spins during t_1 under concomitant decoupling of the protons, coherence is efficiently transferred through the system of J-coupled ribose ^{13}C and ^{1}H spins on account of the large J-couplings involved ($J_{CH} \approx 145$ Hz, $J_{CC} \approx 40$ Hz), so that at the ^{1}H$_i$ (i = 1'-5'/5'') frequency a cross peak appears, not only at the directly bonded ^{13}C but also at the frequencies of the other ^{13}C nuclei of the ribose ring. Since of all sugar proton resonances the H1' resonances overlap least, we use the TOCSY ladders in the H1' spectral region to unequivocally establish the C4'(n) to C5'(n) connection (Fig. 5).

The sequential step in the assignment is subsequently performed by finding, in the HCP spectrum, the C4'(n + 1) plane, i.e. a C4' plane with a P(n + 1) cross peak at the resonance position of the P(n) cross peak of residue n. As an example we consider the sequential assignment of residues A8 and G9 (Fig. 3). The C4'$_{A8}$-plane (Fig. 3, second right, middle panel) shows two cross peaks at the H4'$_{A8}$ resonance frequency. The cross peak with the upfield ^{31}P resonance is the P(n=8) cross peak as follows from the C5'$_{A8}$ plane (Fig. 3, right, middle panel). The P(n=9) cross peak is connected with the downfield-shifted ^{31}P resonance. It provides the ^{31}P frequency of the P(n=9) cross peak of G9, visible in the C4'$_{G9}$ plane (Fig. 3, left, bottom panel). Via this approach a complete sequential analysis could be performed for the hairpin formed by r(pppGGGC-CAAA-GCCU). No break in the sequential walk occurs in the loop region as would be encountered if assignments were based on sequential NOE contacts.

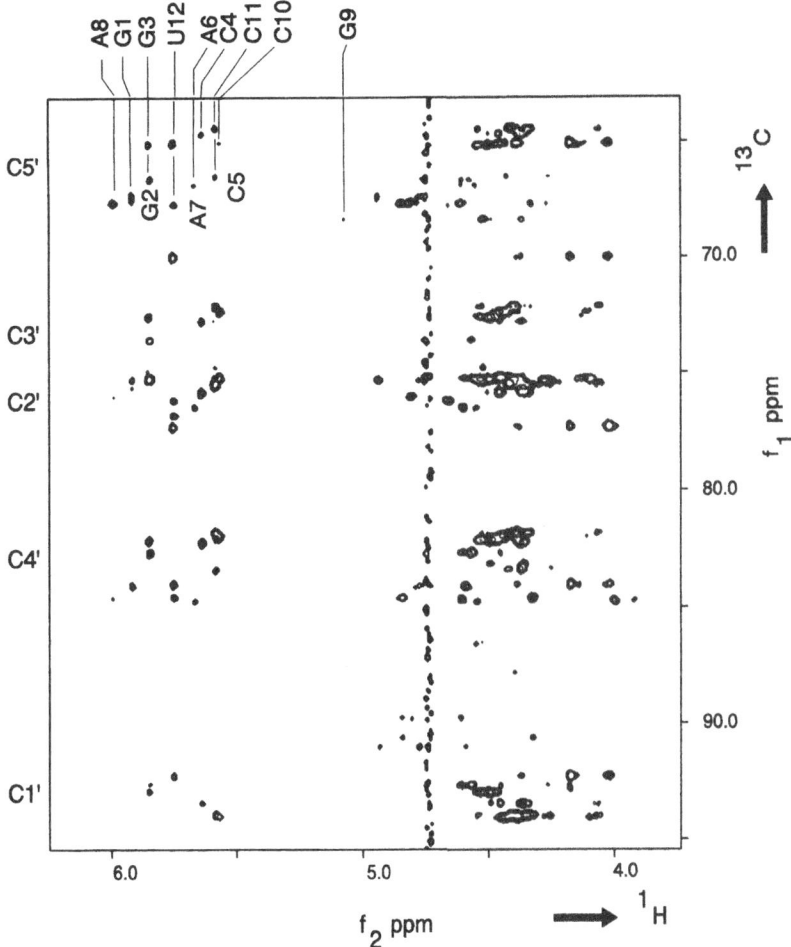

Figure 5. The constant time CCH-TOCSY spectrum (next page), showing ribose region (C1' to C5'). The assignments are indicated in the H1' region. The experimental settings are as given in the legend to Figure 4.

A method complementary to the approach, outlined above, makes use of the higher resolution in the H1' region. A 2D P(CC)H-TOCSY experiment, which combines the PCH and CCH-TOCSY experiment into one triple resonance experiment, allows a sequential backbone assignment via the H1' resonances in a 2D ^{1}H-^{31}P spectrum. The pulse sequence for this experiment is presented in Fig. 6. The transfer of coherence takes place from ^{31}P to ^{1}H through ^{13}C via the coherence pathways sketched in Figs. 7.

The pulse sequence commences with a series of 180° pulses to saturate the ^{1}H spin system and create NOE transfer to the ^{31}P spin system. Subsequently, a 90°$_x$ on ^{31}P generates P_y coherence which is labeled during the t_1 period, while simultaneously the ^{1}H and ^{13}C are decoupled. After the ^{31}P evolution time the in-phase ^{31}P coherence, P_y, is converted into anti-phase coherence, P_xC_z, during the two delays τ. The second 90°-^{31}P pulse converts this anti-phase coherence into zz-order, P_zC_z. Then the gradient pulse removes all unwanted transverse coherence, after which anti-phase P_zC_x coherence is generated by the first 90°$_y$-^{13}C pulse. During the delays τ', anti-phase P_zC_x-coherence is refocused into in-phase C_y coherence. In turn, after application of a ^{13}C trim pulse along the y-axis to remove unwanted ^{13}C coherence, the in-phase C_y-coherence is

Figure 6. Pulse sequence of the P(CC)H- TOCSY experiment. The thick and thin bars indicate 90° and 180° pulses, respectively. The dotted bar indicates before the DIPSI sequence indicates a ^{13}C trim pulse. Phase cycling was as follows: ϕ_1 = x, x, -x, -x (+TPPI(t1)); ϕ_2 = y; ϕ_3 = y, -y; ϕ_4 = x, -x, -x, x. Details concerning delays are given in the legend of Figure 7.

transferred through the network of J-coupled ^{13}C spins of the sugar ring by means of a homonuclear ^{13}C-^{13}C mixing sequence (DIPSI 3). In the final step, the C_y coherence is transferred to the directly bound ^1H via cross polarization by applying a DIPSI3 mixing sequence on both ^1H and ^{13}C simultaneously. Subsequently, the ^1H signal is acquired during t_2 with simultaneous GARP decoupling of ^{13}C. The results obtained by this experiments, when the pulse sequence is applied to the hairpin formed by 5'pppGpGpGpCpCpApApApGpCpCpU3', are shown in Fig. 8. Of the three pathways shown in Figs. 7, the C4' routes (Fig. 7A), i.e. $P_i \Rightarrow C4'_i \Rightarrow C1'_i \Rightarrow H1'_i$ and $P_i \Rightarrow C4'_{i-1} \Rightarrow C1'_{i-1} \Rightarrow H1'_{i-1}$, are the most efficient contributors to the crosspeak intensities in Fig. 8 [Wijmenga et al., (1995)]. An almost uninterrupted sequential walk can be made from the 5'-terminal residue G1 to the 3'-terminal residue U12. In Fig. 8 only the connectivity between A8pG9 (P9) is absent (a very weak intensity is visible at a lower threshold). This can be attributed to two effects. First, the torsion angle β could be shifted towards the g⁻ or g⁺ region, leading to a smaller value of the $J_{P5'C4'}$-coupling and thus to a decreased transfer efficiency via the C4'-route. This cross peak is present in the HCP spectrum discussed above, although with lower intensity, so that its absence in the P(CCH)TOCSY spectrum cannot be completely attributed to a change in the value of β. Secondly, further examination of the spectra shows that the upfield shifted resonance of G9 (see Fig. 8) is significantly broadened probably as a result of conformational averaging. Such an effect may considerably reduce the transfer efficiency in the particular C4'-pathway. This conclusion is corroborated by the observation that in a HMQC spectrum the H1'/C1' cross peak is also difficult to detect.

The next step of the spectrum interpretation involves the connection of the backbone and sugar resonances to the resonances of the bases. Also here one wants to use coherence transfer methods in order to avoid introducing conformational dependence in the assignment procedure. Several experiments have been published that allow such an approach. Here we introduce the triple resonance HCN experiment through which the $H1'_i \Rightarrow C1'_i \Rightarrow N1_i$ or $N9_i$

A

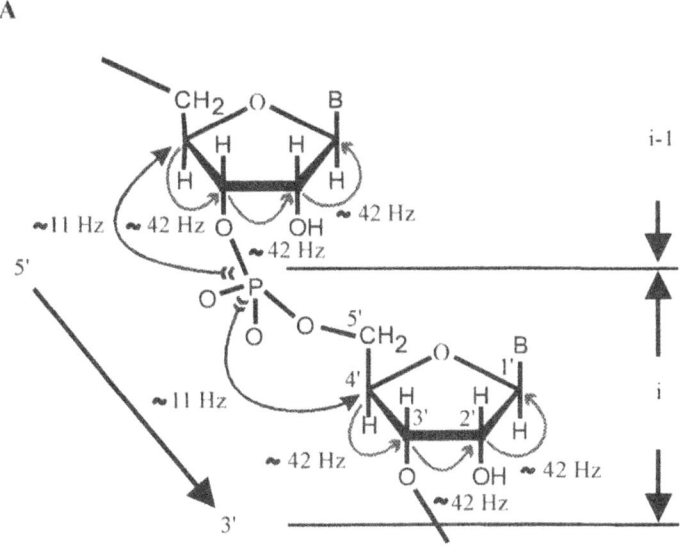

B

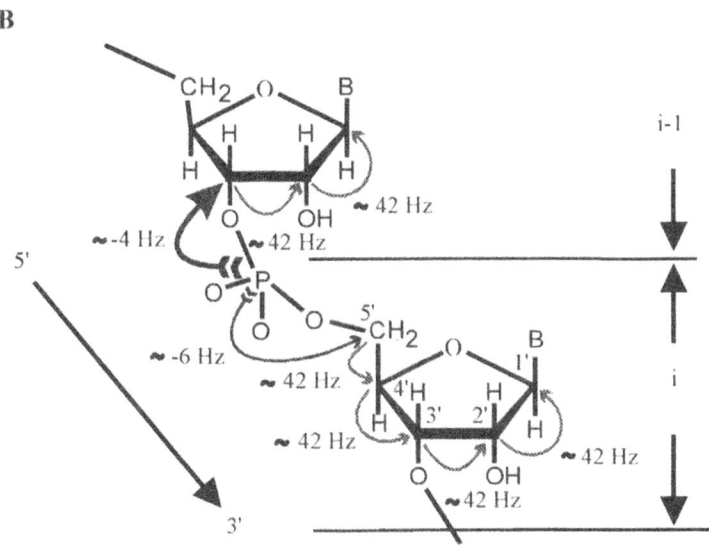

Figure 7. Schematic of the RNA ribose-phosphate backbone, showing the coherence pathways as indicated by arrows in the P(CC)H-TOCSY experiment. (A) The main C4'-route (see text). (B) The C5'-route and C3'-route (see text).

and the $H6_i$ or $H8_i \Rightarrow C6_i$ or $C8_i \Rightarrow N1_i$ or $N9_i$ are connected. The pulse sequence used for this experiment is the same as that used for the HCP experiment (see Fig. 9), but with the ^{31}P pulses replaced by ^{15}N pulses. A cross section in the three-dimensional spectrum, perpendicular to the ^{15}N frequency axis, through the $N1_i$ or $N9_i$ resonance should contain a cross peak between the H1' and C1' resonance as well as between the H6 and C6 or the H8

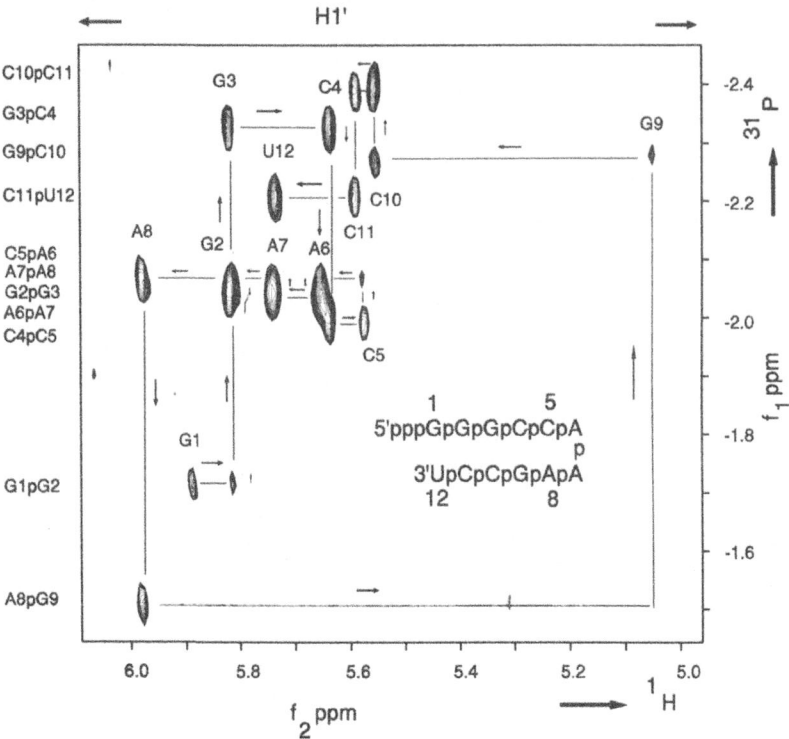

Figure 8. H1' region of the 600 Mhz 2D P(CC)HTOCSY spectrum of ^{13}C labeled RNA hairpin, shown schematically in the inset. The arrows indicate the 5' -> 3' direction of the sequential walk along the sugar-phosphate backbone. The labels at the cross peaks indicate the H1' assignment, while the ^{31}P assignment is given along the left side of the spectrum at the vertical position (^{31}P) of the cross peaks. The spectrum was recorded in approximately 12 hours with the following acquisition settings: 512 scans for each FID of 1024 points (t_2), 64 t_1 values. The delay Δ was set to 25 ms; τ, τ' = 12.5 ms; 1 ms trim pulse; DIPSI3 mixing time for isotropic mixing equal to 13.0 ms and DIPSI3 mixing time for cross-polarization set to 6.5 ms, both with a rf field strength of 8333.3 Hz, 1.0 s relaxation delay, 1 ms sine-shaped 12 G/cm gradient pulse, low power (625 Hz) Garp decoupling of ^{13}C, spectral width 486 Hz, and 5000 Hz for ^{31}P and ^{1}H, respectively; carrier positions at -2.09 ppm, 70 ppm and 4.62 ppm for ^{31}P, ^{13}C and ^{1}H, respectively. Typical processing parameters were: zero-filling twice in t_1 and once in t_2, and 0.4 π shifted sin^2-window multiplication in t_1 and t_2. The final data matrix consisted of 128 x 1024 data points.

and C8 resonance. This is demonstrated in Fig. 10 for a Cytosine monophosphate sample (Note the folding of H6/C6 cross peak).

With these and similar methods, developed by other groups, assignments can be performed in isotope enriched RNA. It is our experience that the current techniques can be successfully applied to molecules of the size of about 45 nucleotides.

HELIX STRUCTURE AROUND TWO CONSECUTIVE GA BASE PAIRS IN AN RNA DOUBLE HELIX

Application of the methods discussed above allows one to study suitable nucleic acid structures in very much detail. This is illustrated here for the duplex (double helix) formed

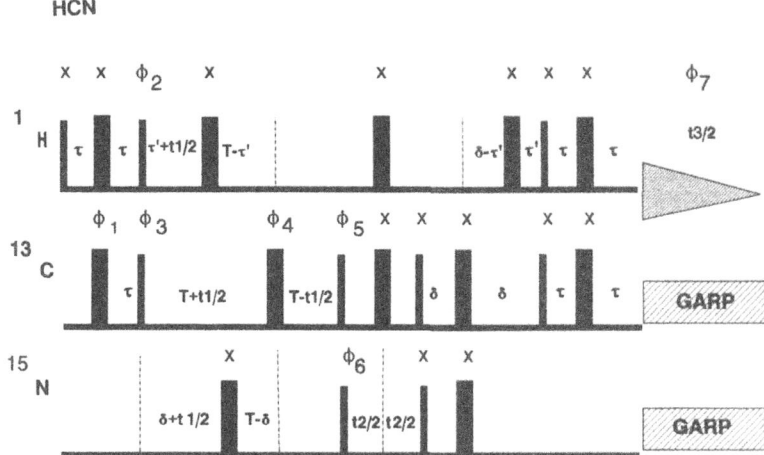

Figure 9. Pulse sequence of the HCN experiment. The phasecycling is as in the HCP experiment (see legend Fig. 2). For the delay settings see legend to Figure 10.

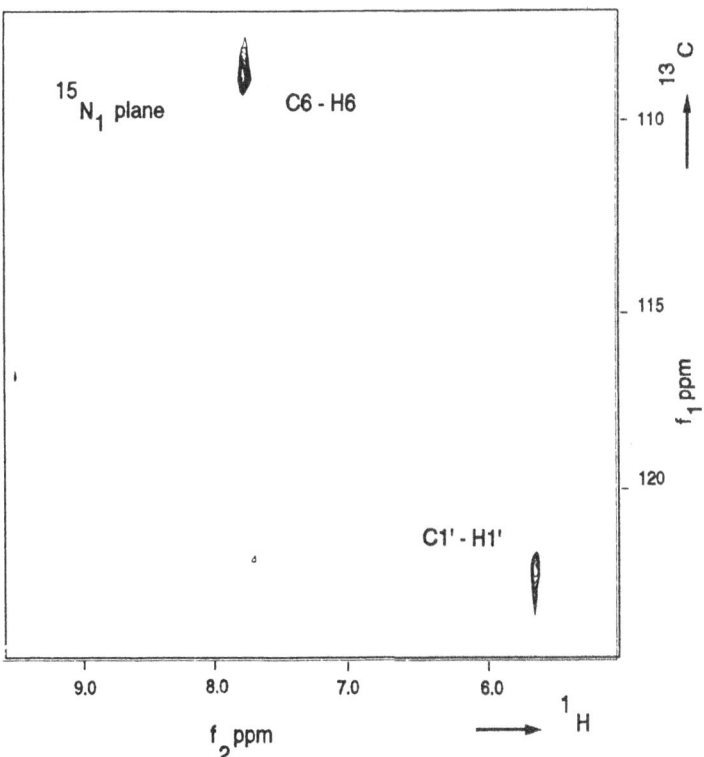

Figure 10. A ^{13}C, ^{1}H plane from the HCN spectrum, taken at the different ^{15}N$_1$ resonance frequency. The HCN was recorded with the following delay settings: T = δ = 12.0 ms, τ = τ' = 1.5 ms (≈ 1/4J$_{CH}$), T-δ = 5 μs, 1 s relaxation delay, low power (3 Watt) GARP decoupling of ^{15}N and ^{13}C, spectral widths, 2941 Hz, 1500 Hz, and 6000 Hz, and carrier positions, 116 ppm, 150 ppm, and 4.75 ppm for ^{13}C, ^{15}N, and ^{1}H, respectively. Note that due to aliasing in the ^{13}C dimension the C1', H1' as well as the C6, H6 cross peak have shifted in the ^{13}C dimension to approximately 125 ppm and 105 ppm, respectively.

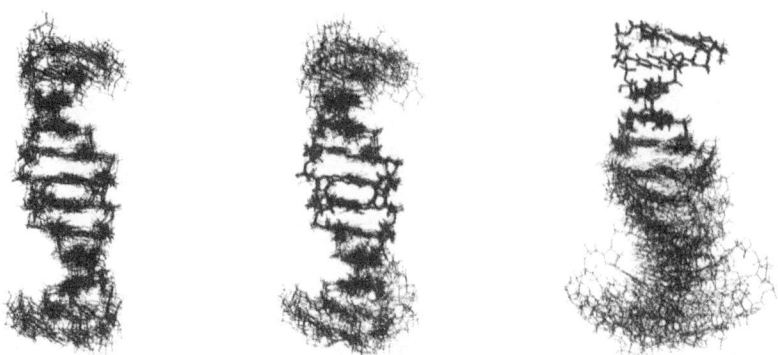

Figure 11. Overlays of a set of 15 XPLOR refined structures of the duplex discussed in text.

by the self-complementary sequence 5'-GGGCUGAAGCCU-3'. In Fig. 11 the fifteen best structures, fulfilling the NMR constraints, have been superimposed in three different ways. In the first representation (Fig. 11, left) the different structures have been superimposed such that a best fit overlay is obtained for all heavy atoms. In the other representations a best fit superposition is obtained for the four central base pairs (Fig. 11, middle) or for the four terminal base pairs (Fig. 11, right). The average number of NMR restraints per residue is twentyfive which means that the structure is quite well-defined. This doesn't show in Fig. 11, left but becomes manifest at the local level in Fig. 11, middle and right. The reason is that in non-global molecules such as this 'long' double helix no long distance constraints are available so that the position of residues far apart from one another is not well-defined.

In the duplex, the central bases form non-Watson-Crick GA base pairs of the type shown in Fig. 12, so-called sheared base pairs, where the amino-group of adenine hydrogen bonds to N3 of its guanine partner and the amino-group of this guanine hydrogen bonds to N7 of the adenine. In the derivation of the structure no assumptions and restraints regarding the hydrogen bonds were introduced for these base pairs.

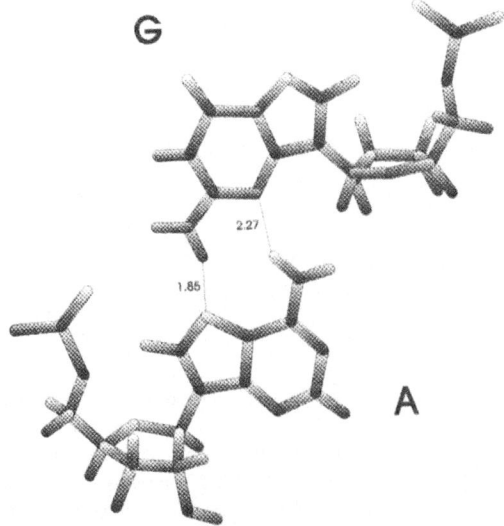

Figure 12. Sheared GA base pair as found in the duplex discussed in the text. The hydrogen bonds are indicated as drawn lines; the bond lengths given are as found in the average refined structure.

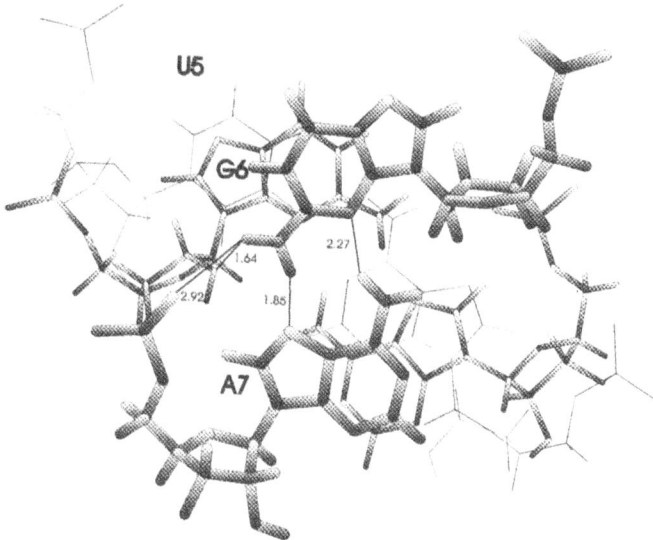

Figure 13. Top view of the UGAA part of the averaged refined structures, showing the cross-strand stacking of the two adenine bases and of the two guanine bases in the tandem GA base pairs; the hydrogen bonds stabilizing the structure are also given.

The presence of the sheared GA base pairs introduces very interesting distortions in the duplex compared with the A-type helix normally found for RNA duplexes. Due to the extreme high helical twist of the GA to GA base pair step (i.e. ~ 70°), the bases of the opposite strands stack, that is on the one hand the adenines on the other hand the guanines (Fig. 13). The sugar ring of the G's in the GA pairs adopts a pure S-conformation in contrast to the normal N-conformation found in standard A-type RNA helices. This allows the guanine of the opposite strand to form, via its amino group a hydrogen bond with the hydroxy group of this sugar (Fig. 13 and 14A). The helical twist between the GA pairs and the neighboring AU pairs is however small (~ 17°) so that the average twist per base pair step in the double helix is that found in the A-type helix.

Comparison with other structures in which two such consecutive GA base pairs are formed, i.e. in a duplex studied by Turner and SantaLucia [SantaLucia Jr. and Turner (1993)] (Fig. 14B) and in the hammerhead studied by McKay and collaborators [Pley et al., (1994)] (Fig. 14C), shows that this detailed structure depends on the identity of the base pairs adjacent to the two GA base pairs. In our molecule the GA pairs are flanked by AU pairs r(A-GA-U), in the duplex studied by SantaLucia and Turner the GA pairs are flanked by GC pairs r(G-GA-C) and in the hammerhead by a GC pair and an AU pair r(G-GA-U). Interestingly, in SantaLucia's duplex all sugars have an N-pucker and now the G's of the GA pairs form, via their amino group, a hydrogen bond with the phosphate group of the opposite strand. This is impossible in our duplex because, as mentioned, the sugar of the central G's has adopted an S-pucker allowing hydrogen bond formation between its hydroxyl group and the amino group of the G's. In the hammerhead a mixed situation obtains. On the side where the GA pair is flanked with a GC pair the sugar of the G in the GA pair maintains its N-conformation as in the SantaLucia duplex. On the other hand, the sugar of the G in the GA pair adjacent to the AU pair has switched to a S-pucker as in our duplex. Hydrogen bonding of the G-amino groups to the phosphate in the opposite strand or to the sugar hydroxyl group of the opposite strand takes place correspondingly. Thus, the admittedly very

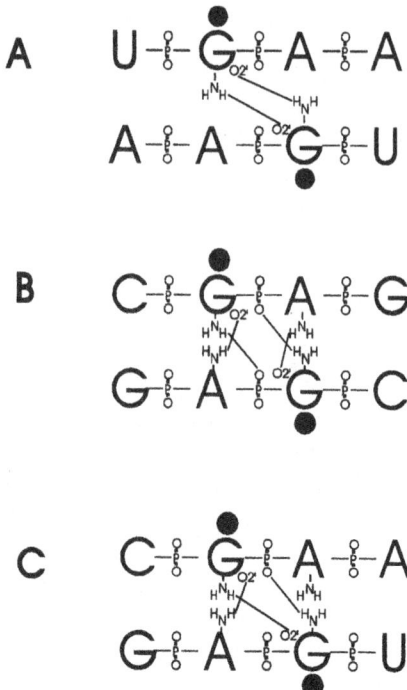

Figure 14. Schematics of the hydrogen bond stabilization of the structure of tandem GA base pairs with different flanking sequences (see text).

limited data set suggests that simple rules govern the detailed conformation of non-canonical double helix regions in these RNA duplexes. The currently available NMR methods allow one to establish the structure in sufficient detail. For instance, in the present example the number of NMR restraints was sufficiently high that no hydrogen bonds needed to be invoked neither between the G and the A bases nor between the G amino groups and the hydroxyl groups of the sugar to derive the structure of the RNA duplex.

REFERENCES

Blommers, M.J.J.; Haasnoot, C.A.G.; Hilbers, C.W.; van Boom, J.H.; van der Marel, G.A. *NMR studies of loopfolding in a DNA hairpin molecule*; in *Structure and dynamics of biopolymers*, ed. Nicolini, C.; NATO ASI Series, Series E: Applied Sciences-No. 133; Martinus Nijhoff Publishers: Dordrecht, pp 78-91 (1987).

Blommers, M.J.J.; van de Ven, F.J.M.; van der Marel, G.A.; van Boom, J.H.; Hilbers, C.W. *Eur. J. Biochem.* *201*, 33-51 (1991).

Grzesiek, S.; Bax, A. *J. Magn. Res. 96*, 432-440 (1992).

Frank-Kamenetskii, M.D.; Mirkin, S.M. *Annu. Rev. Biochem. 64*, 65-95 (1995).

Heus, H.A.; Wijmenga, S.S.; van de Ven, F.J.M.; Hilbers, C.W. *J.Am. Chem. Soc. 116*, 4983-4984 (1994).

Heus, H.A.; Pardi, H.A. *Science 253*, 191-194 (1991).

Hilbers, C.W.; Heus, H.A.; van Dongen, M.J.P.; Wijmenga, S.S. *The hairpin elements of nucleic acid structure: DNA and RNA folding*; in *Nucleic Acids and Molecular Biology*, Vol. 8; eds. Eckstein, F. and Liley, D.M.J.; Springer-Verlag: Berlin, Heidelberg, pp 56-104 (1994).

Mooren, M.M.W.; Pulleyblank, D.E.; Wijmenga, S.S.; van de Ven, F.J.M.; Hilbers, C.W. *Biochemistry 33*, 7315-7325 (1994).

Pleij, C.W.A. *TIBS 15*, 143-147 (1990).

Pley, H.W.; Flaherty, K.M.; McKay, D.B. *Nature 372*, 68-74 (1994).

SantaLucia Jr., J.; Turner, D.H. *Biochemistry 32*, 12612-12623 (1993).

Scott, W.G.; Finch, J.T.; Klug, A. *Cell 81*, 991-1002 (1995).

Wijmenga, S.S.; Mooren, M.M.W.; Hilbers, C.W. *NMR of Nucleic Acids; from spectrum to structure*; in *NMR of Macromolecules*, ed. Roberts, G.C.K.; Oxford University Press: Oxford, pp 217-288 (1993).

Wijmenga, S.S.; Heus, H.A.; Leeuw, H.A.E.; Hoppe, H.; van der Graaf, M.; Hilbers, C.W. *J. Biomol. NMR 5*, 82-86 (1995).

15

SOLUTION DYNAMICS OF THE *TRP*-REPRESSOR STUDIED BY NMR SPECTROSCOPY

Oleg Jardetzky and Zhiwen Zheng

Stanford Magnetic Resonance Laboratory
Stanford University
Stanford, California 94305-5055

Motions in proteins generally reflect the flexibility of the structure, which is important for the function of all but the simplest rigid peptides. In recent years detailed studies of internal motions in proteins by NMR have become possible (Englander & Kallenbach, 1984; Palmer, 1993; Wagner, 1993; Jardetzky & Lefèvre, 1994). High resolution NMR is unique among physical methods, as it can provide information about motions at individual atomic sites, and therefore allows the construction of motional maps (Jardetzky & Roberts, 1981; Ernst et al., 1987). For studies of protein flexibility and stability, NMR also has a significant advantage of detecting dynamic processes over a wide range of timescales. High frequency motions (sub-nanosecond to millisecond) can be investigated rigorously by heteronuclear relaxation in both laboratory and rotating frames. Slow conformational exchange (with lifetime of exchange ranging from millisecond to seconds and hours) can be investigated by techniques such as magnetization transfer, isotope replacement and lineshape analysis. The question therefore arises: How are the motions detected by NMR related to motions that are essential for function?

At least a partial answer to this question has emerged from the detailed studies of the solution structure and dynamics of the *trp*-repressor from *E. coli* carried out in our laboratory over the past several years. The *trp*-repressor is a 25 kDa allosteric protein, responsible for the regulation of transcription initiation in at least five operons (trp, trpEDCBA, aroH, trpR and mtr) involved in tryptophan metabolism (Squires et al., 1975; Gunsalus & Yanofsky, 1980; Zurawski et al., 1981; Klig et al., 1988; Heatwole & Somerville, 1991, 1992; Sarsero et al., 1991). Structurally, the repressor is a homodimer product of the trpR gene, each monomer containing 107 amino acids. The ligand-free protein - the aporepressor - binds two molecules of tryptophan to form the holorepressor, the form activated for binding to the operator-promoter regions of the operon DNA. Crystallographic (Schevitz et al., 1985; Zhang et al., 1987) and NMR (Arrowsmith et al., 1990, 1991a; Zhao et al., 1993) studies have revealed that the repressor consists of three major structural domains. The NH2-terminal arm is unstructured in both crystal and solution structures. The central core domain, formed by helices ABC and F from both polypeptide chains appears rather closely packed

Dynamics and the Problem of Recognition in Biological Macromolecules, edited by Jardetzky and Lefèvre
Plenum Press, New York, 1996

209

in both crystal and solution structures. The DE helix-turn-helix DNA-binding domain extending from the central core is well defined in the crystal. It is less well defined in the solution structures due to the lack of sufficient NOEs which results from rapid amide proton exchange found in this region (Czaplicki et al., 1991). Based on the alpha proton chemical shift and the ^{15}N relaxation findings discussed below, however, the DNA-binding region should not be viewed as unstructured but rather as an unstable region with a mainly helical conformation. The *trp*-repressor solution structures by Zhao et al. are presented in Figure 1.

So far three major NMR techniques have been employed for the studies of backbone dynamics of *trp*-repressor over three distinct time domains. First, amide ^{15}N relaxation (Zheng et al., 1995a) has been measured to characterize internal motions at sub-nanosecond timescale as well as the overall molecular reorientation (nanosecond time scales). Second, saturation transfer and amide proton relaxation experiments have been used to define the amide proton exchange process on the millisecond to second time scale (Gryk et al., 1995; Finucane & Jardetzky, 1995; Gryk & Jardetzky, 1996). Last, deuterium exchange experiments were performed to study the degrees of protection of amide protons from exchange on the timescale of minutes to hours (Czaplicki et al., 1991; Finucane & Jardetzky, 1996).

1. HIGH FREQUENCY MOBILITIES (ON NANO-SECOND TO SUBNANO-SECOND TIMESCALES) BY ^{15}N RELAXATION

It has been known for a long time that NMR spin relaxation, which is related to a set of specific spectral density functions, can provide valuable information on the molecular mobilities on the time scale around the resonance frequencies (Abragam, 1965). Heteronuclear relaxation (i.e., ^{13}C, ^{15}N) is especially well suited for the studies of internal motions of a protein for two reasons. First, the isotope labels can be evenly distributed over the entire polypeptide chain, allowing us to probe mobilities at many points along the backbone. Second, the heteronuclear relaxation mechanisms for these two nuclei are relatively simple, dominated by the dipolar interaction with directly bonded protons, which makes an accurate determination of spectral density functions possible. The development of proton-detected heteronuclear relaxation methods (Kay et al., 1989) has resulted in an increasing number of

A   B

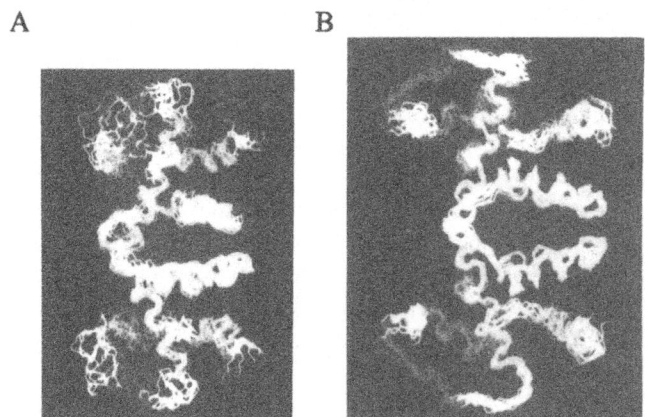

Figure 1. A family of 10 solution structures of (A) aporepressor and (B) holorepressor obtained from calculations using the XPLOR program: only the backbone atoms are displayed. The 6 heleces (A,B,C,D,E,F) are displayed. A color representation of this figure can be found facing p.113.

reports characterizing both internal motions as well as overall molecular motions for small size proteins or domains (Palmer, 1993; Wagner, 1993, review).

In a recent report by Zheng et al. (1995a), backbone ^{15}N T_1, T_2 and NOE for *trp*-repressor have been measured and the results show no significant variations over the entire protein backbone with the exception of both the N and C termini, where longer than average T_1 and T_2 relaxation times, with corresponding lower than average NOE values were observed (see Figure 2). On one hand, this is somewhat surprising since the ^{15}N relaxation data suggests that the DNA binding region is as rigid as the hydrophobic core region, in contrast to the solution structures where the DE region appears to be ill defined. On the other hand, the ^{15}N relaxation results are consistent with the alpha proton chemical shift findings, supporting the argument that the DNA-binding region is ordered, mostly in the helical form in solution. The ^{15}N relaxation data for aporepressor exhibit similar T_1, T_2 and NOE profiles as for the holorepressor (Zheng, unpublished result). No significant differences in the internal motions on the sub-nanosecond time-scale were found in the DNA-binding region between

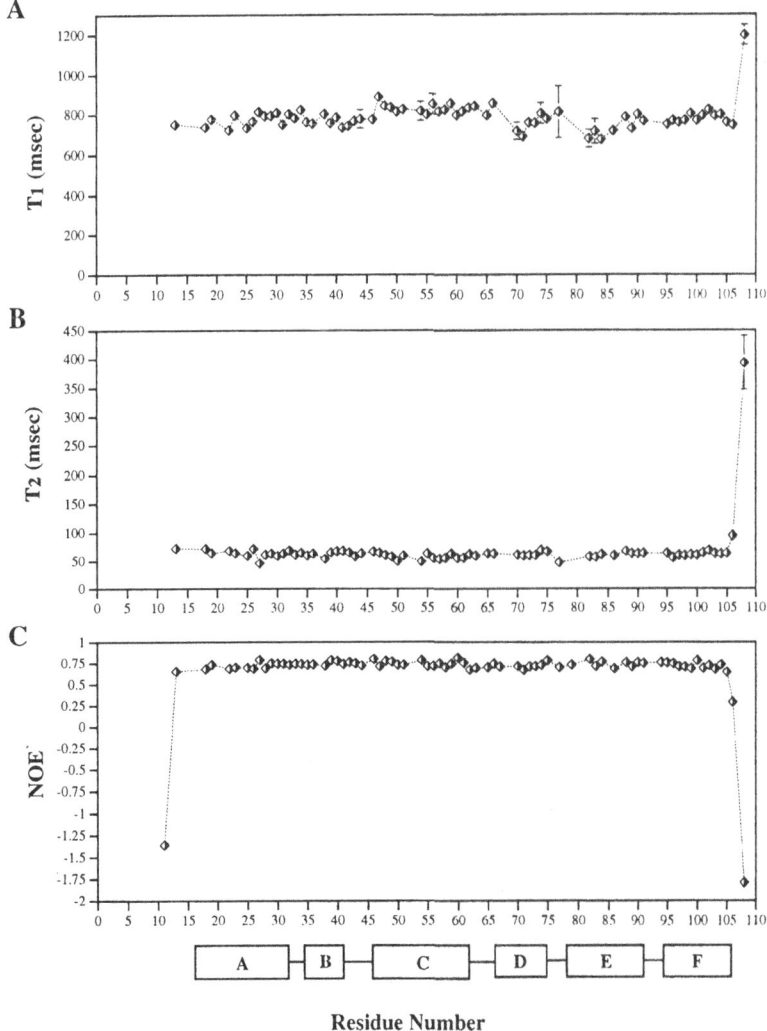

Figure 2. Plots of the measured T_1 (A), T_2 (B), and NOE (C) values vs. residue number. The secondary structure determined in solution for *trp*-repressor (Zhao et al., 1993) is indicated with boxes.

the two forms of repressors, indicating that no local folding process is involved upon complex formation with tryptophan. The higher degree of mobility found in both N-and C-termini is consistent with both solution and crystal structures.

The ratio of T_1 and T_2 for residues with relatively large NOE values are commonly used to estimate overall molecular tumbling rates (Kay et al., 1989). In most ^{15}N relaxation studies isotropic rotation is assumed for the simplification of data analysis. It is interesting to note that for the *trp*-repressor small variations in the T_1, T_2 ratio over the protein backbone were found to lie well beyond experimental uncertainty. These variations were correlated with the orientation of the NH vectors, or the orientation of each helix, indicating that the anisotropy of molecular reorientation in this system is detectable by ^{15}N relaxation parameters. Zheng et al. (1995a) have determined the rotational diffusion coefficients along different orientations of the molecule based on the T_1, T_2 ratio. It was found that the ratio in the diffusion coefficients between the parallel and perpendicular orientations of the approximately axially symmetric rotamer is 1.2:1 with the unique axis closely along helix C of each polypetide chain as shown in Figure 3. The internal motions of the NH vectors could be adequately described by Howarth's "wobble-in-a-cone" model, with a relatively small average half-angle of the wobble cone of 22°. The conclusion of this study is that the repressor molecule displays no flexibility on a nanosecond time scale, except for its N- and C-terminal ends.

2. LOW FREQUENCY STRUCTURAL FLUCTUATION (ON A TIME SCALE OF MILLISECONDS TO HOURS) THROUGH AMIDE HYDROGEN EXCHANGE

Since the pioneer work by Linderstrøm-Lang (1955), amide hydrogen exchange has been appreciated for its dependence on the structural fluctuation of a protein (Hvidt & Nilsen,

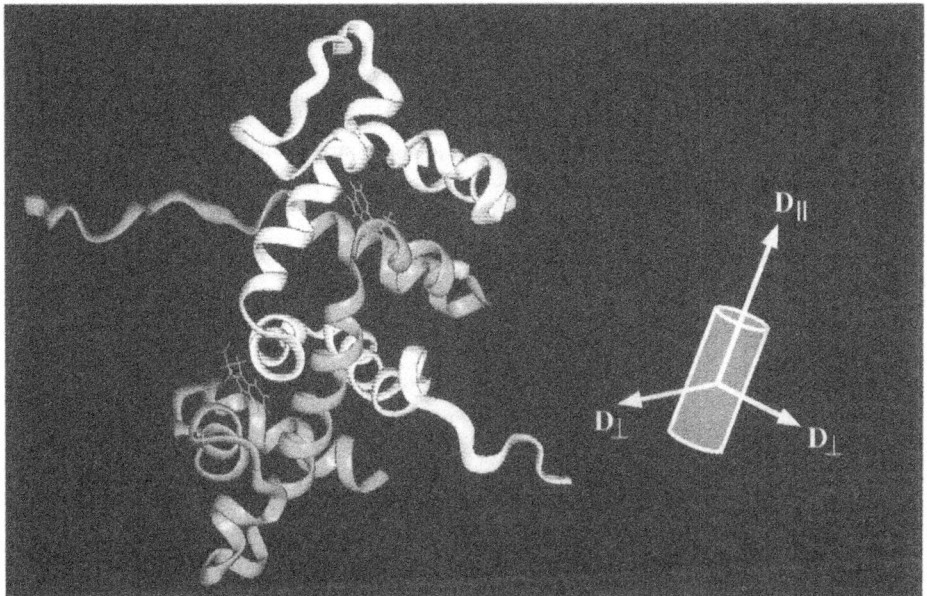

Figure 3. Ribbon diagram of one of the *trp*-repressor solution structures (Zhao et al., 1993). The axes of the axially symmetric rotational diffusion tensor (D∥ and D) are given schematically in a cylindrical diagram. The unique axis of the diffusion tensor is oriented closely along the two C helices of the dimeric molecule. A color representation of this figure can be found facing p. 113.

1966). The use of hydrogen exchange for the investigation of protein segmental flexibility has been greatly augmented in the past twenty years by the emergence of two-dimensional NMR spectroscopy so that most target protons can be resolved in the spectra. As in the case of ^{15}N relaxation experiments, the hydrogen exchange behavior over the entire protein backbone becomes amenable to study (Englander & Kallenbach, 1984, review). The exchange rates for fully extended peptides are very fast, and can be predicted with a reasonable accuracy based on their primary sequences (Molday et al., 1972; Bai et al., 1993). The fact that amide protons buried inside a highly packed protein molecule may still exchange at a slow but finite rate with solvent is a convincing evidence of some sort of internal structural fluctuation. The interpretation of hydrogen exchange in terms of molecular motions, however, is not as straight-forward as for the ^{15}N relaxation. Two major models have been proposed to interpret hydrogen exchange processes. The first model involves the exposure of the amide protons to the bulk solvent through a local unfolding of a secondary structure (Linderstrøm-Lang, 1955; Englander et al., 1980). The second model postulates solvent penetration into the protein interior to allow hydrogen exchange occur (Woodward & Hilton, 1980).

2.1. Rapid Hydrogen Exchange in the DNA-Binding Region and at the Two Termini

The profile of hydrogen exchange for the DNA-binding region of the *trp*-repressor is quite different from that of ^{15}N relaxation data. Residues in both the DNA binding region and at the two termini have very fast exchange rates at relatively high pH (pH 7.6), in contrast with the core region which is highly protected from exchange (Arrowsmith et al., 1991b; Czaplicki et al., 1991). Gryk et al. (1995) have carefully measured amide proton exchange rates for the fast-exchanging amides in the holorepressor using a combination of T_1 and saturation transfer (Waelder et al., 1975) experiments. The new method, as reported by Zheng et al. (1995b), not only measures the intensity attenuation of amide proton resonances due to the exchange with saturated solvent but also allows for the detection of possible hindered process occurred during saturation transfer, which could result in biphasic magnetization recovery profiles. The authors have found that the amide proton relaxation profiles observed for *trp*-repressor can be divided into the following three patterns. (1) pH-independent single exponential, a pattern for residues in the hydrophobic core indicating no exchange on the relaxation timescale is observable. The much slower exchange in this region will be discussed later. (2) pH-dependent single exponential, a pattern characteristic of residues on both terminal ends. (3) pH-dependent bi-exponential, This is the pattern observed for residues in DNA-binding region suggesting a hindered process with rates on the same order of magnitude as the relaxation rate may occur.

Gryk et al. (1995) have analyzed the experimental data using a general solution of the McConnell equation (1965) for the two-side exchange model of Linderstrøm-Lang (1955) as shown in Figure 4.

It becomes apparent that the usual special case of the Linderstrøm-Lang model assuming fast exchange between a closed and an open form cannot account for the observed experimental results since an one-step or pseudo one-step dynamical process predicts only

$$NH_{closed} \underset{k_{close}}{\overset{k_{open}}{\rightleftharpoons}} NH_{open} \underset{}{\overset{k_{ex}}{\rightleftharpoons}} H_2O$$

Figure 4. The Linderstrøm-Lang model for amide proton exchange.

a single-exponential recovery profile (Krishna et al., 1979; Spera et al., 1991). In the *trp*-repressor case it was found that amide protons on both termini are in their open states most of the time, but those in the DNA-binding region are in slow exchange between the open and closed forms with an approximately equal population in each state. The exchange rates (k_{ex}) were found to be in agreement with values expected for random-coil peptides (Bai et al., 1993) only for the very fast case at both termini, but slower by two orders of magnitude for the residues in the DNA-binding region as shown in Figure 5. Gryk et al. (1995) have postulated two possible models to explain the disagreement found in the DNA-binding region. In the first model, the "open" state in the Linderstøm-Lang model is simply not completely open and has an intrinsic exchange rate different from that of fully extended peptide. The second model postulates two "closed" states to be in slow exchange with each other, one of the them in rapid exchange with the open state, from which solvent exchange is assumed to proceed at a rate consistent with the literature values. The two models are indistinguishable by kinetic experiments. In the light of chemical shift and ^{15}N relaxation findings, the authors have drawn a picture of molecular motions for the DNA-binding domain. The helix-turn-helix motif involves the opening of the individual hydrogen-bonds, one at a time, without complete unraveling of the whole helix. At any given time only half of the hydrogen bonds in the helix may break, but the two helices in the DNA-binding region remain in a helical conformation all the time.

Finucane and Jardetzky (1995) have conducted similar experiments for *trp* aporepressor. In comparison with holorepressor, no significant difference in the exchange rates were found for residues in the N-terminal region and for C-terminal residue 108 where exchange rates are similar to those for fully exposed amide protons. The amide protons in the hydrophobic core are still too slow to be detectable in the saturation transfer experiments. The exchange rates (k_{ex}) for most part in the DNA-binding region, however, are increased by a factor of 2 to 5 in aporepressor, compared with holorepressor. The stabilization of amide protons upon binding with tryptophan ligand occurs throughout the region, on all faces of helices, both solvent exposed and interior. This damping at a distance led the authors to conclude that a concerted dynamic process is involved in the protein backbone which is modulated by ligand binding and in turn affects the observed backbone proton exchange.

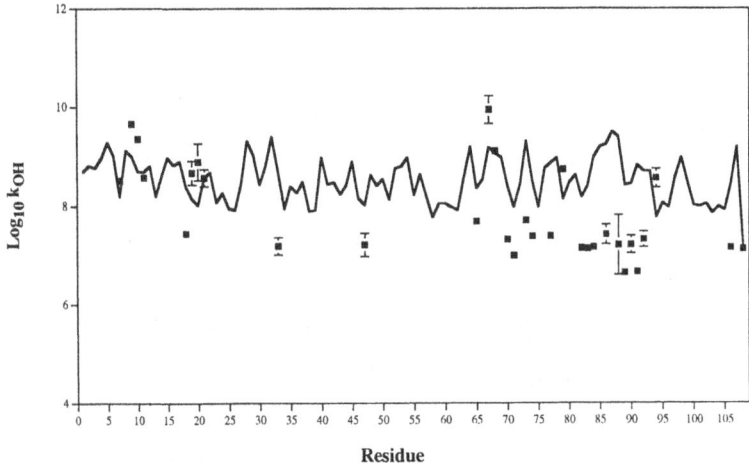

Figure 5. Comparison of calculated (—) and observed (•) k_{ex} values for *trp*-repressor. The calculated values were obtained using the method of Bai et al. (1993).

The cited experiments indicate that the DNA-binding region of the *trp*-repressor is quite fragile on a millisecond time scale, although it retains a predominantly helical conformation and appears stable on the nanosecond time scale. Inequality of the protection factors along each of the helices excludes the possibility that the helices unravel in a cooperative fashion in the exchange process. This is in contrast to the core of the molecule, in which exchange is on a time scale of hours or days under the same experimental conditions. The functional significance of this difference is discussed below.

2.2. Slow Hydrogen Exchange Found In Hydrophobic Core Region

The exchange rates with lifetimes longer than 10 minutes can be directly measured by observing the decay of amide proton peak intensities in an isotope replacement experiment. Czaplicki et al. (1993) have measured amide proton exchange rates for *trp*-repressor in both holo and apo forms. In the first spectra right after addition of D_2O into the lyophilized protein sample, only amide protons found in the hydrophobic core region (A,B,C and F) appear, while amide protons in the remainder of the molecule exchange too fast to be seen. As shown in Figure 6, the exchange lifetimes range from tens of minutes to above 100 hours for two forms of repressor. It is noteworthy, that L-tryptophan binding causes the exchange time to be longer in all parts of the holorepressor as compared to the aporepressor, by a factor of 2 to 10 in the hydrophobic core, and by a factor of 2 - 5 in the DNA binding region. The stabilization of the repressor molecule by ligand binding is thus propagated throughout the entire molecule and can thus be said to involve an action at a distance.

Recently, Finucane and Jardetzky (1996) have examined the pH-dependence of the slowly exchanging amide protons for *trp*-repressor. A very interesting observation has been

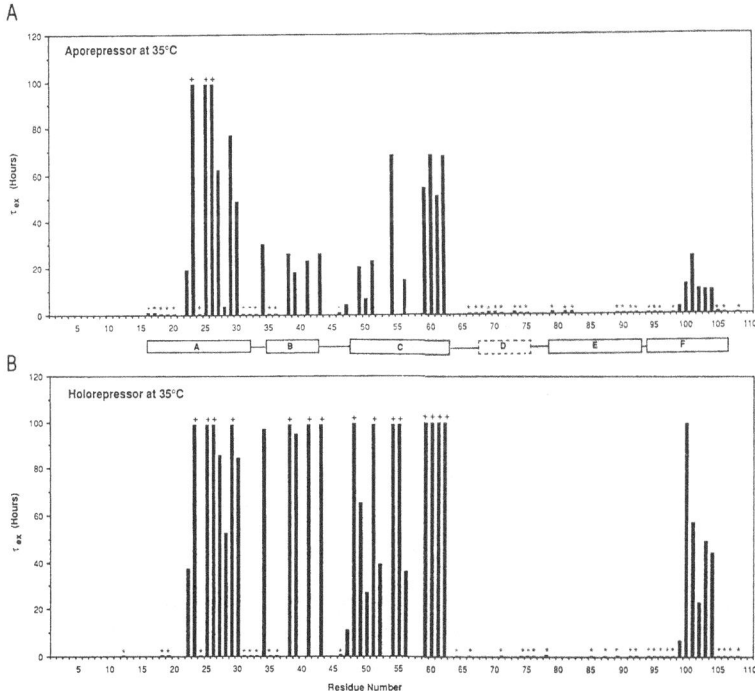

Figure 6. Lifetimes ($\tau_{ex}=1/k_{ex}$) of amide protons vs. residue number for (A) aporepressor and (B) holorepressor at 35°C.

made: the exchange rates for helices A and F show linearity with hydrogen concentration. However, those for the more buried helix B and C, whose amide proton exchange rates are among the slowest at pH 7.6 have a minimum rate at around pH 6. Normally the minimum points of exchange rates for extended peptides appear at around pH 3.5 (Wüthrich, 1986). The authors have attributed the abnormal behavior found for helices B and C to the local unfolding of the tertiary structure of the protein. The dynamic picture of the repressor molecule which emerges from the hydrogen exchange studies differs from that which emerged from the ^{15}N relaxation study. On the longer time scale the molecule consists of a stable core with a well structured, but fragile Helix-turn-Helix DNA binding region.

2.3. Kinetics of *trp* Ligand Exchange

An additional important exchange process that occurs in the *trp*-repressor system results from the reversible binding of the corepressor tryptophan. This has been studied in detail by Schmitt et al. (1995). The salient finding is that at temperatures comparable to those at which relaxation and proton exchange studies were carried out (45° C) the rate of exchange between free and bound ligand (~ 2000 s-1) is about 10 - 100 times faster than the backbone proton exchange. Thus the rate of hydrogen bond breakage in the Helix-turn-Helix is not correlated with ligand binding. The latter may however be correlated with the rate of the en-bloc conformational shift of the Helix-turn-Helix, which is known to occur from *trp*-binding from both crystallographic (Schevitz et al., 1985; Lawson et al., 1988) and NMR studies (Zhao et al., 1993). In the ternary complex (*trp*+repressor+DNA) the ligand exchange is slowed down by a factor of approximately 1000 (Arrowsmith et al., 1991b).

2.4. Amide Proton Exchange for AV77 Mutant

A further illustration of the significance of the structural fluctuation of *trp*-repressor has been given recently by Gryk and Jardetzky (1996) who studied the structure and dynamics of a hinge mutant, the AV77 repressor. Ala77 is located in the turn between helices D and E of the DNA-binding region. Alanine 77 to valine substitution results in a repressor with enhanced activity, a "super-repressor" whose binding with DNA-operators no longer depends on tryptophan concentration (Kelly & Yanofsky, 1985). At first one would expect some sort of structural change occurring in the mutant molecule to be responsible for its altered biological function. However, this is not the case. Based on the comparison of NMR NOE and chemical shift data, Gryk and Jardetzky (1996) have found that no substantial difference in the tertiary structure exists between AV-77 and wild-type for both holo- and aporepressors (such findings were also found to be consistent with unpublished crystallographic results). The similarity in the chemical shifts between wild-type and AV77 holorepressors are shown in Figure 7 in the amide region of a ^{1}H-^{15}N two-dimensional HMQC spectrum.

On the other hand, the hydrogen exchange measurements by Gryk and Jardetzky (1996) have revealed that the exchange rates for hydrogen-bonded amide protons in the DNA-binding domain are decreased for the most part by at least one order of magnitude in both the apo and the holo forms of the mutant. The point mutation has a slightly higher stabilization effect compared to a 2-5 fold stabilization of holorepressor over aporepressor. The relative amide proton intensities without and with solvent saturation for both wild-type and the mutant, apo and holorepressors are shown in Figure 8 as a function of residue number. Unlike the effect of ligand-binding which results in an increase in the stability of amide protons throughout the entire protein, the authors found that the alanine to valine mutation leaves the stability of the amide protons for the central core unchanged. The experiment gives a clear indication that protein dynamics can affect function. Accompanying stabiliza-

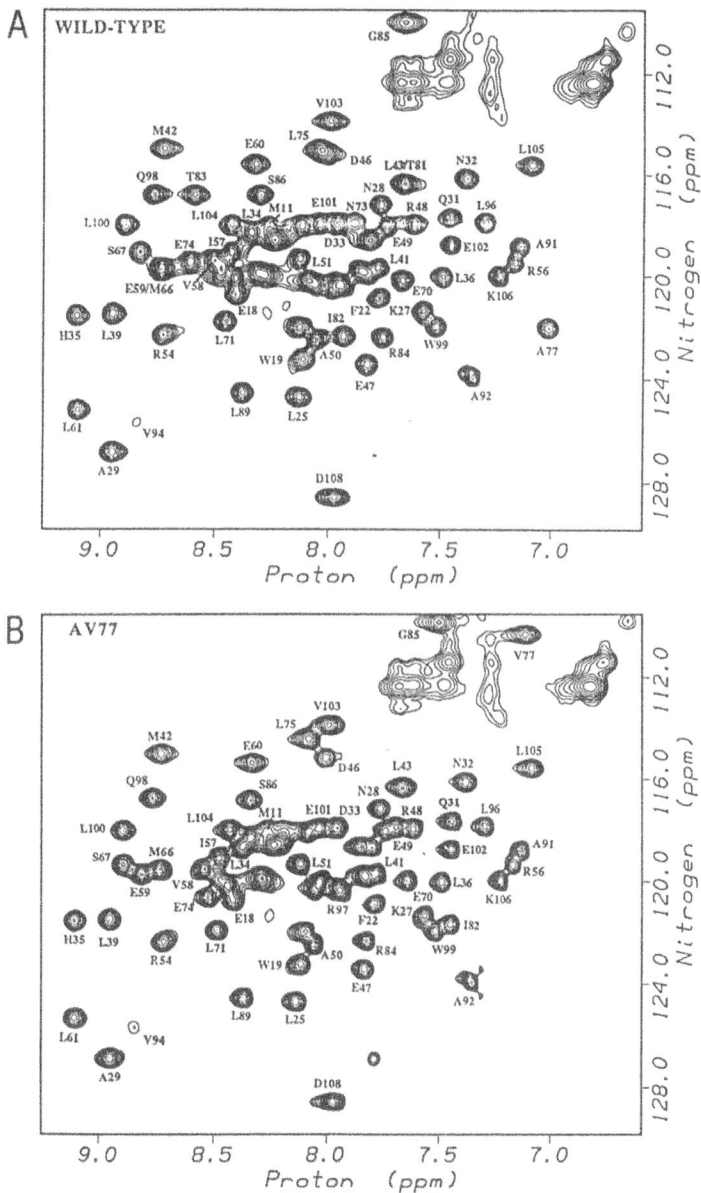

Figure 7. Comparison of the most densely packed region of the two-dimensional HMQC spectra of (a) wild-type and (b) AV77 holorepressor.

tion of the hinge by mutation are both loss of *trp* sensitivity and of the versatility in binding to different DNA sequences (Yanofsky, 1994, personal communication), which is further discussed in the next section. The shifts in the time scales of different motions which result from ligand binding or mutation are summarized in Figure 9. Changes in the Ramachandran angles at the hinges flanking the DNA-binding region are summarized in Table 1.

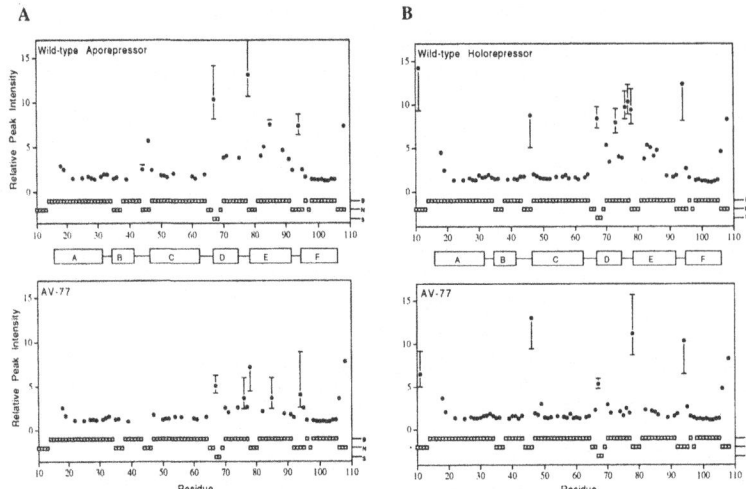

Figure 8. Relative peak intensity without and with solvent saturation vs. residue for the wild-type and AV77 proteins. A) Comparison of the aporepressor. B) Comparison of the holorepressor.

3. FUNCTIONAL RELEVANCE OF *TRP*-REPRESSOR INTERNAL MOTIONS

Our current understanding of the dynamics of the *trp*-repressor molecule can be summarized as follows:

1. On the nanosecond time scale the molecule diffuses in solution as a single rigid unit with no significant regional differences in internal motions, except for the N- and C- termini. The internal motion in the peptide backbone is severely restricted to excursions of the order of 20-30° for the NH vector wobbling in a cone.

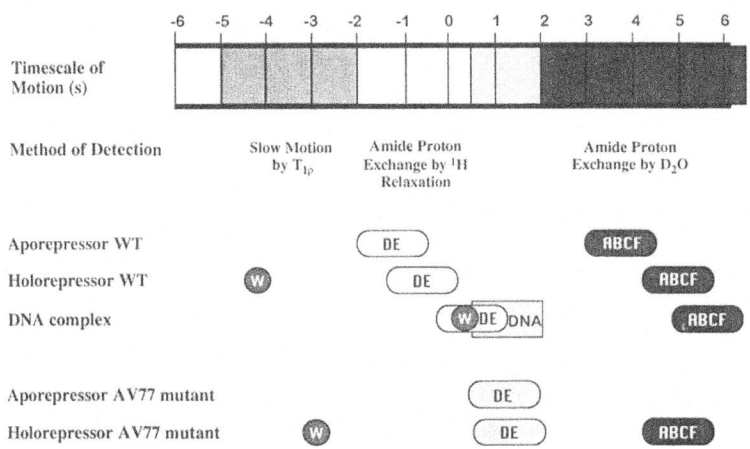

Figure 9. Time scale of dynamic processes in *trp* repressor. W represents tryptophan-*trp* repressor exchange frequency, DE represents amide proton exchange frequency in the D and E helices and ABCF represents the amide proton exchange frequency in the A, B, C and F helices. DNA indicates the lifetime of the *trp* repressor-DNA complex.

Table I. Phi (φ) and psi (ψ) angles obtained from the crystal structures of residues in the hinges for the different forms of *trp* repressor.

	Aporepressor (orthorhombic form)[a]	Holorepressor (orthorhombic form)[b]	Holorepressor (trigonal form)[c]	Holorepressor complexed with DNA[d]	Difference Apo-Holorepressor (orthorhombic)	Difference Apo-Holorepressor (trigonal)	Difference Holo (o) - Holo complexed with DNA	Difference Holo (t) - Holo complexed with DNA
Hinge 1								
Asn 32 φ	-101.58	-106.15	-107.28	-106.46	4.57	5.70	0.31	-0.82
ψ	-11.40	8.87	2.25	4.48	-20.27	-13.65	4.39	-2.23
Asp 33 φ	69.78	62.78	65.18	64.03	7.00	4.60	-1.24	1.15
ψ	35.09	30.57	34.87	29.16	4.52	0.22	1.41	5.71
Leu 34 φ	-114.52	-111.46	-119.37	-112.50	-3.06	4.85	1.04	-6.87
ψ	18.84	13.01	16.64	19.35	5.83	2.20	-6.34	-2.71
Hinge 2								
Leu43 φ	-116.39	-113.81	-120.69	-118.62	-2.58	4.30	4.81	-2.07
ψ	154.62	158.14	148.63	159.74	-3.52	5.99	-1.60	-11.11
Thr44 φ	-85.12	-88.04	-81.57	-93.69	2.92	-3.55	5.65	12.12
ψ	165.28	162.51	157.05	158.54	2.77	8.23	3.96	-1.49
Pro45 φ	-62.18	-54.50	-50.06	-49.23	-7.68	-12.12	-5.27	-0.83
ψ	-46.88	-42.42	-45.83	-48.41	-4.46	-1.05	5.99	2.58
Hinge 3								
Gly64 φ	66.64	73.17	76.02	64.83	-6.53	-9.38	8.34	11.19
ψ	7.64	21.10	24.36	30.82	-13.46	-16.72	-9.72	-6.46
Glu65 φ	-81.14	-84.50	-102.86	-99.32	3.36	21.72	14.82	-3.54
ψ	-35.10	-43.51	-47.90	-19.82	8.41	12.80	-23.69	-28.08
Met66 φ	-96.29	-93.78	-67.80	-115.91	-2.51	-28.49	22.13	48.11
ψ	147.85	143.79	137.25	152.85	4.06	10.60	-9.07	-15.60
Ser67 φ	-83.87	-79.52	-92.27	-80.99	-4.35	8.40	1.47	-11.28
ψ	155.83	164.51	168.13	166.56	-8.68	-12.30	-2.05	1.57
Hinge 4								
Leu75 φ	-107.58	-95.91	-112.60	-90.33	-11.67	5.02	-5.57	-22.26
ψ	-22.93	-8.98	-6.20	-14.82	-13.95	-16.73	5.84	8.62
Gly76 φ	74.26	83.23	78.32	65.78	-8.97	-4.06	17.45	12.54
ψ	25.33	-5.13	27.05	35.34	30.46	-1.72	-40.47	-8.29
Ala77 φ	-100.54	-74.92	-108.66	-102.85	-25.62	8.12	27.93	-5.81
ψ	133.26	137.65	105.18	163.32	-4.39	28.08	-25.67	-58.14
Gly78 φ	-56.99	-72.09	-95.60	-81.73	15.10	38.61	9.64	-13.87
ψ	144.97	162.45	168.81	144.08	-17.48	-23.84	18.37	24.73
Hinge 5								
Ala91 φ	-74.61	-82.68	-77.01	-82.46	8.07	2.40	-0.22	5.45
ψ	-10.11	-12.93	-19.48	-6.47	2.82	9.37	-6.46	-13.01
Ala92 φ	-69.57	-74.11	-72.91	-81.16	4.54	3.34	7.05	8.25
ψ	140.60	157.52	142.08	150.26	-16.92	-1.48	7.26	-8.18
Pro93 φ	-60.30	-58.50	-62.98	-59.57	-1.80	2.68	1.07	-3.41
ψ	139.14	139.42	151.95	149.01	-0.28	-12.81	-9.59	2.94
Val94 φ	-48.15	-54.43	-58.48	-56.40	6.28	10.33	1.97	-2.08
ψ	-38.56	-40.59	-24.25	-43.64	2.03	-14.31	3.05	19.39

[a] Zhang et al., 1987; [b] Lawson et al., 1988; [c] Schevitz et al., 1985; [d] Average between four values given in the Protein Data Bank file by Otwinowski et al., 1988.

2. The en-bloc displacement of the D and E helices around hinges at R63 and A92, observed upon tryptophan binding in both the x-ray and NMR structures is very likely correlated in time with the exchange rate of the ligand tryptophan on a sub-millisecond time scale.

3. The fluctuation (opening and closing) of H-bonds in the DNA binding domain is superimposed on the preceding and occur with frequencies of 20-50 s-1.

4. The observation that an increase in the stability occurs for the wild-type repressor upon the binding of L-tryptophan supports the views in the literature that the tryptophan ligand acts as a key that locks the DNA-binding domain to fluctuate among only a small number of favorable conformations so that it binds specifically to an operator-DNA sequence (Schevitz et al., 1985; Zhao et al., 1993).

It appears that the flexibility in the DNA-binding domain is required by the ligand to regulate the repressor-DNA-binding process. First, flexibility is required to optimize contacts with the operator DNA. Second, flexibility permits a balance between the formation (stabilization) and the breakage (destabilization) of contacts which is necessary if the molecule is to be susceptible to regulation by ligand binding. Third, flexibility makes it possible for the molecule to optimize its contacts in complexes to several different operator DNAs. The finding by Gryk and Jardetzky (1996) that a point mutation causes changes in both dynamics and function, but not in its structure is a clear indication that dynamics of a protein play an important role in governing its function. The correlation of this finding with the finding of Klig and Yanofsky that the stabilized AV77 mutant binds equally well to the consensus, but less well to the trpR operator than the wild type signifies a loss of versatility resulting from the stabilization.

The dynamics of the *trp*-repressor represent an interesting example of the role protein flexibility can play in the function of an allosteric mechanism. While in this case high frequency motions are highly restricted, flexibility reflected in low frequency motions plays an important role in optimizing contacts ("induced fit"), providing for versatility in reading different DNA sequences and allowing reversible binding of the corepressor, essential for regulation. We may expect in other systems however to find a variety of other patterns in the relations between the flexibility detectable by NMR and the flexibility essential for function.

4. REFERENCES

Arrowsmith, C.H., Pachter, R., Altman, R., Iyer, S. & Jardetzky, O. (1990). Sequence-specific ^{1}H NMR assignments and secondary structure in solution of *Escherichia coli trp*-repressor. *Biochemistry 29*, 6332-6341.

Arrowsmith, C., Pachter, R., Altman, R. & Jardetzky, O. (1991a). The solution structures of *Escherichia coli trp*-repressor and *trp* aporepressor at an intermediate resolution. *Eur. J. Biochem. 202*, 53-66.

Arrowsmith, C.H., Czaplicki, J., Iyer, S.B. & Jardetzky, O. (1991b). Unusual dynamic features of the *trp*-repressor from *Escherichia coli. J. Am. Chem. Soc. 113*, 4020-4022.

Arvidson, D.N., Pfau, J., Hatt, J.K., Shapiro, M., Pecoraro, F.S. & Youderian, P. (1993). Tryptophan super-repressors with alanine 77 changes. *J. Biol. Chem. 268*, 4362-4369.

Arvidson, D.N., Arvidson, C.G., Lawson, C.L., Miner, J., Adams, C. & Youderian, P. (1994). The tryptophan repressor sequence is highly conserved among the *Enterobacteriaceae. Nuc. Acids. Res. 22*, 1821-1829.

Bai, Y., Milne, J.S., Mayne, L. & Englander, S.W. (1993). Primary structure effects on peptide group hydrogen exchange. *Proteins: Structure, Function, and Genetics 17*, 75-86.

Carey, J., Combatti, N., Lewis, D.E.A. & Lawson, C.L. (1993). Cocrystals of *Escherichia coli trp* repressor bound to an alternative operator DNA sequence. *J. Mol. Biol. 234*, 496-498.

Czaplicki, J., Arrowsmith, C.H. & Jardezky, O. (1991). Segmental differences in the stability of the *trp*-repressor peptide backbone. *J. Biomolecular NMR 1*, 349-361.

Ernst, R. R., Bodenhausen, G. & Wokaun, A. (1989). Principles of nuclear magnetic resonance in one and two dimensions. Clarendon Press, Oxford.

Englander, S.W. & Kallenbach, N.R. (1984). Hydrogen exchange and structural dynamics of proteins and nucleic acids. *Quart. Rev. Biophys. 16*, 521-655.

Finucane, M.D. & Jardetzky, O. (1995). Mechanism of hydrogen-deuterium exchange in *trp*-repressor studied by ^1H-^{15}N NMR. *J. Mol. Biol. 253*, 576-589.

Finucane, M.D. & Jardetzky, O. (1996). The pH-dependence of hydrogen-deuterium exchange in *trp*-repressor. *Protein Science 5, in press.*

Gryk, M.R., Finucane, M.D., Zheng, Z. & Jardetzky, O. (1995). Solution dynamics of the *trp* repressor: A study of amide proton exchange by T_1 Relaxation. *J. Mol. Biol. 246*, 618-627.

Gryk, M.R. & Jardetzky, O. (1996). AV77 hinge mutation stabilizes the helix-turn-helix domain of *trp* repressor. *J. Mol. Biol. 255*, 204-214.

Guenot, J., Fletterick, R.J. & Kollman, P.A. (1994). A negative electrostatic determinant mediates the association between the *Escherichia coli trp* repressor and its operator DNA. *Protein Science 3*, 1276-1285.

Harrison, S. & Aggarwal, A. K. (1990). DNA recognition by proteins with the helix-turn-helix motif. *Annu. Rev. Biochem. 59*, 933-969

Heatwole, V.M. & Somerville, R.L. (1991). The tryptophan-specific permease gene, *mtr*, is differentially regulated by the tryptophan and tyrosine repressors in *Escherichia coli* K-12. *J. Bacteriology 173*, 3601-3604.

Heatwole, V.M. & Somerville, R.L. (1992). Synergism between the *trp* repressor and tyr repressor in repression of the *aroL* promoter of *Escherichia coli* K-12. *J. Bacteriology 174*, 331-335.

Hvidt, A. & Nilsen, S.O. (1966). Hydrogen exchange in proteins. *Adv. Prot. Chem. 21*, 287-386.

Jardetzky, O. & Roberts, G. C. K. (1981). NMR in Molecular Biology. Academic Press, New York.

Jardetzky, O. & Lefèvre, J-F. (1994). Protein Dynamics. *FEBS Lett. 338*, 246-250.

Kay, L.E., Torchia, D.A. & Bax, A. (1989). Backbone dynamics of proteins as studied by ^{15}N inverse detected heteronuclear NMR spectroscopy: Application to staphylococcal nuclease. *Biochemistry 28*, 8972-8979.

Kelley, R.L. & Yanofsky, C. (1985). Mutational studies with the *trp* repressor of *Escherichia coli* support the helix-turn-helix model of repressor recognition of operator DNA. *Proc. Natl. Acad. Sci. USA 82*, 483-487.

Kim, K,-S. & Woodward, C. (1993). Protein internal flexibility and global stability: Effect of urea on hydrogen exchange rates of bovine pancreatic trypsin inhibitor. *Biochemistry 32*, 9009-9013.

King, R., Maas, R., Gassner, M., Nanda, P.K., Conover, W.W. & Jardetzky, O. (1978). Magnetic relaxation analysis of dynamic processes in macromolecules in the pico- to microsecond range. *Biophys. J. 24*, 103-117.

Krishna, N.R., Huang, D.H., Glickson, J.D., Rowan III, R. & Walter, R. (1979). Amide hydrogen exchange rates of peptides in H_2O solution by ^1H nuclear magnetic resonance transfer of solvent saturation method. *Biophys. J. 26*, 345-366.

Lawson, C.L., Zhang, R.-G., Schevitz, R.W., Otwinowski, Z., Joachimiak, A. & Sigler, P.B. (1988). Flexibility of the DNA-binding domains of *trp* repressor. *Proteins: Structure, Function and Genetics 3*, 18-31.

Lipari, G. & Szabo, A. (1982). Model-free approach to the interpretation of nuclear magnetic resonance relaxation in macromolecules: 1. Theory and range of validity. *J. Am. Chem. Soc. 104*, 4546-4559.

Luisi, B.F. & Sigler, P.B. (1990). The stereochemistry and biochemistry of the *trp* repressor-operator complex. *Biochemica et Biophysica Acta 1048*, 113-126.

Marmorstein, R.Q., Sprinzl, M. & Sigler, P.B. (1991). An alkaline phosphotase protection assay to investigate *trp* repressor/operator interactions. *Biochemistry 30*, 1141-1148.

Molday, R.S., Englander, S.W. & Kallen, R.G. (1972). Primary structure effects on peptide group hydrogen exchange. *Biochemistry 11*, 150-158.

Otwinowski, Z., Schevitz, R.W., Zhang, R.-G., Lawson, C.L., Joachimiak, A., Marmorstein, R.Q., Luisi, B.F. & Sigler, P.B. (1988). Crystal structure of *trp* repressor/operator complex at atomic resolution. *Nature 335*, 321-329.

Palmer, A.G., III (1993). Dynamics properties of proteins from NMR spectroscopy. *Curr. Opin Biotech. 4*, 395-391.

Saunders, M. & Wishnia, A. (1958). Nuclear magnetic resonance spectra of proteins. *Ann. N. Y. Acad. Sci. 70*, 870-874.

Schevitz, R.W., Otwinowski, Z., Joachimiak, A., Lawson, C.L. & Sigler, P.B. (1985). The three-dimensional structure of *trp* repressor. *Nature 317*, 782-786.

Schmitt, T.H., Zheng, Z. & Jardetzky, O. (1995). Dynamics of tryptophan binding to *Escherichia coli trp* repressor wild type and AV77 mutant: an NMR study. *Biochemistry 34*, 13183-13189.

Spera, S., Ikura, M. & Bax, A. (1991). Measurement of the exchange rates of rapidly exchanging amide protons: Application to the study of calmodulin and its complex with a myosin light chain kinase fragment. *J. Biomol. NMR 1*, 155-165.

Spolar R.S. & Record, Jr., M.T. (1994). Coupling of local folding to site-specific binding of proteins to DNA. *Science 263*, 777-784.

Staacke, D., Walter, B., Kisters-Woike, B., Wilcken-Bergmann, B.v. & Müller-Hill, B. (1990). How *Trp* repressor binds to its operator. *EMBO J. 9*, 1963-1967. Corrigendum published on page 3023.

Waelder, S., Lee, L. & Redfield, A.G. (1975). Nuclear magnetic resonance studies of exchangeable protons. I. Fourier transform saturation-recovery and transfer of saturation of the tryptophan indole nitrogen proton. *J. Am. Chem. Soc. 97*, 2927-2928.

Wagner, G. & Wüthrich, K. (1982). Amide proton exchange and surface conformation of the basic pancreatic trypsin inhibitor in solution. *J. Mol. Biol. 160*, 343-361.

Wagner, G. (1993). NMR relaxation and protein mobility. *Curr. Opin. Struct. Biol. 3*, 748-754.

Zhang, H. (1993). Structural investigation of *trp*-repressor-DNA complex by NMR. Ph.D. thesis, Stanford University, Stanford, CA.

Zhang, H., Zhao, D., Revington, M., Lee, W., Jia, X., Arrowsmith, C. & Jardetzky, O. (1994). The solution structures of the *trp* repressor-operator DNA complex. *J. Mol. Biol. 238*, 592-614.

Zhang, R.-G., Joachimiak, A., Lawson, C.L., Schevitz, R.W., Otwinowski, Z., Sigler, P.B. (1987). The crystal structure of *trp* aporepressor at 1.8Å shows how binding tryptophan enhances DNA affinity. *Nature 327*, 591-597.

Zhao, D., Arrowsmith, C.H., Jia, X. & Jardetzky, O. (1993). Refined solution structures of the *Escherichia coli trp* Holo- and Aporepressor. *J. Mol. Biol. 229*, 735-746.

Zheng, Z., Czaplicki, J. & Jardetzky, O. (1995a). Backbone dynamics of *trp*-repressor studied by ^{15}N NMR relaxation. *Biochemistry, 34*, 5212-5223.

Zheng, Z., Gryk, M., Finucane, M. & Jardetzky, O. (1995b). Investigation of protein amide proton exchange by ^{1}H longitudinal spin relaxation. *J. Magn. Reson., Ser. B 108*, 220-234.

16

A REFINED NMR SOLUTION STRUCTURE OF THE POU-SPECIFIC DOMAIN OF THE HUMAN OCT-1 PROTEIN

Michel Cox,[1][*] Niek Dekker,[1] Rolf Boelens,[1] Hans C. van Leeuwen,[2] Peter C. van der Vliet,[2] and Robert Kaptein[1][†]

[1] Bijvoet Center for Biomolecular Research
Utrecht University
Padualaan 8, 3584 CH Utrecht, The Netherlands
[2] Laboratory for Physiological Chemistry
Utrecht University
Stratenum, P.O. Box 80042, 3508 TA Utrecht, The Netherlands

ABSTRACT

The 76 residue POU-specific domain (POU_s) of the human transcription factor Oct-1 was studied by two- and three-dimensional homo- and heteronuclear NMR techniques. By means of ($^{13}C,^{1}H$) HMQC experiments on a fractionally ^{13}C-labelled sample, the methyl groups of 10 out of 12 leucine residues could be assigned stereospecifically. This allowed the use of much tighter distance constraints, involving these methyls, in structure calculations. The three-dimensional structure was determined by using a total of 1361 distance constraints and 93 dihedral constraints, derived from the NMR spectra, as input for distance geometry calculations. The resulting 47 conformations were refined by subjecting them to a restrained molecular dynamics protocol, including a total of 15 ps *in vacuo* simulation. A total of 27 structures, that fitted the experimental data best, was selected and found to superimpose on the backbone atoms of residues 1 to 71 with a root mean square deviation of the average co-ordinates of 0.45 Å. The four α-helices, connected with irregular loops, are oriented in much the same way as in a recently published X-ray structure of the POU-domain (consisting of a POU_s and a POU-homeodomain) bound to its target DNA sequence. In the orientation of the recognition helix a deviation of about 22° was observed between X-ray and NMR structure.

---

[*] Michel Cox died on April 26, 1995.
[†] To whom correspondence should be addressed.

Dynamics and the Problem of Recognition in Biological Macromolecules, edited by Jardetzky and Lefèvre
Plenum Press, New York, 1996

INTRODUCTION

POU-proteins are important transcription factors that are involved in transcriptional regulation, development and cell differentation and that can be found in eukaryotic organisms ranging from *Drosophila* to humans (reviewed in: Verrijzer *et al.*, 1993). The architecture of some POU-proteins is depicted in Figure 1.

In general the POU transcription factors consist of a highly homologous region of 150-170 amino acids, the POU-domain, in which the DNA-binding ability of the protein resides, and variable domains, that are rich in Gln, or Ser/Thr and Pro and that are thought to function as transactivation domains. The POU-domain consists of two independently folded modules, as was deduced from several lines of evidence (Verrijzer *et al.*, 1992; Botfield *et al.*, 1992). The C-terminal part, which is homologous and structurally similar to classical homeodomains (Morita *et al.*, 1993), is referred to as POU-homeodomain (POU$_{hd}$). For the N-terminal part, initially no sequence homology with other proteins could be found, hence its name POU-specific domain (POU$_s$). The two subdomains are connected via a linker, variable in both length (14-27 residues) and amino acid sequence.

The POU transcription factor studied by us, Oct-1, is a vertebrate class II POU-protein of ca. 90 kD, and is unique in the family, in that it is expressed in all tissue types. It is involved in the control of the expression of immunoglobulin heavy and light chains, histone 2B and some snRNAs. Class II POU-proteins recognize the octamer DNA sequence (5'-ATGCAAAT-3'). Biochemical studies showed, that both POU$_s$ and POU$_{hd}$ are necessary to recognise the octamer sequence with high specificity and affinity (Verrijzer *et al.*, 1990). In fact, POU$_s$ recognises the left half of this octamer sequence, and POU$_{hd}$ the right half (Verrijzer *et al.*, 1992).

For the NMR studies of the Oct-1 POU domain both subdomains have been expressed in *E. coli* and purified. This has resulted in an initial structure of the POU$_s$ domain (Assa-Munt *et al.*, 1993; Dekker *et al.*, 1993), and in the structure of the Oct-1 POU$_{hd}$ domain (Cox *et al.*, 1995). It was concluded, that the structure of the POU$_s$ domain had high structural homology with that of the helix-turn-helix DNA-binding proteins λ repressor and 434 repressor. On basis of this structural homology Assa-Munt *et al.*, (1993) proposed a similar mode of DNA-binding for the Oct-1 POU$_s$ domain as observed for these repressors in the cocrystal structures. Indeed NMR experiments on a POU$_s$-DNA complex (Dekker *et al.*, 1993) showed, that the helix-turn-helix motif was involved in DNA-binding. Also binding studies of Oct-2 mutants demonstrated the role of this motif for POU$_s$ (Jancso *et al.*, 1994).

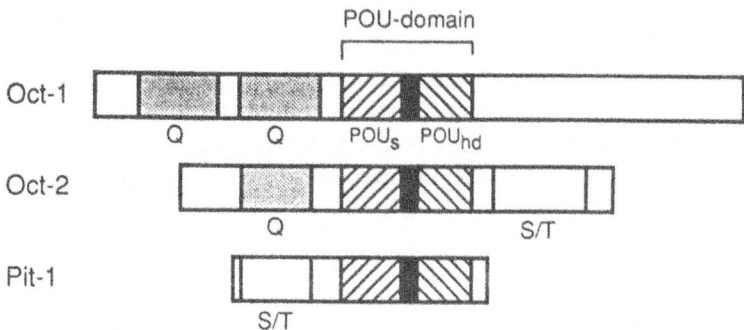

Figure 1. Outline of some POU proteins. The subdomain of the POU domain, POU-specific (POU$_s$) and POU-homeodomain (POU$_{hd}$) are hatched, the linker is black. The transactivation domains (Q: glutamine-rich; S/T: serine/threonine rich) are shaded.

The solution structure of the Oct-1 POU$_{hd}$ domain is very similar to that of the canonical homeodomains (Cox *et al.*, 1995), as had been anticipated on the basis of the 30% homology. Recently, the crystal structure of the complex of an intact POU domain with a DNA octamer was solved (Klemm *et al.*, 1994). A structural comparision of POU$_{hd}$ free in solution and bound to DNA in the complex, showed that the recognition helix of POU$_{hd}$ is elongated in the complex with 6 residues. Apparently this recognition helix is stabilised in the POU complex by additional DNA contacts, a behaviour which is reminiscent of the complex formation of the *Antennopedia* homeodomain (Qian *et al.*, 1993).

The present study reports a refined solution structure of the POU$_s$ domain from human Oct-1. Based on NOE and J-coupling data from homo- and heteronuclear NMR experiments, constraints were obtained which were used as input for distance geometry calculations, followed by a Restrained Molecular Dynamics protocol. The refined structure of POU$_s$ was compared to the one in the X-ray structure of the complex of the POU-domain with its cognate DNA sequence (Klemm *et al.*, 1994). It is found, that that the core of free and bound POU$_s$ is very similar, but that the orientation of the recognition helix of POU$_s$ differs substantially in the two structures.

MATERIALS AND METHODS

Sample Preparation

The POU$_s$ domain of human Oct-1 was obtained from the corresponding DNA sequence cloned into an expression vector and overexpressed in *E. coli*. Production and purification of the protein were performed using methods described previously (Cox *et al.*, 1993). The fractionally ^{13}C labelled sample was prepared by growing *E. coli* cells in 5 l minimal medium which contained 10 g of unlabelled glucose as the sole source of carbon. One hour before induction with IPTG, 1 g of ^{13}C glucose was added (Neri *et al.*, 1989). Protein samples were concentrated to 2-4 mM typically, in 100 mM NaCl and 5 mM DTT. D$_2$O samples were prepared by lyophilising H$_2$O samples. The pH was set to 5.0 by adding small aliquots of DCl or NaOD.

NMR Spectroscopy

The assignment of ^{1}H and ^{15}N resonances was accomplished by means of homonu-clear COSY, TOCSY and NOESY spectra, as well as by heteronuclear (^{15}N, ^{1}H)-HMQC experiments and a 3D (^{15}N, ^{1}H) NOESY-HSQC experiment as described elsewhere (Cox *et al.*, 1993). For the determination of distance constraints NOESY experiments were per-formed on H$_2$O samples recorded on a Bruker AMXT 600 spectrometer with mixing times of 75 and 150 ms, as well as a NOESY experiment on a D$_2$O sample with a mixing time of 75 ms, recorded on a Bruker AMX 500 spectrometer. A (^{13}C, ^{1}H) HMQC experiment (Bax *et al.*, 1990) as well as a TOCSY-relayed (^{13}C, ^{1}H) HMQC (Otting & Wüthrich, 1988) experiment was recorded on the fractionally ^{13}C-labelled POU$_s$ sample on a Bruker AMX 500 spectrometer. To facilitate the assignment of these spectra, a (^{13}C, ^{1}H)-HMQC spectrum was recorded on a sample with natural abundance of ^{13}C. ^{1}H-^{13}C decoupling in $\omega 2$ was achieved by a GARP sequence during acquisition (Shaka *et al.*, 1985). All spectra were recorded at 298 K. The reported ^{13}C chemical shifts are calibrated to the Cα resonance of glucose at 76.536 ppm. The (^{13}C, ^{1}H) spectra were processed to 1Kx1K data points, applying square sine windows in both domains prior to Fourier transformation.

NMR Experimental Data

To obtain inter proton distance information, NOESY spectra both of an H_2O and a D_2O sample, with a mixing time of 75 ms were recorded. A short mixing time is needed to avoid spin diffusion from being an important factor in the intensities of the cross-peaks. Cross-peak intensities, classified as strong, medium/strong, medium, medium/weak or weak, were translated to upper bound distances of 2.5, 2.9, 3.3, 3.7 and 4.2 Å, respectively. In addition, weak cross-peaks from a NOESY spectrum with a 150 ms mixing time, recorded on an H_2O sample, were implemented as distance constraints with an upper bound of 5.5 Å. All lower bounds were set to 2.0 Å. From these spectra a total of 1361 non-redundant distance constraints were derived. Distance-angle constraints were obtained in an iterative procedure. With preliminary distance constraints, sets of 15 structures were calculated with distance geometry, which allowed the assignment of previously ambiguous cross-peaks. Table 1 lists the constraints according to their type. From previous amide exchange experiments a total of 41 H-bonds were determined in the helices (Cox et al., 1993), and used as input for the structure generation with distance geometry. Each H-bond, between the amide proton and the carbonyl oxygen of the i+4th residue, was accounted for by two constraints of $2.8 < d_{NO} < 3.2$ Å and $4.0 < d_{NC} < 4.3$ Å. The H-bond constraints were not used in the restrained molecular dynamics calculations.

A difficulty with leucine residues in protein structure determination by NMR lies in the stereospecific assignment of the two methyl groups. With the regular NMR techniques like COSY-type experiments to derive coupling constants, and NOESY experiments to obtain intra-residual distances, it is possible to determine the prochirality of the methyls in valine residues. However, it is difficult to stereospecifically assign methyl groups for the longer side chain of leucine residues on the base of COSY and NOESY experiments only. An elegant way to obtain these assignments, proposed by Neri and co-workers (Neri et al., 1989), involves fractional ^{13}C labelling of the protein, using a mixture of 10% uniformly ^{13}C labelled glucose and 90% unlabelled glucose as the sole carbon source. In a $(^{13}C, {}^1H)$ HMQC experiment, the pro-R methyl of leucine will appear as a doublet in the ^{13}C dimension due to a $^1J13_C\text{-}13_C$ coupling of circa 33 Hz between $C\delta$ and $C\gamma$. The pro-S methyl will appear as a singlet, since, in that case, the $C\gamma$ is an unlabelled carbon. In this way, for 10 out of the 12 leucines present in POU_s, the methyls could be assigned stereospecifically. For the methyl groups of the other leucines the standard pseudo-atom correction of 2.4 Å was used; for the ring protons of Tyr and Phe and for methylene protons pseudo-atom corrections were used

Table 1. Summation of constraints used as input
for structure calculations

Distance constraints[a]	
Intraresidual	168
Sequential	522
Medium range	386
Long range	285
Total	1361
Dihedral constraints	
ϕ-angle	74
χ_1-angle	19
total	93

[a]In addition 41 hydrogen bonds, implemented as
82 constraints, were used in DG, but not in REM
and RMD.

of 2.0 and 1.0 Å, respectively (Wüthrich *et al.*, 1983). Cross-peaks involving methyl groups were corrected for the three proton intensity and a pseudo-atom correction of 0.3 Å was added (Koning *et al.*, 1990).

A number of dihedral constraints were derived for ϕ- and χ_1-angles. A (^{15}N, ^{1}H) HMQC spectrum, recorded with a high resolution in the ^{15}N direction, provided data on the ϕ-angles. The $^3J_{HN\alpha}$ coupling constant causes a splitting in the signals of this spectrum, and the measured value of the constant can be translated to a torsion angle. In total, 73 ϕ-angles were constrained with upper and lower margins of 10° for $\phi = -65°$, and margins of 30° for $\phi = -120°$. $^3J_{\alpha\beta}$ coupling-constants, obtained from DQF-COSY experiments, combined with NOE intensities of N-Hβ and Hα-Hβ cross-peaks, gave stereospecific assignment of β-methylene protons. From this, the χ_1-angle of 10 residues could be constrained with an upper- and lower margin of 30°, selecting one of the three favourable rotamers. Additionally, for 6 residues, one of the three rotamers could be excluded. This was implemented by setting a larger margin of $\pm 90°$ in the dihedral constraint. Table 2 lists the assignment of the χ_1-angles.

Structure Calculations

A total of 47 structures was generated with distance geometry calculations using the DGII package implemented in the Biosym software (Havel, 1991). This consisted of a triangulation step, an embedding step and finally an optimisation step, employing a simplified simulated annealing procedure. The 47 conformations were subsequently subjected to the following refinement protocol. A restrained energy minimisation (REM) was carried out to obtain structures with an acceptable energy to perform a restrained molecular dynamics (RMD) run. The RMD calculation consisted of a 5 ps Simulated Annealing (SA) run, in which the temperature was decreased gradually from 800 K to 300 K, followed by a 10 ps RMD run at 300 K. The time step used was 2 fs. Finally, an extra Restrained Energy Minimisation was performed. All REM and RMD calculations were done with the GROMOS force field and software (van Gunsteren *et al.*, 1987). All calculations were carried out on Silicon Graphics Personal Iris 4D/35 and 4D/120 GTX workstations.

RESULTS AND DISCUSSION

NMR Spectroscopy

Figure 2 shows part of the (^{13}C, ^{1}H) HMQC spectrum of the fractionally ^{13}C labelled POU$_s$ sample. In the crowded part of the spectrum, between 0.8 and 1.0 ppm, there is severe proton overlap. It is in some cases difficult to identify signals as being a doublet or two

Table 2. Assignment of χ_1-angles

Residue	Conformation	Residue	Conformation
Asn1	gg/tg	Asn50	gg
Phe8	gt	Phe53	gt/tg
Phe12	gt	Phe55	tg
Phe21	tg	Cys57	tg
Asp25	tg	Trp66	gt/tg
Asn28	gg/tg	Asn68	tg
Phe38	gt/tg	Asp69	tg
Phe46	gt	Asn72	gg/tg

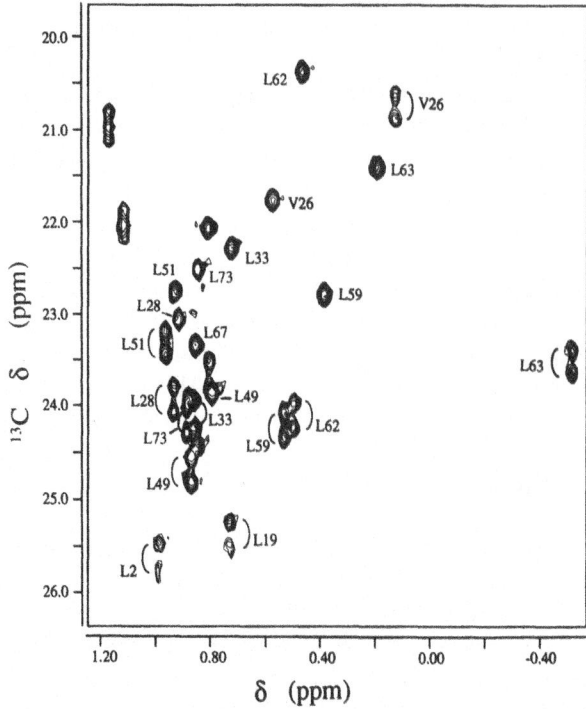

Figure 2. Part of the methyl region of the (^{13}C, ^{1}H) HMQC-spectrum of 10% ^{13}C-labelled POU$_s$, recorded on a Bruker AMX 500 at 298 K. Stereospecifically assigned methyl signals are indicated.

singlets. In the (^{13}C, ^{1}H) HMQC spectrum on a sample with natural abundance in ^{13}C (not shown), all methyls appear as singlets. By comparing this spectrum with the (^{13}C, ^{1}H) HMQC spectrum of the fractionally ^{13}C labelled POU$_s$, the doublet/singlet ambiguity was relieved. The assignment of proton resonances was done by comparing the resonances with homonuclear spectra. In the TOCSY-relayed (^{13}C, ^{1}H) HMQC spectrum, the methyl resonances could be linked with the corresponding Hγ. This extra frequency made it possible to stereospecifically assign the methyls of 10 leucine residues. The ^{13}C chemical shifts and the prochirality of the assigned methyls are given in Table 3. In an analogous way, the spectra enabled us to stereospecifically assign the methyls of the only valine residue in POU$_s$. As was previously determined on basis of NOE intensities and J-coupling data, the experiment confirmed that the conformation of Val26 is g$^+$, which is the preferred rotamer for valines (Janin *et al.*, 1978).

Structure Determination

For structure analysis, a selection from the 47 conformations was made. Based on low constraint energy and few constraint violations, 27 structures were selected. Figure 3A shows the 27 conformations, superimposed onto each other. It can be seen that after the refinement the structures are well converged. For the five residues at the C-terminus, Asn72-Asp76, no long range NOEs were identified, and only few medium range NOEs. This is reflected in the considerable freedom of this part of the molecule. When superimposing all atoms of residues 1-71 to the average co-ordinates, the r.m.s.d. is 1.03 Å, and for the backbone atoms only the r.m.s.d. is 0.45 Å (Table 4). A way of detecting relative conformational freedom is by inspection of the r.m.s.d. of the atom positions per residue. Low r.m.s.d.

Table 3. Assignment of leucine methyls

Residue	Upfield methyl 1H		Downfield methyl 1H	
	Prochirality	13C δ (ppm)	Prochirality	13C δ (ppm)
Leu2	S	—	R	25.63
Leu19	R	25.40	S	—
Leu33	S	22.28	R	24.07
Leu49	S	23.83	R	24.65
Leu51	S	22.74	R	23.30
Leu59	S	22.76	R	24.20
Leu62	R	24.07	S	20.38
Leu63	R	23.50	S	21.40
Leu67	R	—	S	23.30
Leu73	S	22.50	R	24.15

values of the Cα atom coordinates, displayed in Figure 4A, are correlated with a large number of constraints for these residues, shown in Figure 4B. In general, the core residues exhibit low r.m.s.d. values, while the residues at the surface, as well as in the C-terminus, have higher r.m.s.d. values. In general there is a tendency in protein structures determined by NMR for loops connecting the helices to have more freedom. Strikingly, this is not the case for this structure, presumably because in each of the three loops there is a residue with a relatively large number of constraints, which fixes the loop to some extent (Phe21, Tyr34 and Phe53).

For the 27 refined POU$_s$ structures, a number of side chain and backbone torsion angles were compared with ideal values, with the program PROCHECK (Morris *et al.*, 1992). As can be seen in Table 5, the data of the POU$_s$ structures fits the ideal data very well. The average deviation of the bond lengths after refinement from ideal geometry is 0.007 Å and the average deviation of bond angles is 1.14 degrees.

It can be concluded that the fit to the experimental data, as checked by inspection of distance and dihedral constraint violations, is good. As listed in Table 6, there are on average 1.9 distance constraint violations per structure larger than 0.6 Å, and 6.5 violations larger than 0.5 Å. For dihedral constraints, 1.1 violation per structure larger than 15.0° was found, and 11.9 violations larger than 5.0°.

The structure of POU$_s$ consists of four helices, connected with short loops. All helices are found to be α-helical, as the observation of $i,i+4$ contacts indicate. The two larger helices 1 (residues 1-19) and 4 (residues 54-71) are positioned parallel to each other and form a scaffold on which the shorter helices 2 (residues 23-32) and 3 (residues 40-49) are lying. The proline at position 61 in helix 4 induces a kink of 22 ± 2°. This value is in agreement

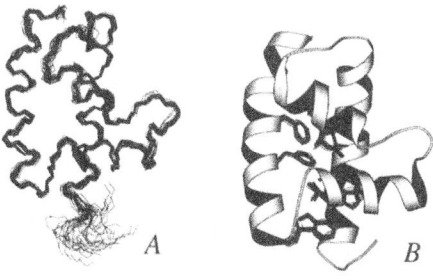

Figure 3. Refined structure of POU$_s$. *(A)* The 27 conformations superimposed on the backbone atoms. *(B)* Ribbon representation of the average structure. The side chains which constitute the core of the protein are shown for all 27 structures.

Table 4. R.m.s.d. values (Å) for the 27 refined structures with respect to the average structure

	Backbone atoms	All atoms
Residues 1–71	0.45	1.03
Helix 1 (1–19)	0.35	1.20
Helix 2 (23–34)	0.27	0.75
Helix 3 (40–49)	0.23	0.85
Helix 4 (54–71)	0.22	0.72

with the average angle found in 9 other proline containing α-helical proteins (Barlow & Thornton, 1988) of $26 \pm 5°$. The kink in helix 4 and a bend in helix 1 results in a close packing of side chains all along the two helices. The second and third helix form a helix-turn-helix like motif. The inner side of all helices have hydrophobic residues, which constitute the core of the protein. Many of these residues are highly conserved among POU-domains, such as residues Phe8, Phe12, Leu19, Phe21, Val26, Ile43, Phe46, Leu63, Trp66 and Leu67. Figure 3B displays the side chains of these residues, as well as the backbone of the average structure. It is clear that these residues on the inside of the protein have a well-defined position; the r.m.s.d. of the ten residues is 0.50 Å for the heavy atoms. On the other hand, the side chains at the surface of the protein tend to be much more flexible, as is usually seen in globular proteins. This is visualised in Figure 5. The heavy atoms of all four helices are displayed with the side facing the protein surface pointing to the left. Note that the interior side chains are in general better defined than those at the surface of the protein. As has been noted before (Assa-Munt *et al.*, 1993; Dekker *et al.*, 1993), the overall fold of POU$_s$ is very similar to the structure of the first four helices of λ repressor. While the sequence homology is only 22%, the r.m.s.d. of the Cα atoms of 59 corresponding residues is as low as 2.2 Å.

The refined structures were checked for hydrogen bonds, by searching for donor-acceptor pairs with a distance shorter than 2.8 Å and an angle larger than 135° to account for a proper H-bond geometry. Many hydrogen bonds are of the C'O$_i$-NH$_{i+4}$ type, characteristic for α-helices. Most of these coincide very well with the presence of slowly or intermediately exchanging amide protons, as shown in Table 7. Apart from this, a number of hydrogen bonds were found involving side chain atoms. The Hε or Hη protons of Arg15 are hydrogen bonded

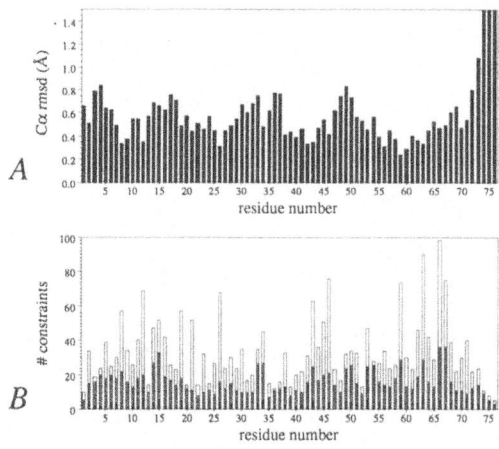

Figure 4. *(A)* The average r.m.s.d. of the atomic co-ordinates per residue, after superimposing on all atoms. *(B)* Plot of the number of distance constraints per residue. Three types of constraints are indicated: black, intra-residual; grey, sequential; white, medium- and long-range.

Table 5. Stereochemical quality of the 27 refined conformations

	Refined structures	Ideal values
Pro ϕ	-64.6 ± 1.4	-65.4 ± 11.2
Helix ϕ	-63.1 ± 14.4	-65.3 ± 11.9
Helix ψ	-45.5 ± 16.2	-39.4 ± 11.3
χ_1 (g-)	60.0 ± 17.4	64.1 ± 15.7
χ_1 (trans)	183.9 ± 17.7	183.6 ± 16.8
χ_1 (g+)	-68.4 ± 16.0	-66.7 ± 15.0
χ_2	173.8 ± 19.6	177.4 ± 18.5
Peptide ω	179.0 ± 3.2	180.0 ± 5.8
H-bond energy	-2.4 ± 0.9	-2.0 ± 0.8
Cα chirality	33.8 ± 2.6	33.9 ± 3.5

with the carboxyl group of Glu71 in 16 out of 27 structures. This fixes the C-terminal ends of helices 1 and 4 with respect to each other. Residue Gln40 donates an Hϵ proton to the carboxyl group of Gln23, which is an interesting part of the molecule, since it is involved in DNA-binding by the protein (Dekker *et al.*, 1993). In the crystal structure of the Oct-1 POU-domain complexed with its cognate DNA sequence (Klemm *et al.*, 1994), these two residues are the main participants in an extensive hydrogen bond network between POU$_s$ and the DNA. Apparently, part of this network is already present in the free protein. Other hydrogen bonds are found in the loops between helices, e.g. the amide proton of Met56 with the carbonyl oxygen of Leu51, and the amide proton of Phe21 with the carbonyl oxygen of Arg16. The irregular loop between helices 2 and 3 is defined by two H-bonds between the amide of Tyr34 and the carbonyl of Met30, and between the amide of Phe38 and the carbonyl of Tyr34. In none of the interhelical loops however, could any regular turn structure be determined.

Independently from our work, another structure determination by NMR of Oct-1 POU$_s$ has been carried out (Assa-Munt *et al.*, 1993). When superimposing the backbone atoms of these two structures, an r.m.s.d. value of 2.3 Å is obtained, suggesting a similar three-dimensional structure. Moreover, superpositioning of the seperate helices yields values of 1.34, 0.41, 0.92 and 0.69 Å, for the four consecutive helices. These figures indicate that the helices as such are quite similar in the two structures, but the orientation of them differs somewhat, or the connecting loops have different orientations. On inspection of the structures, this is indeed the case. The fourth helix has a very similar bend in both cases. The first helix in our structure has a slightly stronger overall bend, which explains the higher r.m.s.d. value for this helix. In the overall structure, the third helix, as well as the loop between the two helices of the HTH-motif, displays most of the differences. Still, many residues in the hydrophobic core are oriented in the same way. The χ_1 angles of residues F8, F12, F21, V26,

Table 6. Average number of constraint violations per structure in the 27 refined structures

Distance constraints	
Number > 0.40 Å	22.7
Number > 0.50 Å	6.5
Number > 0.60 Å	1.9
Number > 0.70 Å	1.1
Dihedral constraints	
Number > 5.0 degrees	11.9
Number > 10.0 degrees	2.6
Number > 15.0 degrees	1.1

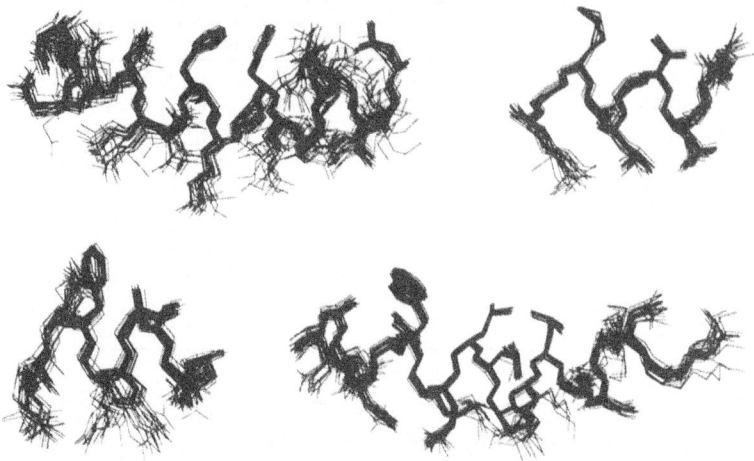

Figure 5. Secondary structure elements in POU$_s$. A superposition of the backbone atoms (N, Cα, C') was performed on each of the four helices, for the 27 structures. The orientation of the helices is such that the solvent exposed side faces down.

F46, L63, W66 and L67 differ within only 23 degrees in the two structures, whereas the side chains at the surface display large variations since they are poorly constrained.

The Helix-Turn-Helix Motif of POU$_s$

The second and third helix of POU$_s$ form a helix-turn-helix (HTH) like motif, which differs from the canonical HTH-motifs found in for instance the λ and 434 repressors. The sequence alignment of a number of HTH-motifs is given in Figure 6. There is a strong conservation of the lengths of the helices in the canonical HTH-motifs. Furthermore, the turn of three residues starts in general with a glycine. In the case of POU$_s$, in between the two helices an extra six residues are present. This results in an extension of the first helix with one helical turn and an elongation of the turn in the HTH-motif of 3 residues. The effect of the six extra residues on the structure of the HTH-motif is visualised in Figure 7, depicting the Cα-traces of the HTH-motifs of POU$_s$ and a number of the canonical HTH-motifs. A one residue insert has also been reported for the turn in the HTH-motif of LexA (Fogh *et al.*, 1994) and the third repeat of c-Myb (Ogata *et al.*, 1992). Furthermore, six and eight residue inserts occur in the turn of the HTH-motifs of GH5 (Ramakrishnan *et al.*, 1993) and HNF-3γ (Clark *et al.*, 1993).

In all these cases, the inserts have only little effect on the relative orientation of the two helices in the motif. Mainly the length of the helices and turns is affected in the case of the six and eight residue insertions, which are pointing away from the DNA and the rest of the protein. An extreme case is the HTH-motif of LFB1/HNF1. A total of 23 residues, inserted in between the two helices, adopt a poorly defined loop region. Furthermore, there is some sequence homology among the HTH-motifs in amino acids that contact the DNA. At position 4 and 5 there is a strongly conserved long aliphatic side chain and an alanine or glycine residue, respectively. Also, in the second helix, at position 12, there is an almost absolutely conserved long aliphatic side chain. This demonstrates that the HTH-motif is a stable structure element that can have several variations in amino acid sequence without affecting the overall fold. Therefore, in an evolutionary sense, it is a very successful motif, occurring in many, divergent organisms (reviewed in: Pabo & Sauer, 1992). Interestingly, the glutamine

Table 7. Hydrogen bonds observed in the α-helices of
27 POU$_s$ conformations

Donor HN	Acceptor C=O	Presence (%)	NH exchange rate[a]
Leu5	Asp1	74.1	m
Ala9	Leu5	100	s
Lys10	Glu6	100	s
Thr11	Gln7	100	m
Phe12	Phe8	100	s
Lys13	Ala9	100	m
Gln14	Lys10	96.3	m
Arg15	Thr11	100	m
Arg16	Phe12	100	m
Ile17	Lys13	100	s
Lys18	Gln14	100	m
Leu19	Arg15	100	s
Val26	Thr22	100	s
Gly27	Gln23	100	m
Leu28	Gly24	100	m
Ala29	Asp25	100	m
Met30	Val26	100	m
Gly31	Gly27	88.9	m
Lys32	Leu28	85.2	n.d.
Ser44	Gln40	100	m
Arg45	Thr41	100	m
Phe46	Thr42	100	s
Glu47	Ile43	96.3	s
Ala48	Ser44	25.9	s
Leu49	Arg45	100	s
Met56	Ser52	100	m
Cys57	Phe53	100	m
Leu59	Asn55	100	s
Lys60	Met56	100	s
Leu62	Lys58	100	s
Leu63	Leu59	100	s
Glu64	Lys60	100	s
Lys65	Pro61	100	s
Trp66	Leu62	100	s
Leu67	Leu63	100	s
Asn68	Glu64	100	s
Asp69	Lys65	100	m
Ala70	Trp66	100	m
Glu71	Leu67	100	m
Leu73	Asp69	51.9	n.d.

[a]The amide proton exchange rate is presented in this column; s denotes slow
exchange rate; m denotes intermediate exchange rate; n.d. denotes not
determined, due to overlap.

residue at the start of either helix of the POU$_s$ HTH-motif is absolutely conserved in the λ and 434 repressor and cro proteins, and in all POU-proteins as well. This implies that these proteins all share the same mode of anchoring to the DNA, namely by a concerted interaction of the two glutamines to the DNA backbone, as is supported by structure determination studies of co-crystals (Klemm *et al.*, 1994; Beamer & Pabo, 1992; Rodgers & Harrison, 1993; Mondragón & Harrison, 1991).

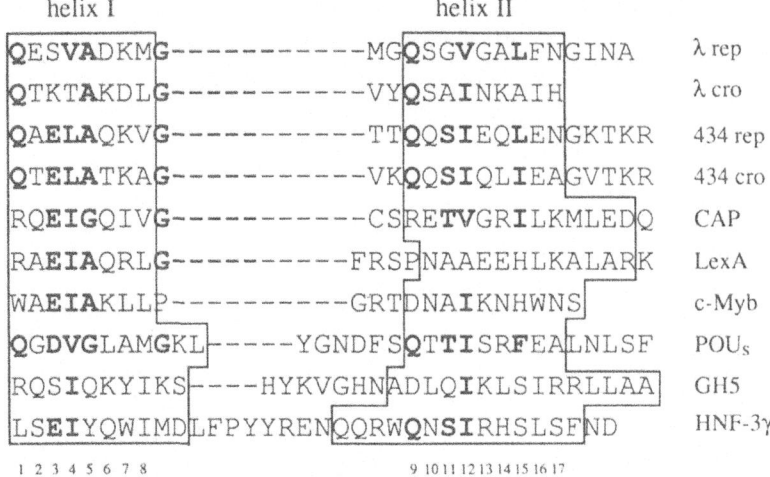

<pre>
        helix I   helix II
    QESVADKMG-----------MGQSGVGALFNGINA   λ rep
    QTKTAKDLG-----------VYQSAINKAIH   λ cro
    QAELAQKVG-----------TTQQSIEQLENGKTKR   434 rep
    QTELATKAG-----------VKQQSIQLIEAGVTKR   434 cro
    RQEIGQIVG-----------CSRETVGRILKMLEDQ   CAP
    RAEIAQRLG-----------FRSPNAAEEHLKALARK   LexA
    WAEIAKLLP-----------GRTDNAIKNHWNS   c-Myb
    QGDVGLAMGKL-----YGNDFSQTTISRFEALNLSF   POUₛ
    RQSIQKYIKS----HYKVGHNADLQIKLSIRRLLAA   GH5
    LSEIYQWIMDLFPYYRENQQRWQNSIRHSLSFND   HNF-3γ

    1 2 3 4 5 6 7 8   9 10 11 12 13 14 15 16 17
</pre>

Figure 6. Alignment of a number of helix-turn-helix motif sequences. Residues in the two helices are boxed. In bold face the functionally conserved residues are indicated.

NMR Structure versus X-ray Structure of the Complex

Recently the crystal structure of a POU-domain bound to its cognate DNA sequence has been determined at 2.4 Å resolution (Klemm *et al.*, 1994). Although this X-ray structure does not have a high resolution, it is interesting to compare the POUₛ domain of this complex with the NMR-structure of the free protein. On the whole the structures of the POUₛ domains are very similar. The backbone atoms of residues 1-71 superimpose with an r.m.s.d. of 2.1 Å. The core of the two structures is quite similar. The most notable difference lies in the orientation of the third or recognition helix, which deviates by approximately 22°, as shown

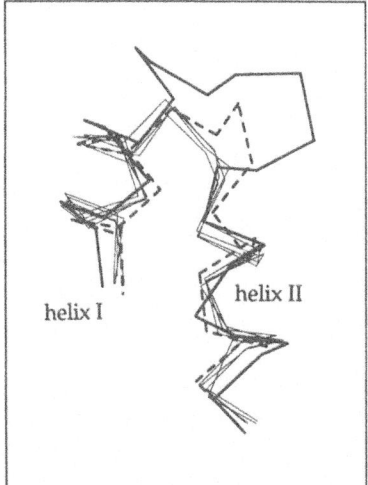
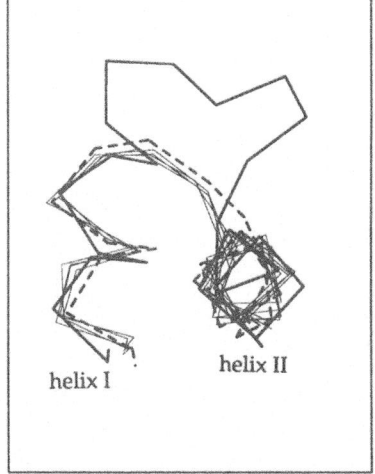

Figure 7. Two views of the Cα traces of a number of helix-turn-helix motifs. Thin traces: λ repressor, lac headpiece, 434 repressor and 434 cro; dashed trace: LexA; thicktrace: POUₛ.

POUs NMR

POUs Xray

Figure 8. Helix-turn-helix motifs in POU-specific domains. Grey: NMR structure of the free protein; black: crystal structure of the helix-turn-helix motif of POU$_s$ in the structure of the complex of the POU-domain with its cognate DNA sequence. The superposition is done on the 72 Cα atoms of the POU$_s$ domain.

in Figure 8. Also the positioning of the loop between helices 2 and 3 differs in the two structures. The significance of this difference is yet unclear. Apparently the protein structure changes upon binding to the DNA, presumably in the region contacting the DNA. On the other hand there could be crystal packing effects, especially at the outside of the molecule. The loop between helices 2 and 3 is indeed solvent exposed and could therefore be suffering from these effects.

CONCLUSION

We present here the refined structure of POU$_s$ characterised with a high resolution. Both the fit to the experimental data, as well as the stereochemical parameters of the structures are good. This allows for a detailed comparison with other protein structures. The four helices of POU$_s$ form a globular protein with a topology similar to the first four helices of λ repressor. Some structural features emerging from this study are the presence of a proline-induced kink in the fourth helix, and an interaction between the side chains of Gln23 and Gln40. A comparison of the helix-turn-helix motif of POU$_s$ with a number of canonical HTH-motifs, shows that although POU$_s$ has a six-residue insert in the motif, the extra residues can be accommodated very easily, resulting in a motif in which the helices have almost an identical orientation as in the canonical HTH-motifs. Moreover, the amino acids at positions 23 and 40, two glutamines, are conserved in the λ and 434 repressor and cro proteins, suggesting an identical structural frame-work for DNA-binding. This clearly shows that although the HTH-motif displays variations in amino acid constitution and size, it has a constant overall fold.

ACKNOWLEDGMENTS

We thank M.J.J. Strating for purifying NMR samples. The investigations were supported by the Netherlands Foundation for Chemical Research (SON) with financial aid from the Netherlands Organisation for Scientific Research (NWO).

REFERENCES

Assa-Munt, N., Mortishire-Smith, R.J., Aurora, R., Herr, W. and Wright, P.E. (1993) Cell 73, 193-205.

Beamer, L.J. and Pabo, C.O. (1992) J. Mol. Biol. 227, 177-196.

Barlow, D.J. and Thornton, J.M. (1988) J. Mol. Biol. 201, 601-619.

Bax, A., Ikura, M., Kay, L., Torchia, D.A. and Tschudin, R (1990), J. Magn. Reson. 86, 304-318.

Botfield, M.C., Jancso, A. and Weiss, M. A. (1992) Biochemistry 31, 5841-5848.

Clark, K.L., Halay, E.D., Lai, E. and Burley, S.K. (1993) Nature 364, 412-420.

Cox, M., Dekker, N., Boelens, R., Verrijzer, C.P.,van der Vliet, P.C. and Kaptein, R. (1993) Biochemistry 32, 6032-6040.

Cox, M., van Tilborg, P.J.A., de Laat, W., Boelens, R., van Leeuwen, H.C.,van der Vliet, P.C. and Kaptein, R. (1995) J. Biomol. NMR, in press.

Dekker, N., Cox, M., Boelens, R., Verrijzer, C.P., van der Vliet, P.C. and Kaptein, R. (1993) Nature 362, 852-854.

Fogh, R.H., Ottleben, G., Rüterjans, H., Schnarr, M., Boelens, R. and Kaptein, R. (1994) EMBO J. 13, 3936-3944

van Gunsteren, W.F. and Berendsen, H.J.C. (1987). Groningen Molecular Simulation (GROMOS) Library Manual. Biomos BV, Nijenborgh 16, 9747 AG Groningen, The Netherlands.

Havel, T.F. (1991), Progr. Biophys. Molec. Biol. 56, 43-78.

Janin, J., Wodak, S., Levitt, M. and Maigrait, B. (1978), J. Mol. Biol. 125, 357-386.

Jancso, A., Botfield, M.C., Sowers, L.C. and Weiss, M.A. (1994) Proc. Natl. Acad. Sci. USA, 91, 3887-3891.

Klemm, J.D., Rould, M.A., Aurora, R., Herr, W. and Pabo C.O. (1994) Cell, 77, 21-32.

Koning, M.M.G., Boelens, R. and Kaptein, R. (1990), J. Magn. Reson. 90, 111-123.

Mondragón, A. and Harrison, S.C. (1991) J. Mol. Biol. 219, 321-334.

Morita, E.H., Shirakawa, M., Hayashi, F., Imagawa, M. and Kyogoku, Y. (1993) FEBS Letters 321, 107-110.

Morris, A.L., MacArthur, M.W., Hutchinson, E.G. and Thornton, J.M. (1992) Proteins: Struct., Funct. & Genetics 12, 345-364.

Neri, D., Szyperski, T., Otting, G. Senn, H. and Wüthrich, K. (1989), Biochemistry 28, 7510-7516.

Ogata, K., Hojo, H., Aimoto, S., Nakai, T., Nakamura, H., Sarai, A., Ishii, S. and Nishimura, Y. (1992) Proc. Natl. Acad. Sci. USA 89, 6428-6432.

Otting, G. and Wüthrich, K. (1988), J. Magn. Reson. 76, 569-574.

Pabo, C.O. and Sauer, R.T. (1992) Annu. Rev. Biochem. 61, 1053-1095.

Qian, Y.C., Otting, G., Billeter, M., Müller M., Gehring, W.J. and Wüthrich, K. (1993), J.Mol. Biol. 238, 333-345

Ramakrishnan, V., Finch, J.T., Graziano, V, Lee, P.L. and Sweet, R.M. (1993) Nature 362, 219-223.

Rodgers, D.W. and Harrison, S.C. (1993) Structure 1, 227-240.

Shaka, A.J., Barker, P.B. and Freeman, R.J. (1985) J. Magn. Reson. 64, 547-552.

Verrijzer, C.P., Alkema, M.J., van Weperen, W.W., van Leeuwen, H.C., Strating, M.J.J. and van der Vliet, P.C. (1992) EMBO J. 11, 4993-5003.

Verrijzer, C.P., Kal, A.J. and van der Vliet, P.C. (1990) Genes Dev. 4, 1964-1974.

Verrijzer, C.P. and van der Vliet, P.C. (1993) Biochim. Biophys. Acta 1173, 1-21.

Wüthrich, K., Billeter, M. and Braun, W. (1983), J. Mol. Biol. 169, 949-961.

17

APPLICATIONS OF MULTIDIMENSIONAL SOLID-STATE NMR SPECTROSCOPY TO MEMBRANE PROTEINS

A. Ramamoorthy, F. M. Marassi, and S. J. Opella

Department of Chemistry
University of Pennsylvania
Philadelphia, Pennsylvania 19104

I. INTRODUCTION

A singular challenge in structural biology is the experimental determination of the structures of membrane proteins. Because these proteins are difficult to crystallize there are few examples with structures determined by X-ray diffraction (Deisenhofer et al., 1985; Weiss et al. 1991; Iwata et al., 1995). Multidimensional solution NMR methods are difficult to apply to membrane proteins because of the slow reorientation rates and broad resonance linewidths that accompany solubilisation in detergent micelles. Proteins in the other well characterized model membrane environment of lipid bilayers are even less well suited for solution NMR methods because the individual protein molecules are effectively immobilized when complexed with phospholipids. However, it is useful to keep in mind that NMR studies of membrane proteins are formidable because of the motional properties of the samples rather than any intrinsic properties of the proteins themselves and that solid state NMR spectroscopy is fully capable of overcoming the difficulties resulting from the very slow reorientation rates (Opella, 1994).

Solid state NMR methods are appropriate for studying membrane proteins in both oriented and unoriented bilayer samples, as illustrated schematically in Figure 1. Since membrane proteins are immobilized by the ordered lipid environment of the bilayers, on the relevant NMR time scales, they behave spectroscopically like solids. Thus, solid state NMR methods are applicable to these samples even though they are fully hydrated and exhibit some properties generally associated with liquids or liquid crystals. While most of the residues in a membrane protein are involved in stable secondary structural elements, primarily α-helices, and are immobile in lipid environments, some individual segments of the protein do undergo motions that affect their NMR spectral properties. Because effects of overall reorientation are absent in bilayer samples, it is particularly straightforward to describe the local dynamics of membrane proteins by solid-state NMR spectroscopy. Information about dynamics assists in the determination of structure for these proteins

Dynamics and the Problem of Recognition in Biological Macromolecules, edited by Jardetzky and Lefèvre
Plenum Press, New York, 1996

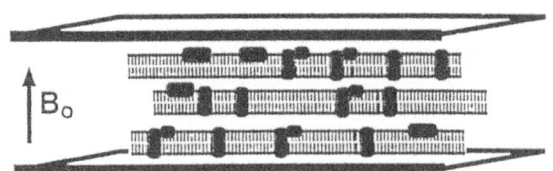

Figure 1. Model of peptides and proteins in oriented lipid bilayers.

because they typically consist of rigid structured helices separated by relatively mobile loop regions, with mobile N- and C-terminal regions (Shon et al., 1991).

The approach to structure determination described in this chapter takes advantage of the ability to specifically, selectively and uniformly label proteins with ^{15}N in amide backbone sites, and to orient them along with the phospholipid bilayers in the magnetic field of the spectrometer. This has a number of benefits which include the placement of the two directly bonded spin S=1/2 nuclei, ^{1}H and ^{15}N, in the peptide bonds of greatest interest. It also means that single crystal samples of ^{15}N labeled synthetic peptides provide valuable model systems for the development of new spectroscopic experiments.

Solid-state NMR spectroscopy remains separate from solution NMR spectroscopy largely for historic reasons even though these two types of experiments have a great deal in common. The technical difficulties of solid state NMR experiments, especially the use of very high power radiofrequency irradiations, limit their implementation on the type of commercial spectrometers most commonly available. However, the greater technical difficulty and complexity in solid-state NMR experiments are small trade offs for the drastic steps needed to perform high resolution spectroscopy on proteins in phospholipid bilayers and other immobile systems. The situation is illustrated with the ^{1}H NMR spectrum in Figure 2 of a polycrystalline sample of the model peptide N-acetyl-leucine. It is obviously impossible to extract detailed spectroscopic, much less structural, information from this spectrum. Improvements in instrumentation and increases in field strength are unlikely to improve the situation appreciably. Therefore, the methods of high resolution solid state NMR spectroscopy were developed to overcome the severe broadening present in spectra like that shown in Figure 2 (Waugh et al., 1968).

High resolution solid state NMR spectroscopy consists of a collection of irradiation procedures and sample manipulations that yield high resolution spectra of solid samples. These methods are capable of transforming a spectrum like that of Figure 2 into well resolved multidimensional spectra capable of yielding spectral parameters useful for structure determination. The effects of the application of selected solid state NMR procedures on the ^{15}N NMR spectra of N-acetyl-leucine are illustrated in Figure 3. A key element in the most successful strategy for high resolution solid state NMR spectroscopy is the focus on the multiple roles, including the detection of magnetization under high resolution conditions, from a dilute spin nucleus, in this case ^{15}N at a labeled site, or alternatively ^{13}C in natural

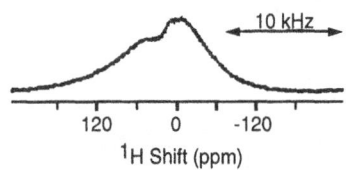

Figure 2. ^{1}H NMR spectrum of a powder sample of N-acetyl-^{15}N-leucine.

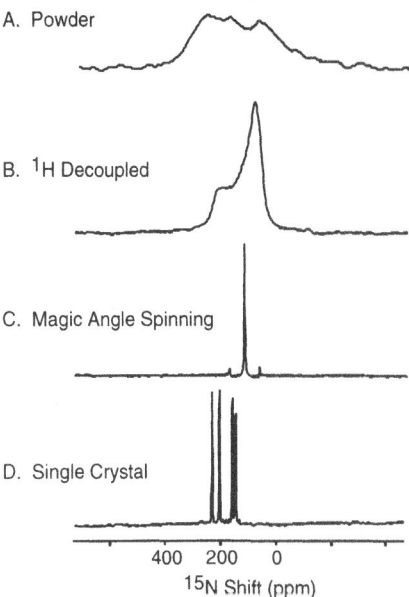

A. Powder

B. ^{1}H Decoupled

C. Magic Angle Spinning

D. Single Crystal

400 200   0
^{15}N Shift (ppm)

Figure 3. ^{15}N NMR spectra of N-acetyl ^{15}N-leucine under various conditions. (A) Spectrum of a powder sample; (B) Same as A with continuous decoupling of protons; (C) Same as B with magic angle spinning at a frequency of 4 kHz; (D) Spectrum of a single crystal sample at an arbitrary orientation with respect to the magnetic field.

abundance or labeled sites (Pines et al., 1973). It is feasible to observe ^{1}H resonances that are substantially narrowed through the application of multiple pulse sequences, however, the applicability of this approach to complex biochemical and chemical problems remains limited and dilute spin approaches have been much more successful. The ^{15}N NMR spectrum in Figure 3A is similar to the ^{1}H spectrum in Figure 2 as starting point since the resonance intensity is very broad due to the effects of dipolar couplings present in a powder sample although, in this case, they are hetero- rather than homo-nuclear couplings. Importantly, heteronuclear dipolar couplings are substantially easier to suppress than homonuclear couplings and this can be accomplished without the scaling (reduction) of chemical shift frequencies that accompanies multiple pulse line narrowing of ^{1}H resonances. This is illustrated in Figure 3B where the characteristic shape of the chemical shift anisotropy powder pattern from the ^{15}N amide site of the model peptide can be observed without severe broadening and distortion from the heteronuclear dipolar couplings to bonded and other nearby hydrogen nuclei. The signals are also much more intense because cross-polarization, which is used to generate the observed ^{15}N magnetization, is integrated with the irradiation of the proton resonances used to suppress the heteronuclear dipolar couplings in the classic pulse sequence for proton enhanced nuclear induction spectroscopy.

Although a chemical shift powder pattern spectrum has a distinctive shape, which is informative about the magnitudes of the principal values of the of the chemical shift tensor, it is still quite broad by high resolution NMR standards. The span of the ^{15}N amide chemical shift tensor is about 170 ppm and when powder patterns from multiple sites are present in the spectrum the possibilities for resolving features from different sites are very limited. Therefore, additional spectroscopic manipulations are needed to obtain chemical shift resolution in solid state NMR spectra. Dramatic improvements in spectral resolution of powder samples can be accomplished with magic sample spinning as shown in the spectrum of Figure 3C where a single line resonance is flanked by small spinning sidebands separated by the spinning frequency of the rotor (Schaefer and Stejskal, 1976). The comparison of the spectra in Figures 3A and 3C clearly illustrates how the combined effects of ^{1}H irradiation for decoupling of the heteronuclear dipolar interactions and magic angle sample spinning to average the chemical shift anisotropy transform the spectra of solids from low resolution "wide lines" into those with high resolution. There is one chemical type of amide nitrogen in the sample, which is why there is only one resonance in the spectrum. If there were multiple labeled residues in a larger peptide, then there would be a corresponding number of resonances in the one-dimensional spectrum.

In addition to magic angle spinning, there is another way to obtain narrow resonances of dilute spin nuclei after the broadening effects of dipolar couplings are removed, and this is sample orientation (Pausak et al., 1973). The best way to orient a molecule is to form a single crystal. The presence of four resolved lines in the spectrum of Figure 3D, obtained from a single crystal sample, versus a single line in the spectrum obtained from a polycrystalline sample with magic angle sample spinning, demonstrates the fundamental principles used as a basis for structure determination by solid state NMR spectroscopy of oriented samples. An oriented sample gives a single line resonance for each unique site. The resonance frequency of the chemical shift interaction, like the measurable spectral parameters for other spin-interactions, depends on the orientation of the relevant bond or atom with respect to the direction of the applied magnetic field. The four resonance frequencies in Figure 3D show that there are four different molecular orientations in the single crystal sample. This is also known from the X-ray crystal structure. When the overall orientation of the crystal is changed, then the resonance frequencies of all four lines change. Thus, the spectrum in Figure 3D illustrates these two points: high resolution can be achieved equally well by sample orientation as magic angle sample spinning, and the observable parameters, here resonance frequencies, reflect the orientations of the groups with respect to the applied magnetic field.

It is possible to utilize the special spectroscopic properties exhibited by uniaxially oriented samples, where the axis of molecular orientation lies parallel to the direction of the applied magnetic field, as the basis for structure determination (Opella and Waugh, 1977). A uniaxially oriented sample has the property that each site on one molecule can be transformed into the identical site on another molecule through a combination of translation, inversion and rotation operations about an axis parallel to the direction of the applied magnetic field. Just as seen for single crystals, the spectra of oriented protein samples are characterized by single line resonances, or multiplets, rather than powder patterns. Since these spectral parameters vary with orientation, and their geometrical dependencies are well understood, they provide a method for structure determination. These properties were first demonstrated on fibers of highly oriented polyethylene and have subsequently been applied to a wide variety of peptides, proteins and other biopolymers (Opella et al., 1996).

The three dimensional structure of the polypeptide backbone of a protein is described equivalently by lines (vectors) representing bonds between non-hydrogen atoms, as in Figure 4A, and by the rectangular outlines of the peptide planes, as in Figure 4B. The peptide bond, together with its directly bonded atoms, forms a rigid planar unit that is the focal point in the determination as well as the presentation and analysis of protein backbone structures by solid-state NMR spectroscopy. The orientation of a peptide plane with respect to the direction of the applied magnetic field, Bo, is described by two polar angles, α and β, defined in Figure 4C. These angles can be determined from the spectral parameters measured in solid state NMR experiments on oriented samples. Once the orientations of all the individual peptide planes in a protein are determined, then they can be assembled into a complete protein structure because they are all related by the common axis system defined by Bo (Opella et al., 1987).

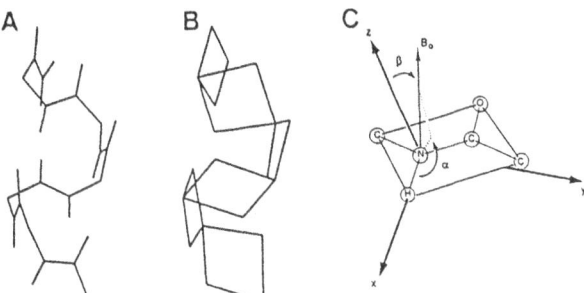

Figure 4. Representations of the structure of the polypeptide backbone with the helix axis aligned vertically. (A) Vector drawing with a line between each non-hydrogen atom; (B) planar representation with connected peptide planes; (C) definition of the axis system used to relate the measured NMR spectral parameters to the peptide plane.

A single measurement rarely leads to a unique peptide plane orientation. This is why it is not sufficient to obtain high resolution chemical shift spectra like that shown in Figure 3D. Although resolution may be adequate, especially in specifically or selectively labeled samples, the amount of structural information is not, unless supplemented by results from other methods. Thus, the need arises to measure two, or preferably three, spectral parameters for each site or peptide plane in order to determine the complete three-dimensional structure.

There are no fundamental size limitations to the polypeptides that can be studied by solid state NMR spectroscopy, since both small and large proteins are equally immobilized within the lipid bilayer. However, since the methodology is still at an early stage in its development, applications to smaller, more tractable peptides and proteins are of immediate interest. The 23 residue magainin antibiotic peptide (Bechinger et al., 1991; Ramamoorthy et al., 1995) and the membrane bound form of the 50 residue filamentous fd bacteriophage coat protein (McDonnell et al., 1993) provide the examples discussed in this chapter. They also serve to demonstrate two different methods of sample preparation, since the magainin peptides were made by solid phase peptide synthesis, and the coat proteins were prepared by expression in bacteria. Chemical synthesis facilitates the incorporation of isotopic labels at specific sites, while expression in bacteria lends itself to selective and uniform labeling of the polypeptides. It is possible to prepare very highly oriented samples for the NMR experiments from both types of samples by reconstituting the peptides into lipid bilayers and spreading these between glass plates under controlled conditions of humidity.

Magainin antibiotic peptides were originally identified in frog skin because of their ability to protect against wound infections (Zasloff, 1987). Magainin2, GIGKFLHSAKKFGKAFVGEIMNS-NH2, is a typical member of this family of peptides, and has a broad range of protective activities. It has similarities to other highly charged antibiotic peptides, such as cecropins, bombolitins, mastoparans and melittin, found in other organisms. Magainin has been shown to be an amphipathic α-helix in membrane environments (Marion et al., 1988; Opella et al, 1993), and our solid state NMR experiments show that its helix axis lies in the plane of the phospholipid bilayers, suggesting that its mechanism of action may be fundamentally different from other amphipathic peptides generally recognized as channel forming (Bechinger et al., 1991).

The major coat protein of a filamentous bacteriophage is stored as a membrane protein in the host prior to being assembled into the intact virus particles as they are extruded through the cell membrane out into the periplasm Makowski, 1984). The coat protein is small and can be prepared in large quantities and labeled with stable isotopes. It is a valuable model systems for developing methods for studying membrane proteins since it contains the principal features of the major class of membrane proteins: a long hydrophobic membrane spanning helix, a shorter amphipathic helix in the plane of the membrane, a loop connecting the two helices, and mobile N- and C- terminal regions (McDonnell et al., 1993).

II. MULTIDIMENSIONAL SOLID STATE NMR SPECTROSCOPY

There are only a few nuclear spin interactions to be concerned with in proteins labeled with ^{15}N in backbone amide sites. These include the ^{1}H chemical shift of the amide hydrogen, the ^{15}N chemical shift of the amide nitrogen, the heteronuclear dipolar coupling of these directly bonded ^{1}H and ^{15}N nuclei, the strong homonuclear couplings among all the ^{1}H nuclei in the sample, and the weak homonuclear couplings among labeled ^{15}N sites in nearby residues. The effects of most of these interactions can be seen in the example spectra in Figures 2 and 3. In Figure 3, in particular, the results of applying the most direct methods of controlling the spectral manifestations of these spin interactions through radiofrequency irradiations and sample manipulations are shown. In this section, we discuss the spectroscopy involved in separating the effects of individual spin interactions so that spectral parameters that represent only a single well defined interaction can be resolved and measured to provide the basis for high resolution spectroscopy as well as structure determination. This is in contrast to the procedures illustrated with the spectra in Figure 3 which, with the exception of sample orientation, simply removed the effects of the anisotropic spin interactions from the spectra.

A. Homonuclear Dipolar Interactions

The dominant spin interaction in essentially all samples of interest in chemistry and biochemistry is the homonuclear dipolar coupling among the abundant hydrogens. The effects of this can be seen in the very broad ^{1}H NMR spectrum of a polycrystalline sample of N-acetyl-leucine in Figure 2. This spectrum is typical of what can be obtained using conventional NMR approaches, such as those used in solution NMR spectroscopy, on not only polycrystalline samples of peptides and proteins, but also proteins embedded in phospholipid bilayers or associated in supramolecular structures. There are simply too many strong interactions characterized by a range of frequencies to obtain high resolution in conventional ^{1}H NMR spectra of solids. However, the proton-proton homonuclear dipolar coupling did provide one of the first opportunities for high resolution solid state NMR spectroscopy. Pake demonstrated that, in favorable cases, molecular dilution by itself was sufficient to isolate the couplings between pairs of hydrogens so as to resolve individual splitting in single crystal samples or to characterize the underlying powder pattern (Pake, 1948). Since the distance and angular dependencies of the dipolar coupling between two spin 1/2 nuclei is so simple and well established, without any need to invoke electronic theory, this seminal experiment maintains its profound influence to this day. A simple statement of the goals of much of the spectroscopic development for structure determination of proteins is that one would like to be able to replicate these results in much more complex molecular systems than water in gypsum. However, little more than the spectrum in Figure 2 is needed as a reminder that this is a very difficult undertaking.

B. Chemical Shift Interactions

The first approach to high resolution solid state NMR spectroscopy utilized pulse sequences to selectively suppress the strong homonuclear dipolar couplings among hydrogens while preserving their chemical shifts. This remains one of the most remarkable achievements in NMR spectroscopy, and led to a family of multiple pulse methods and coherent averaging theory (Waugh et al., 1968). It also led to a substantial body of research; chemical applications of solid state NMR spectroscopy flourished for a while based on the novelty of being able to record and characterize chemical shifts of selected ^{1}H nuclei in powder and single crystal samples (Mehring, 1976). However, these pulse sequences scaled the chemical shift frequencies and given the low fields of the magnets available at the time, it was difficult to perform experiments on samples of even moderate complexity. Thus, there were essentially no applications to biopolymers. Also the early multiple-pulse sequences did not lend themselves to the measurement of individual homonuclear dipolar coupling in solids.

It was the development of proton enhanced nuclear induction spectroscopy (Pines et al., 1973) that enabled solid state NMR to be applied to a broad range of chemical and biochemical applications. Remarkably, this approach takes advantage of the properties of dilute spins, such as ^{13}C and ^{15}N, in the presence of the abundant ^{1}H spins in double resonance experiments. It is then possible to greatly enhance sensitivity through cross polarization and to obtain high resolution by decoupling the heteronuclear dipolar interactions through irradiation at the ^{1}H resonance frequency. The homonuclear dipolar interactions among the dilute spins are weak, by definition, and do not affect the resolution of the spectra very much, while the strong homonuclear dipolar interactions among the hydrogens are suppressed along with the heteronuclear dipolar interactions by the irradiation of the ^{1}H resonance frequency. High resolution can be obtained routinely on single crystal samples (Pausak et al., 1973), as shown in Figure 3D, and oriented samples of membrane proteins in lipid bilayers (Bechinger et al., 1991).

Figure 5 compares the one-dimensional ^{15}N NMR spectra obtained from two different samples of specifically ^{15}N labeled magainin peptides in oriented bilayers (Ramamoorthy et

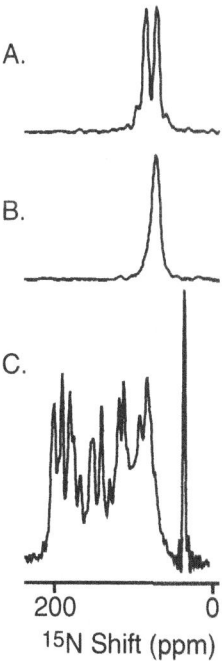

Figure 5. ^{15}N NMR spectra of ^{15}N-labeled peptides and proteins in oriented lipid bilayers. (A) Mixture of ^{15}N Phe 16 and ^{15}N Val 17 labeled magainin peptides; (B) ^{15}N Phe 16 labeled magainin peptide; (C) uniformly ^{15}N labeled fd bacteriophage coat protein.

al., 1995a). There are two single line resonances in the spectrum of Figure 5A because it was obtained from a sample containing a mixture of magainin peptides labeled at two different residues, Phe 16 and Val 17. In contrast the spectrum in Figure 5B, which was obtained on a sample containing only ^{15}N Phe 16 labeled magainin, has one single resonance at 73.8 ppm. The resonance at 83.8 ppm in the spectrum in Figure 5A can be assigned by difference to Val 17. The spectrum in Figure 5C was obtained from a sample of uniformly ^{15}N labeled fd coat protein in oriented bilayers (Marassi et al., 1996). The 48 amide ^{15}N resonances in the spectrum have frequencies along the full range available for these sites, with values between 36 and 211 ppm reflecting the range of orientations for the different peptide planes.

The spectra in Figure 5 illustrate two important points. First that sufficiently high resolution is available to resolve many of the resonances in a uniformly labeled 50-residue membrane protein. Second, that it is possible to utilize the ^{15}N chemical shift frequency to identify residues in trans-membrane versus in-plane helices. This is possible because the amide ^{15}N chemical shift tensor has its principal elements aligned approximately along the N-H bond axis (Wu et al., 1995). For example, oriented magainin-2 whose α-helical structure has been established by solution NMR in micelles has an ^{15}N chemical shift spectrum with resonance frequencies near the principal element of the amide ^{15}N chemical shift tensor associated with N-H bond orientations perpendicular to the direction of the magnetic field. This is consistent with in-plane orientation of the helix. The oriented coat protein, on the other hand, gives a spectrum where both in-plane and trans-membrane residues can be identified. Amide nitrogen resonances from residues in the hydrophobic membrane spanning helix have ^{15}N chemical shift frequencies near that of principal element of the tensor aligned approximately along the N-H bond, while residues in the shorter amphipathic helix, which lies in the plane of the membrane, have ^{15}N chemical shift frequencies near the principal element of the tensor associated with N-H bond orientations perpendicular to the direction of the magnetic field. The amino groups from the five lysines and the N-terminal alanine contribute to the peak at 36 ppm.

Chemical shift frequencies are not sufficient to determine three-dimensional structure without additional information such as the secondary structure of the residues of interest. Therefore, for solid state NMR spectroscopy to be an independent method of structure determination an additional development was needed in order to take advantage of the high resolution available from spectra like those in Figure 5 and measure dipolar couplings associated with individual sites. Thus, it was with the advent of pulse sequences capable of separating dipolar couplings from chemical shifts that solid state NMR spectroscopy began to emerge as a method for determining molecular structure (Waugh, 1976).

C. Heteronuclear Dipolar Couplings

The heteronuclear dipole-dipole interaction is almost always the major factor in determining the lineshapes, linewidths and relaxation parameters in NMR spectra of dilute spin 1/2 nuclei in solids, for example the spectrum in Figure 3A. It was the development of two-dimensional separated local field spectroscopy that enabled heteronuclear dipole-dipole couplings to be resolved and measured in single crystals and oriented samples. The chemical shift dimension in separated local field spectra has intrinsically high resolution because of the effectiveness of continuous ^{1}H irradiation, in suppressing heteronuclear dipolar couplings (Hester et al., 1975). However, even with the use of homonuclear multiple pulse decoupling, the doublets or multiplets observed in the dipolar dimension are quite broad. We developed the first of a new family of pulse sequences in order to further narrow the resonances observed in the dipolar dimension, to enhance both the resolution and the precision of measurements of heteronuclear dipolar couplings (Wu et al., 1994).

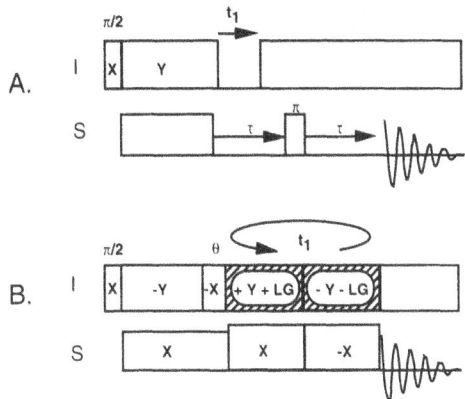

Figure 6. (A) Two-dimensional pulse sequence for separated-local-field (SLF) spectroscopy. (B) Two-dimensional PISEMA pulse sequence for the high resolution dipolar spectroscopy. PISEMA employs the flip-flop Lee-Goldburg (FFLG-2) pulse sequence designated in the hatched areas.

The simplest and most widely used separated local field pulse sequence is diagramed in Figure 6A. It relies on the relatively large magnitude of the heteronuclear dipolar coupling between directly bonded nuclei to overwhelm the effects of homonuclear and longer range heteronuclear dipolar couplings during the t_1 interval. This pulse sequence is generally successful in associating a defined dipolar coupling with each of the resonances resolved on the basis of their chemical shifts as shown in the spectrum in Figure 7A. Measurements have been made on many peptides and proteins, including in oriented bilayers using this approach (Cross and Opella, 1994). However, much better results can be achieved using the PISEMA (polarization inversion spin exchange at the magic angle) pulse sequence diagramed in Figure 6B (Wu et al., 1994). It gives the spectrum in Figure 7B with dipolar linewidths of

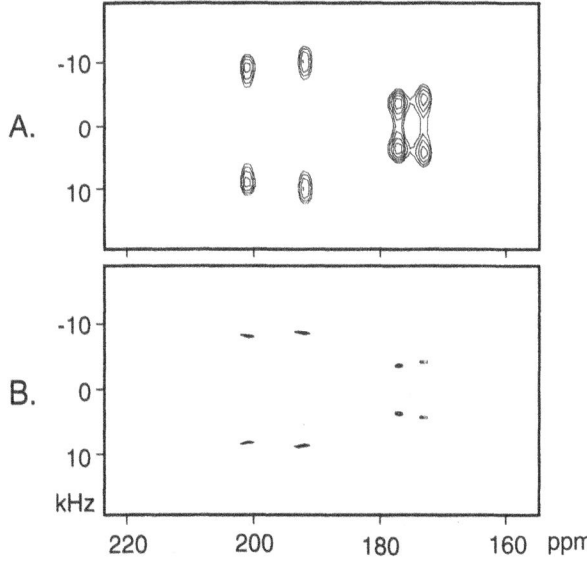

Figure 7. Two-dimensional separated local field spectra of a single crystal sample of N-acetyl ^{15}N-leucine at an arbitrary orientation relative to the external magnetic field. (A) Spectrum obtained using the pulse sequence in Figure 6 A; (B) Spectrum obtained using the pulse sequence in Figure 6 B.

less than 200 Hz, more than one order of magnitude narrower than those observed with the conventional separated local field experiment. Moreover, the scale factor is a very favorable 0.82. The combination of substantial line narrowing and minimal scaling results in a striking improvement in the resolution of the spectra along the dipolar frequency dimension.

The PISEMA experiment is based on the flip-flop (Bielecki et al., 1990) Lee-Goldburg (Lee and Goldburg, 1965) pulse sequence, which is used to spin lock the ^{1}H magnetization along the magic angle. The spin lock field applied to the ^{15}N spins is phase alternated synchronously with the flip-flop Lee-Goldburg procedure to enable coherent exchange of magnetization by cross polarization. The application of the flip-flop Lee-Goldburg sequence greatly extends the oscillation time of the dilute spin magnetization, which results in narrower linewidths, as it seeks to regain equilibrium with the abundant ^{1}H spin magnetization through the phase reversals.

A useful index of resolution in NMR spectroscopy is the ratio of the total spectral range available, in this case the span of the heteronuclear dipolar coupling, to the linewidths of the resonances. This ratio is about 50 for the PISEMA spectrum in contrast to a ratio of about 2 for the conventional separated local field experiment. For comparison, the ratio is about 40 in the ^{15}N chemical shift dimension. This indicates that it is feasible to distinguish among molecular sites on the basis of small variations in the frequencies of their heteronuclear dipolar couplings, in addition to chemical shift frequencies. It also makes very precise measurements of the dipolar coupling frequencies available for analysis in structure determination.

Two-dimensional heteronuclear correlation spectroscopy has proved to be a useful and popular method for determining spin connectivities in biological molecules. In solids, the correlation of chemical shift frequencies is obtained through mixing procedures involving the heteronuclear dipolar couplings (Caravatti et al., 1982; Roberts et al., 1984). This greatly simplifies the detection of the chemical shift spectrum by avoiding the sampling of proton magnetization within the restrictions of multiple pulse windows. Two-dimensional experiments that correlate the chemical shifts of ^{1}H and ^{15}N nuclei are especially useful for ^{15}N labeled peptides and proteins. The chemical shift of amide protons, along with the ^{15}N chemical shift and the ^{1}H-^{15}N dipolar coupling, are valuable in the precise determination of the orientation of individual peptide planes of a protein relative to the external magnetic field. The main advantage of heteronuclear correlation is that it separates the proton resonances over the much wider ^{15}N chemical shift range. Conversely, the spectral range of amide proton chemical shift, about 16 ppm, is useful for resolving among overlapping ^{15}N spectral lines in proteins (Gerald et al., 1993; Wu et al., 1995). This experiment, as many other two-dimensional methods, has preparation, evolution, mixing and detection periods. The preparation consists of a 90° pulse following which both homonuclear ^{1}H-^{1}H and heteronuclear ^{1}H-^{15}N dipolar decoupling is applied while the protons undergo chemical shift evolution. After the first evolution period, in the mixing period, the ^{1}H magnetization is selectively transferred to the directly bonded ^{15}N via the ^{1}H-^{15}N dipolar couplings using spin exchange at the magic angle (SEMA). Finally the ^{15}N magnetization is acquired in the presence of continuous ^{1}H irradiation. High resolution in the ^{1}H chemical shift spectra is achieved with the efficient suppression of ^{1}H-^{1}H homonuclear dipolar couplings by the flip-flop Lee-Goldburg pulse sequence. Although this procedure has a scaling factor of 0.58, generally it is well suited for the study of biological molecules because it performs well with relatively low rf power and is tolerant to pulse imperfections.

The two-dimensional PISEMA and HETCOR spectra in Figures 8 and 9, for magainin and fd coat protein samples illustrate these advantages. The spectra in Figure 8 were obtained from the same sample used for the spectrum in Figure 5A which contains magainin-2 peptides ^{15}N-labeled at phenylalanine 16 and valine 17. In both spectra, each resonance is characterized by two frequencies: the ^{15}N chemical shift and ^{1}H chemical shift in the HETCOR

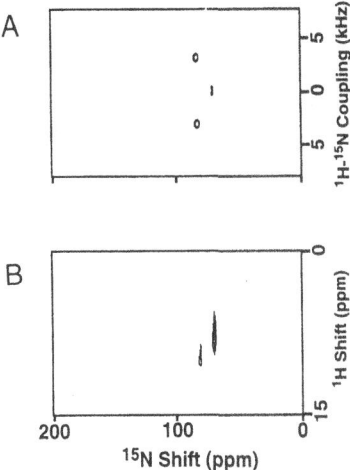

Figure 8. (A) Two-dimensional PISEMA spectrum of a mixture of [15]N Phe 16 and [15]N Val 17 labeled magainin peptides in oriented lipid bilayers. (B) Two-dimensional [1]H /^{15}N chemical shift correlation spectrum of the same sample.

spectrum, and the ^{15}N chemical shift and [1]H-^{15}N dipolar coupling in the PISEMA spectrum. For magainin, the two sites are already well resolved in one dimension, and the two dimensional spectra provide the simultaneous, accurate measurement of two parameters for each of the residues. Taken together, they provide the three frequencies needed for determination of the peptide plane orientation. These experiments demonstrate the feasibility of using multidimensional solid state NMR for structure determination of membrane proteins which are isotopically labeled at all residues. The two-dimensional PISEMA spectrum in Figure 9 was obtained from the 50 residue major coat protein of fd filamentous bacteriophage, uniformly labeled with ^{15}N and reconstituted in phospholipid bilayers oriented between glass plates in a flat coil probe. For each residue, the peptide plane orientation relative to the magnetic field and to the membrane plane yields specific values for the ^{15}N chemical shift and for the [1]H-^{15}N dipolar coupling. Residues contained in the transmembrane segment yield ^{15}N chemical shifts and [1]H-^{15}N dipolar couplings near the $\sigma_\parallel$ and $\nu_\parallel$ principal

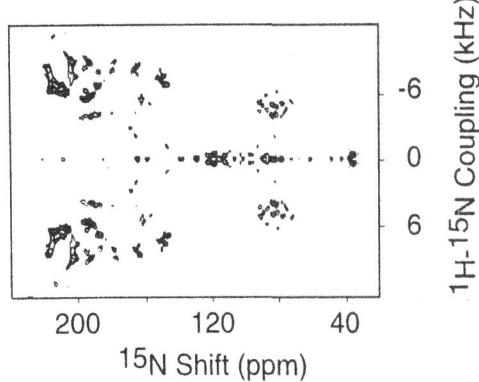

Figure 9. Two-dimensional PISEMA spectrum of uniformly ^{15}N labeled fd bacteriophage coat protein in oriented lipid bilayers.

elements of the tensors, consistent with a peptide plane orientation which places the N-H bond parallel to the magnetic field. On the other hand, residues contained in the in-plane helix have ^{15}N chemical shifts and ^{1}H-^{15}N dipolar couplings near the $\sigma_\perp$ and $\nu_\perp$ principal tensor elements, as expected for an alignment of the N-H bond perpendicular to the magnetic field. Thus the two-dimensional PISEMA spectrum of a uniformly labeled, oriented membrane protein provide much of the resolution and structural information necessary for determining both structure and topology of membrane insertion. Three-dimensional experiments with the additional dimension of the ^{1}H chemical shift further enhance resolution and place tighter restrictions on the peptide plane orientations.

It is somewhat surprising that there have been so few solid-state NMR experiments with dimensionality greater than two. This is in contrast to solution NMR spectroscopy where three- and four-dimensional experiments are routine. Because of the limitations imposed by somewhat broader linewidths, higher dimensional experiments are needed even more in solid-state than in solution NMR spectroscopy. Not only do higher dimensional experiments improve resolution by spreading the resonances along multiple frequency axes, they can also be used to segregate the spectral parameters associated with each of the operative spin interactions by placing them along a unique frequency axis. This is well illustrated with the first of the three-dimensional experiments developed from the original two-dimensional PISEMA experiment (Ramamoorthy et al., 1995b).

The three-dimensional pulse sequence diagramed in Figure 10 accomplishes its goals of providing high spectral resolution, separation and correlation of the spectral parameters associated with both the abundant (^{1}H) and dilute (^{15}N) spins through the incorporation of efficient homonuclear and heteronuclear decoupling schemes. It is a combination of flip-flop Lee-Goldburg decoupling of the ^{1}H spins and continuous irradiation of the ^{15}N spins that suppresses both homo- and heteronuclear dipolar couplings. This allows the ^{1}H spin magnetization to evolve during t_1 under the sole influence of its chemical shifts. During t_2, spin exchange at the magic angle (SEMA) suppresses the unwanted homonuclear dipolar coupling and eliminates the effects of the ^{1}H and ^{15}N chemical shifts, while effecting in-phase coherence transfer between the ^{1}H and ^{15}N spins. Finally, during t_3 high resolution ^{15}N chemical shift spectra are acquired.

Experimental three-dimensional correlation spectra of the model peptide ^{15}N-acetyl-leucine are shown in Figure 11. The three-dimensional powder pattern in Figure 11A encompasses all frequencies available to the ^{1}H chemical shift, ^{1}H-^{15}N dipolar, and ^{15}N chemical shift interactions in the amide group of the model peptide ^{15}N acetyl-leucine. The single crystal spectrum in Figure 11B has four signals, since there are four unique molecules in the unit cell of N-acetyl-leucine. The overall spectral resolution is very high with all four

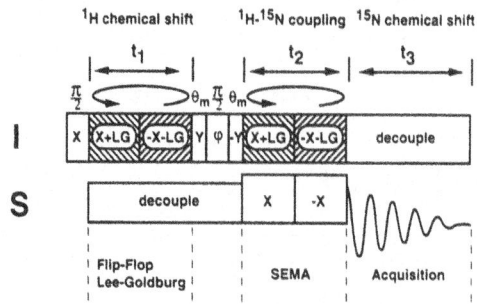

Figure 10. Three-dimensional pulse sequence for I spin chemical shift, I-S dipolar coupling, and S spin chemical shift correlation spectroscopy. In these examples the I spin is ^{1}H and the S spin is ^{15}N.

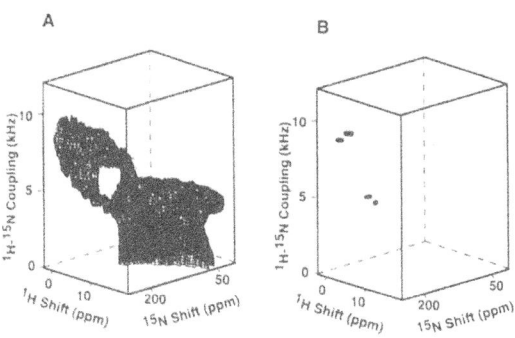

Figure 11. Three-dimensional ^{1}H chemical shift / ^{1}H-^{15}N dipolar coupling / ^{15}N chemical shift correlation spectra of ^{15}N-labeled N-acetyl-leucine obtained using the pulse sequence shown in Figure 10. (A) Polycrystalline sample. (B) Single crystal sample.

resonances resolved along each of the three frequency axes. There are many possible application of this pulse sequence that can be used to obtain high-resolution spectra from complex molecules in single crystal and uniaxially oriented samples. The direct measurement of three spectral parameters for each amide nitrogen site is sufficient in nearly all cases to fully determine the orientation of a peptide plane in a uniaxially oriented sample.

Figure 12A shows a three-dimensional spectrum of magainin-2 peptide incorporated in phospholipid bilayers oriented between glass plates in a flat coil probe (Ramamoorthy et al., 1995a). The spectrum was obtained from the same sample used for the spectrum in Figure 5A which contains magainin peptides ^{15}N-labeled at Phe 16 and Val 17. Each resonance in the three-dimensional spectrum is characterized by the three frequencies, the ^{15}N chemical shift, ^{1}H chemical shift, and ^{1}H-^{15}N dipolar coupling, that can be analyzed by the graphical restriction plot method (Opella et al., 1987) in order to extract the orientations of the individual peptide planes. The linewidths observed for magainin-2 in oriented bilayers were about 1.3 ppm for the ^{1}H resonances, 10 ppm for the ^{15}N resonances, and about 400 Hz for the ^{1}H-^{15}N dipolar couplings. These are quite favorable and can be further improved with additional development.

The graphical restriction plot method for determining peptide plane orientation from the spectral parameters is illustrated in Figure 12B for the peptide planes from Phe 16 and Val 17 in magainin using the data obtained from the three-dimensional spectrum. The dark areas correspond to those pairs of α and β angles, defined in Figure 4, that are consistent with the experimental measurements. The full ranges of angles reflect the fundamental geometrical dependencies of the various spin interactions as well as experimental errors, including uncertainty in the magnitudes and principal values of the spin interaction tensors (Wu et al., 1995). Each experimental parameter restricts the number of possible orientations for an individual peptide plane. Generally, the intersection of the restriction plots from three measurements will suffice to determine the peptide plane orientation unambiguously. The four symmetry-related orientations for a peptide plane reflect the symmetry of the spin interactions of spin S=1/2 nuclei. Further reductions are possible with the interactions of spin S=1 nuclei (McNamara et al., 1995). However, the additional general and chemical requirements inherent in polypeptides generally provide sufficient additional information for assembling of peptide planes into a satisfactory low-energy structure. The orientations for Phe16 and Val17 derived from the restriction plots in Figure 12B are shown in Figure 12C. They provide the input to a program that assembles protein structures from the peptide

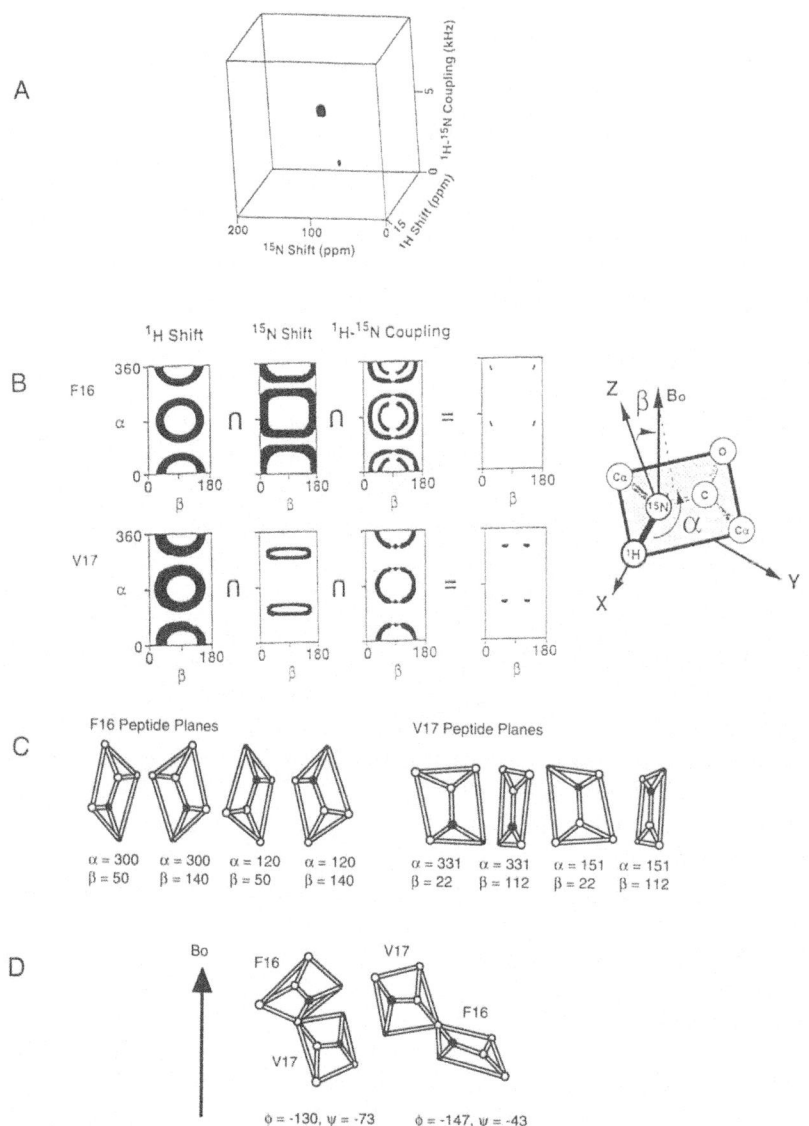

Figure 12. (A) Three-dimensional correlation spectrum obtained on a mixture of ^{15}N Phe 16 and ^{15}N Val 17 magainin peptides in oriented bilayers, using the pulse sequence shown in Figure 10. (B) Restriction plots derived from the NMR parameters measured from the spectrum given in A. These reflect estimates of error and uncertainty of ± 0.2 kHz for the dipolar coupling. The orientations of the principal elements of the tensors were derived from the Ala-Leu dipeptide. The magnitudes were measured from unoriented magainin samples. (C) The four possible peptide planes orientations consistent with the experimental data for ^{15}N-labeled Phe 16 and Val 17 of magainin-2 in oriented bilayers. In this representation the external magnetic field of the spectrometer is directly out of the page toward the viewer.

planes (Opella and Stewart, 1989). The peptide plane orientations are further assembled into the two low energy helical structures shown in the Figure.

E. Spin-Exchange Spectroscopy

Although strong homonuclear couplings are difficult to deal with in complex biological molecules, weak homonuclear dipolar couplings provide a mechanism for spin-exchange. Homonuclear spin-exchange is an important part of NMR spectroscopy. It can be effected by spin-diffusion or cross-relaxation processes originating in the dipole-dipole interactions (Jener et al., 1979). Therefore, it discriminates toward pairs of nuclei in close proximity. It can also be used quantitatively to measure distances and angles.

Two-dimensional homonuclear ^{15}N spin-exchange experiments have been applied to selectively and uniformly ^{15}N labeled proteins and nucleic acids. In these systems, all of the homonuclear dipolar couplings between neighboring ^{15}N nuclei are weak due to their relatively long internuclear distances and low gyromagnetic ratio. There are a variety of situations where dilute spin-exchange experiments are extremely valuable, for example in four-dimensional experiments (Ramamoorthy et al., 1995c).

Two-dimensional ^{15}N spin-exchange spectra of N-acetyl-L-valyl-L-leucine single crystal are shown in Figure 13 for various mixing times. Four resonances are from the two amide nitrogens in each of the two magnetically inequivalent dipeptide molecules in the unit cell. Cross peaks in the two-dimensional ^{15}N spin-exchange spectrum indicate that the respective diagonal resonances exchanged their magnetization during the mixing time and the nuclei corresponding to these diagonal peaks are closer in space. The same experiment with a mixing time of 3s resulted in the spectrum, of Figure 13A, with four diagonal peaks and no cross peaks. Two-dimensional experiments performed as a function of mixing time resulted in the first set of cross peaks only for a mixing time of 4.5s (Figure 13B) for this particular orientation of the single crystal relative to the external magnetic field. This pair of nuclei must be from the same molecule since the intramolecular ^{15}N nuclei are separated by 3.47A° and the closest intermolecular ^{15}N nuclei are separated by 4.99 A°. A cross peak between the intramolecular ^{15}N nuclei in another molecule appears only after 4.7 s mixing time. This difference between the molecules could be attributed to their different orientations with respect to the magnetic field. At longer mixing times, all the resonances are correlated indicating that the ^{15}N spin-exchange reveals the ^{15}N nuclei separated as much as 6 A°. In addition, the powerful ^{1}H spin diffusion phenomenon due to the homonuclear ^{1}H-^{1}H dipolar couplings can be controlled in a useful way to achieve ^{1}H spin-exchange among the nearby amide protons.

Since the three-dimensional experiments resolve individual resonances in uniformly ^{15}N labeled proteins, the next step is to assign the resonances to specific molecular sites. An important potential application of homonuclear spin-exchange experiments is the mapping of the sequential order of amino acids in polypeptides, providing a method for assigning the resonances in spectra obtained from uniformly ^{15}N labeled protein samples. A four-dimensional experiments that correlates the chemical shifts and dipolar coupling frequencies of two heteronuclei with the ^{15}N spin-exchange has been successfully demonstrated on a single crystal sample of uniformly ^{15}N labeled N-acetyl-L-valyl-L-leucine (Ramamoorthy et al., 1995c). Application of just such a single four-dimensional experiment to a protein should enable the assignment of the resonances resolved in three-dimensional spectra. The distances between amide sites on adjacent residues in helical secondary structure (2.75 - 2.95 A°) are substantially shorter than those giving spin-exchange cross peaks in Figure 13. Spin-exchange has been previously detected between ^{15}N-labeled amide sites in a helical protein with two-dimensional experiments (Cross et al., 1983). Since membrane proteins are dominated by alpha helical secondary structure, four-dimensional experiments may provide

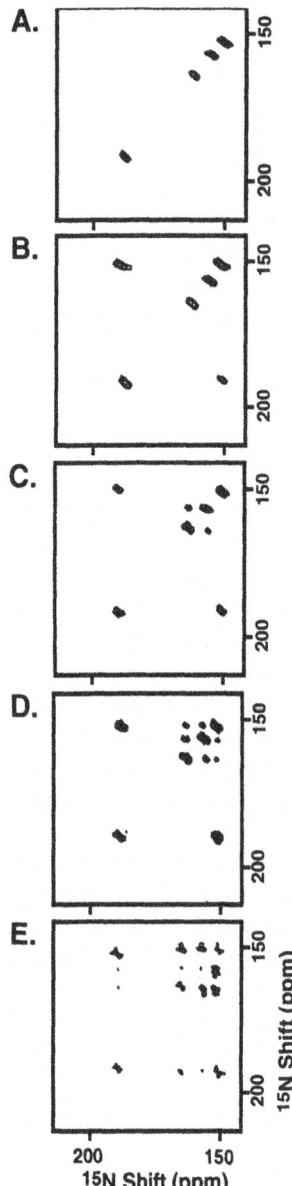

Figure 13. Two-dimensional ^{15}N spin-exchange spectra of a single crystal sample of ^{15}N-acetyl-L-valyl-L-^{15}N-leucine. At short mix times there are no cross-peaks. At longer mix times an increasing number of peaks appear. The mix times for the various experiments are. A. 3.0 sec. B. 4.5 sec. C. 4.7 sec. D. 8.5 sec. E. 10.0 sec.

a systematic method of assigning resonances observed in high resolution three-dimensional correlation spectra.

III. PRACTICAL ASPECTS OF SOLID-STATE NMR STUDIES OF MEMBRANE PROTEINS

The sample with two differently labeled magainin was prepared by mixing 7 mg of each of the labeled peptides with sufficient palmitoyl-oleoyl-phosphatidylcholine (POPC)

(80%) and palmitoyl-oleoyl-phosphatidylglycerol (POPG) (20%) to obtain a final molar ratio of 4% peptide in the lipids. The peptide and lipids were cosolubilised in chloroform with a trace amount of trifluoroethanol, and then spread onto the surface of twenty 18 by 18 mm glass plates whose thickness was reduced from 0.1 mm by etching in a solution of 8% by volume HF in ethanol (Prosser et al., 1995) to give the highest possible filling factor within the basic geometry of a flat-coil probe, actually a square coil. After air drying the sample, residual solvent was removed under vacuum for two hours. The plates were then stacked and the sample hydrated at 45°C for 24 hours. After hydration the sample was equilibrated at 93% relative humidity for an extended period of time in a closed chamber. The sample was wrapped with plastic film before insertion into the RF coil of the probe.

For the uniformly ^{15}N-labeled fd coat protein experiments, the protein was expressed by infecting *E.coli* bacteria, grown on ^{15}N-labeled minimal media, with fd bacteriophage. In order to perform the experiments it is essential to obtain high levels of protein reconstitution into lipid bilayers, and this was achieved by the method of Bayer and Feigenson (1985). Vesicles composed of POPC and POPG (80/20 molar ratio) were prepared by sonication and coat protein solubilised in 10 mM cholate was added to the vesicles suspension to obtain a final molar ratio of 1% protein in the lipid. After diluting by half, the mixture was quickly frozen in liquid nitrogen and allowed to thaw at room temperature. The cholate was subsequently removed by dialysis over a period of 72 hours. Following dialysis the reconstituted vesicles were concentrated and spread on the surface of 44 glass plates of dimensions 11 x 20 x 0.05 mm. The water was evaporated, and the plates were stacked and hydrated at 45°C for 24 hours. The sample was wrapped in plastic and then heat sealed before insertion into the RF coil of the probe.

The major instrumentation requirement for high resolution solid-state NMR experiments is a spectrometer capable of delivering high power rf pulses with sophisticated phase and frequency switching capabilities, and a high power double resonance probe. In the probes, high power Polyflon capacitors are used to tune the resonance circuit to ^{15}N resonance frequency whereas Voltronics capacitors are used to tune the resonance circuit to ^{1}H resonance frequency. Each probe is carefully checked for stable performance with 1-2 kW rf power in the ^{15}N channel and 0.5-1 kW rf power in the ^{1}H channel. A five turn square coil was made of a flat-wire, with a width of 2 mm and a spacing of 2 mm, to a final dimension of 20 x 11 x 4 mm.

Experimental conditions must be carefully optimized in these experiments. Multiple pulse tune-up procedures are followed by calibrating the phase, field strength of the rf pulses, as well as the stability of rf amplifiers. The ^{1}H signal from water in the samples is used to shim to a line width of about 0.1 ppm, to reference the ^{1}H spectrum of the protein relative to TMS at 0 ppm, and to calibrate the 90° pulse width for various rf power levels. Similarly, the ^{15}N signal from ^{15}N labeled ammonium sulfate powder is used to reference the ^{15}N spectrum relative to liquid ammonia at 0 ppm, to set the Hartmann-Hahn cross-polarization match, and to calibrate the 90° pulse width for various rf power levels. Water and ammonium sulfate are reasonable tune-up samples for solid-state NMR experiments, however, the experimental conditions are not exactly the same for membrane proteins oriented in lipid bilayers. Therefore, experimental conditions are readjusted for each sample of interest. To set the Lee-Goldburg match, an off-resonance spin-lock after the preparation 54.7° pulse is used in the ^{1}H channel and at the same time the rf power level of the spin-lock in the ^{15}N channel is varied to maximize the ^{15}N signal intensity. Typically, an rf field strength of 41.67 kHz is used for the Hartmann-Hahn cross-polarization, 56 kHz is used to decouple protons during ^{15}N signal acquisition, and the rf power level in the ^{15}N channel is increased to 51 kHz during the SEMA sequence to set the Lee-Goldburg match. An off-resonance jump of 29.5 kHz is used to suppress ^{1}H-^{1}H homonuclear dipolar couplings. A one dimensional ^{15}N spectrum of the oriented membrane protein is used to set up the optimized two- and three-dimensional experiments. Dry nitrogen or air is used to cool the probe elements from heating during the experiments.

IV. FUTURE PROSPECTS

Solid-state NMR spectroscopy has been utilized in many studies of membrane proteins. These studies have yield much valuable information about the structure and, especially, the dynamics of these proteins, which are so difficult to study by the conventional methods of structural biology, X-ray crystallography and multidimensional solution NMR spectroscopy. The issue of current interest is how close is solid-state NMR spectroscopy to being able to determine the three-dimensional structure of a membrane protein in lipid bilayers.

Since its inception, solid-state NMR spectroscopy was recognized as a method capable of yielding structural information with atomic resolution (Pake, 1948). However, it has taken extensive development of methods, instrumentation, and samples for solid-state NMR to applicable to molecules with the complexity of proteins. The development has come very far, and the pace has accelerated in the past few years.

This chapter describes progress in the development of the approach for determining protein structures using solid-state NMR of oriented samples. Important recent developments include the implementation of higher dimensional experiments, the use of high field spectrometers, and improvements in the orientation of protein containing bilayers. These development have brought the spectroscopy to the point where it is possible to resolve resonances from individual residues in three-dimensional spectra obtained from samples of uniformly ^{15}N labeled proteins. Moreover, each resonance in a three-dimensional correlation spectrum is characterized by three frequencies, which can be used to determine the orientation of individual peptide planes. Related pulse sequences provide systematic methods for assigning the resonances. And there are methods for assembling the structures of the proteins from the orientation of the peptide planes relative to the same external axis defined by the direction of the applied magnetic field. Thus, solid-state NMR spectroscopy is on the verge of being able to determine the structures of membrane proteins in lipid bilayers. Further developments, including bacterial expression systems for membrane proteins, methods for orienting samples routinely, as well as the availability of very high field spectrometers, in the offing provide the basis for optimism about the general applicability of solid-state NMR spectroscopy to protein structure determination.

ACKNOWLEDGMENTS

This research was supported by Grants RO1GM29754 and RO1AI20770 from the General Medical Sciences and Allergy and Infectious Disease Institutes, National Institutes of Health, and utilized the Resource for Solid-State NMR of Proteins at the University of Pennsylvania, supported by Grant P41RR09731 from the Biomedical Research Technology Program, Division of Research Resources, National Institutes of Health. F.M.M. was supported by postdoctoral fellowship 930FEN-1004-43344 from the Medical Research Council of Canada.

REFERENCES

Bayer, R. and Feigenson, G. W. (1985) Biochim. Biophys. Acta 815, 369-379.

Bechinger, B., Kim, Y., Chirlian, L. E., Gesell, J., Neumann, J.-M., Montal, M., Tomich, J., Zasloff, M. and
 Opella, S. J. (1991) J. Biomol. NMR 1, 167-173.
Bieleki, A., Kolbert, A. C., de Groot, H. J. M., Griffin, R. G. and Levitt, M. H. (1990) Adv. Magn. Reson. 14,
 111-124.
Caravatti, P., Bodenhousen, G. and Ernst, R. R. (1982) Chem. Phys. Lett. 89, 363-367.
Cross, T. A., Frey, M. H. and Opella, S. J. (1983) J. Am. Chem. Soc. 105, 7471-7473.
Cross, T. A. and Opella, S. J. (1994) Curr. Opin. in Struct. Biol. 4, 574-581.
Deisenhofer, J., Epp, O., Miki, K., Huber, R. and Michel, H., (1985) Nature 318, 618-624.
Gerald, R., Bernhard, T., Haeberlen, U., Rendell, J. and Opella, S. J. (1993) J. Am. Chem. Soc. 115, 777-782.
Hester, R. K., Ackerman, J. L., Neff, B. L. and Waugh, J. S. (1976) Phys. Rev. Lett. 36, 1081-1084.
Iwata, S., Ostermeier, C., Ludwig, B. and Michel, H. (1995) Nature 376, 660-669.
Jeneer, J, Meier, B. H., Bachmann, P. and Ernst, R. R. (1979) J. Chem. Phys. 21, 4546-4553.
Lee, M. and Goldburg, E. J. (1965) Phys. Rev. A 140, 1261-1271.
Makowski, L. (1984) In Biological Macromolecules and Assemblies (McPherson, A., ed.), 203-253, Wiley,
 New York.
Marassi, F. M., Ramamoorthy, A. and Opella, S. J. (1996) unpublished results.
Marion, D., Zasloff, M. and Bax, A. (1988) FEBS Lett. 227, 21-26.
McDonnell, P. A., Shon, K., Kim, Y. and Opella, S. J. (1993) J. Mol. Biol. 233, 447-463.
McNamara, R., Wu, C. H., Chirlian, L. E. and Opella, S. J. (1995) J. Am. Chem. Soc. 117, 7805-7811.
Mehring, M. (1976) High Resolution NMR spectroscopy in Solids 2nd ed, pp. 246, Springer Verlag, Berlin.
Opella, S. J. and Waugh, J. S. (1977) J. Chem. Phys. 66, 4919-4924.
Opella, S. J., Stewart P. L. and Valentine, K. G. (1987) Q. Rev. Biophys. 19, 7-49.
Opella, S. J. and Stewart P. L. (1989) Methods Enzymol., 176, 242-275.
Opella, S. J., Gesell, J. And Bechinger, B. (1993) in The Amphipathic Helix (Epand, R. ed.), 87-106, CRC
 Press, Boca Raton, FL.
Opella, S. J. (1994) Annu. Rev. Phys. Chem. 45, 659-683.
Opella, S. J., Chirlian, L. E. and Bechinger, B. (1996) in Biological NMR Spectroscopy (Markley, J. L. and
 Opella, S. J. eds.), 139-156, Oxford University Press.
Pake, G. E. (1948) J. Chem. Phys. 16, 327-336.
Pausak, S., Pines, A., Gibby, M. G. and Waugh J. S. (1973) J. Chem. Phys. 59, 591-595.
Pines, A., Gibby, M. G. and Waugh, J. S. (1973) J. Chem. Phys. 59, 569-590.
Prosser, R. S., Hunt, S. A., and Vold, R. R. (1995) J. Magn. Reson. B 107, 109-111.
Ramamoorthy, A., Marassi, F. M., Zasloff, M. and Opella, S. J. (1995a) J. Biomol. NMR 6, 329-334.
Ramamoorthy, A., Wu, C. H. and Opella, S. J. (1995b) J. Magn. Reson. B 107, 88-90.
Ramamoorthy, A., Gierasch, L. M. and Opella, S. J. (1995c) J. Magn. Reson. B 109, 112-116.
Roberts, J. E., Vega, S. and Griffin, R. G. (1984) J. Am. Chem. Soc. 106, 2506-2512.
Schaefer, J. and Stejskal, E. O. (1976) J. Am. Chem. Soc. 98, 1031-1032.
Shon, K., Kim, Y., Colnago, L. A. and Opella, S. J. (1991) Science 242, 1303-1305.
Waugh, J. S., Huber, L. M. and Haeberlen, U.(1968) Phys. Rev. Lett. 20. 180-182.
Waugh, J. S. (1976) Proc. Natl. Acad. Sci. USA 73, 1394-1397.
Weiss, M., Abele, U., Weckesser, J., Welte,W., Schiltz, E. and Schulz, G. (1991) Science 254, 1627-1630.
Wu, C. H., Ramamoorthy, A. and Opella, S. J. (1994) J. Magn. Reson. A 109, 270-272.
Wu, C. H., Ramamoorthy, A., Gierasch, L. M. and Opella, S. J. (1995) J. Am. Chem. Soc. 117, 6148-6149.
Zasloff, M. (1987) Proc. Natl. Acad. Sci. USA 84, 5449-5453.

18

CONFORMATION, MOBILITY, AND FUNCTION OF THE N-LINKED GLYCAN IN THE ADHESION DOMAIN OF HUMAN CD2

Gerhard Wagner,[1] Daniel F. Wyss,[2] Johnathan S. Choi,[1] Jing Li,[3]
Alex Smolyar,[3] Antonio R. N. Arulanandam,[2] Maria H. Knoppers,[2]
Kevin J. Willis,[2] and Ellis L. Reinherz[3]

[1] Department of Biological Chemistry and Molecular Pharmacology
Harvard Medical School
240 Longwood Avenue, Boston, Massachusetts 02115
[2] Procept, Inc.
840 Memorial Drive, Cambridge, Massachusetts 02139
[3] Dana Farber Cancer Institute

INTRODUCTION

Human lymphocytes bear an array of surface glycoprotein receptors that interact with ligands on the surface of other immune cells, such as antigen presenting cells, and are involved in processes of T-cell activation. One important player in this team is human CD2 which binds to the counter receptor CD58 (LFA3) of antigen-presenting cells. Human CD2 is a 50-55 kDa surface glycoprotein which is found on virtually all T lymphocytes as well as on natural killer cells. It initiates the adhesion of T lymphocytes to infected target cells and antigen presenting cells (APCs). At the N-terminus, it consists of two extracellular domains followed by a single transmembrane domain and a proline-rich cytoplasmatic tail of 117 residues which is required for CD2-mediated signal transduction. All adhesion function of CD2 is mediated via the N-terminal adhesion domain. CD2 was predicted to be a member of the immunoglobulin superfamily (IgSF)[1] based on sequence homology of its N-terminal domain and its membrane proximal second extracellular domain to variable and constant IgSF domains, respectively. Indeed, NMR solution studies of the adhesion domains of rat CD2[2] and human CD2[3] revealed conformations typical of immunoglobulin variable (V-set) domains. Furthermore, X-ray structures of the whole extracellular portions of rat CD2[4] and human CD2[5] confirmed this observation and also showed that the structures of the second extracellular domains are similar to constant (C2-set) domains of immunoglobulins. The two extracellular domains form a head-to-tail assembly in which the long axes of the two domains form an angle of ca 40°. However, the relative orientation of the two domains differs by ca. 20° between the rat and the human protein[5] indicating that the two domains may have a certain mobility relative to each other in solution. The extracellular portion of

Dynamics and the Problem of Recognition in Biological Macromolecules, edited by Jardetzky and Lefèvre
Plenum Press, New York, 1996

CD2 has similarities to domains 1 and 2 of either chain of the growth hormone receptor[6] domains 1 and 2 of CD4[7,8], and domains 3 and 4 of CD4[9,10].

Three N-linked carbohydrates are found on the extracellular portion of hCD2, one in domain 1 (at Asn65) and two in domain 2 (at Asn117 and Asn126). The carbohydrates of the second domain can be removed without affecting the adhesion function of hCD2. Removal of the single glycan of the adhesion domain, however, either by treatment with Peptide:N-glycosidase F (PNGase F), or by mutation of the N-glycosylation sequence Asn65-Gly66-Thr67 results in loss of adhesion function[11], or at least a severe destabilization of the protein[12]. Here, we have focused on solving the spatial structure of the carbohydrate and investigating its functional significance.

STRUCTURE OF THE POLYPEPTIDE PART OF THE hCD2 ADHESION DOMAIN

Interested in the role of the N-linked glycan in adhesion function we have initially solved the polypeptide structure of the fully glycosylated adhesion domain of human CD2 by NMR spectroscopy[3,13]. This was achieved with standard homonuclear NMR experiments. Because the protein requires the glycan linked to the side chain of Asn65, the protein could not be expressed in a bacterial expression system. Thus, efficient labeling with ^{15}N and ^{13}C was not possible, and the NMR work relied mostly on homonuclear experiments. Fig. 1 shows a stereo diagram of the polypeptide backbone of an ensemble of 18 NMR structures. The polypeptide has a fold typical of a IgSF V-set domain lacking the first half of strand A. The spread of the structures indicate that there is higher mobility in the N-terminal hexapeptide, the two C-terminal residues, and some of the loops connecting the β-strands (Fig. 1). The core of the protein is well defined.

CD58 BINDING SITE ON CD2

The solution structure of human CD2 served as a guide to design mutants for identifying the CD58 binding site of human CD2. Residues with surface exposed side chains were mutated to alanines, and the full-length mutants were tested for adhesion function in a cell-based adhesion assay. The results showed that the binding site is located on a highly

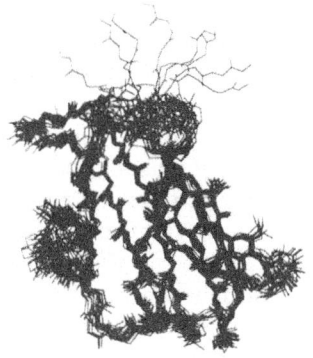 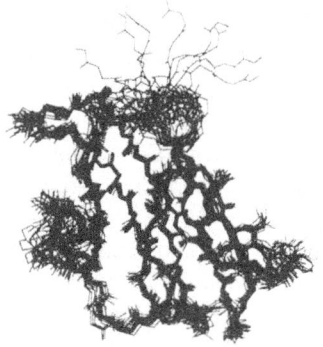

Figure 1. Stereodiagram of an ensemble of NMR solution structures of the adhesion domain of hCD2. Here, only the first three saccharide units are shown (facing to the right).

charged surface area consisting of the GFCC'C" β-sheet[14]. Fig. 2a shows a schematic representation of this face of hCD2. Knock-out mutations are indicated with ellipses, whereas mutations that reduce adhesion are marked with rectangles. Bodian et al.[5] have recently solved the crystal structure of the two-domain extracellular region of human CD2. They have expressed residues 1-182 in Chinese hamster ovary cells in the presence of a glycosidase I inhibitor. In this case, oligosaccharide processing is blocked and glycoproteins are secreted with predominantly Glc_2-$Man_9GlcNAc_2$ oligosaccharides[15].

These preparations are highly sensitive to endoglycosidase H, and treatment with this enzyme yields protein with a single GlcNAc at each glycosidation site. The protein produced this way maintains adhesion function.

CD2 BINDING SITE ON CD58

The counter-receptor CD58 mediates the interactions of antigen presenting cells and target cells, with activated T-cells and thymocytes. Binding of CD58 to CD2 enhances T-cell stimulation. CD58 is homologous to CD2 and CD48. It consists of two extra cellular domains. The N-terminal adhesion domain has homology to variable domains of the immunoglobulin superfamily; the second domain has characteristics of constant domains. CD58 is more extensively glycosylated than CD2 although the extent of glycosylation is not yet fully elucidated.

The adhesion domain of human CD58 is homologous to the adhesion domains of human CD2, rat CD2, CD4 and CD8 for which NMR or X-ray structures are available[2, 3,4,5, 7,8,9,10,16]. Based on the homology, a structure of CD58 has been modeled recently[17]. Based on this modeling, residues with surface exposed side chains were mutated to alanines (E25, K29, K30, K32, D33, K34, E37, E39, R44, K50, R52, D56, E72, D73, E76, E78, D84, K87) and the effect on the adhesion function was tested. The results showed that the CD2 binding

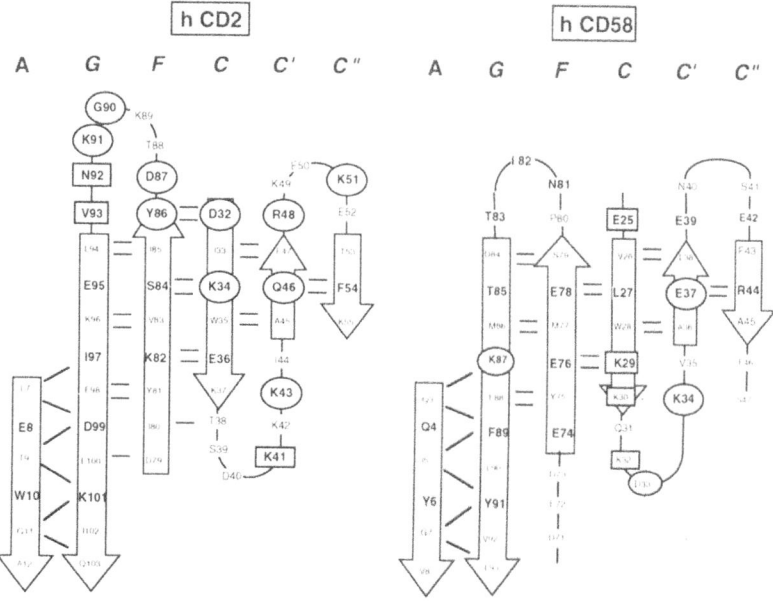

Figure 2. AGFCC'C" faces of CD2 and CD58. Residues with exposed side chains are drawn in bold, knockout Ala mutation sites are indicated by ellipses, Ala-mutations that reduce adhesion are indicated by rectangles.

site includes the C-strand (E25, K29, K30, the CC' loop (K32, D33, K34, the C' strand (E37), and the G-strand (K87).

These mutational studies indicate that the receptor interface between CD2 and CD58 is constituted of two highly charged surfaces. This is surprising considering related interfaces with known structures. These include the interfaces between the variable domains of heavy and light chains of immunoglobulins[18] or the dimer interface of CD8[16]. In these cases, the interface is composed of predominantly apolar side chains. However, one has to consider that the structural model derived by homology modeling from CD2 and CD4 may miss some crucial features of the CD58 structure. For example, there might be a β-bulge in the G-strand, starting with K87 or F88, and a second β-bulge in the C' strand involving residues D33 to V35. These β-bulges could expose several hydrophobic side chains that could be involved in the receptor co-receptor contacts. Bulges in these positions have generally been observed in immunoglobulins and are conserved structural motifs. This face of hCD2 contains two β-bulges as well but they are at different positions than in immunoglobulins[3,13]. If these bulges were present in CD58 they could contribute a number of hydrophobic contacts.

RESONANCE ASSIGNMENTS OF THE hCD2 GLYCAN

Resonance assignments of a carbohydrate in a glycoprotein represent a major challenge. Since glycosylated proteins cannot be produced in *E.coli* expression systems, labeling with stable isotopes (^{15}N and ^{13}C) is prohibitively expensive and all assignments have to rely on homonuclear experiment or on heteronuclear spectra at natural abundance of ^{15}N and ^{13}C. The high mannose glycan in our samples of hCD2 as produced in Chinese hamster ovary (CHO) cells is composed of heterogeneous isomeric structures (glycomers) containing predominantly Man5 (~20%), Man6 (~34%), Man7 (~40%) and Man8 (~6%) glycoforms as determined by electrospray ionization mass spectrometry (ESI-MS)[23]. The glycomers present in our sample arise from trimming the Man$_8$ oligosaccharide (Fig. 3) at different branch points. The assignment of carbohydrate resonances in glycoproteins is complicated by the fact that the vast majority appear in a very small spectral width (~3.5-4.0 ppm) with consequent overlap problems. Exceptions are the anomeric ^1H signals which

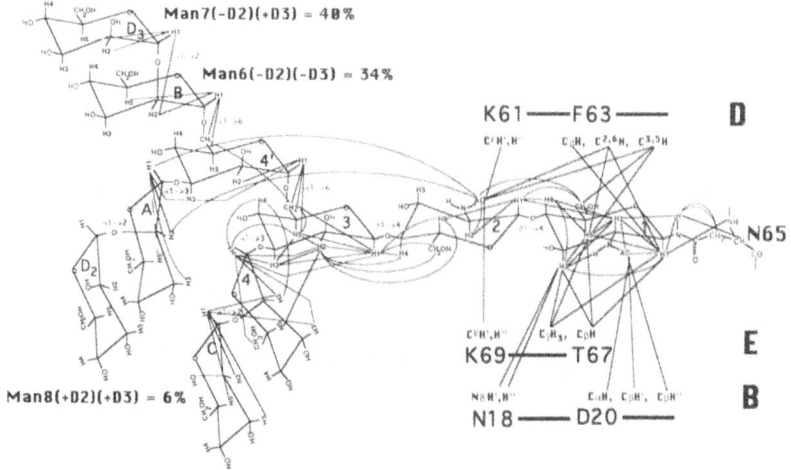

Figure 3. Structure of the N-linked glycan of hCD2. The residues that make contact with the glycan are listed as well as the corresponding strands of the β-sheet. NOEs observed are also indicated.

resonate at lower field (~4.4-5.2 ppm) due to the electron-withdrawing properties of the ring oxygen. As a consequence, the assignment strategy for oligosaccharides is usually based on these resolved anomeric ^{1}H resonances[219]. However, in glycoproteins these anomeric ^{1}H resonances are usually overlapped with H$^\alpha$ resonances from the protein and hence, the anomeric region in 2D ^{1}H-^{1}H spectra is complicated by the appearance of large numbers of cross peaks arising from the spin systems of amino acid residues, such as serines, threonines, and glycines. However, the anomeric carbon resonances of carbohydrates are well separated from all polypeptide carbon signals. Thus, our assignment strategy made extensive use of proton-carbon heteronuclear experiments at natural abundance of ^{13}C[202].

As a key experiment, we recorded ^{1}H-^{13}C HSQC spectra to identify glycan cross peaks, in particular the resonance positions of the anomeric protons. However, most of the other cross peaks of the glycan are also well resolved from polypeptide cross peaks, and reports in the literature on NMR studies of free high-mannose glycans[21] gave some clues for assignments in the glycoprotein. Next, we recorded homonuclear TOCSY experiments with different mixing times between 50 and 250 ms and analyzed the cross peaks lined up at the positions of the anomeric protons identified in the HSQC spectra. With these experiments we could establish connectivities between the anomeric protons and most other protons of a particular monosaccharide. The H^6 protons were identified by selecting for CH$_2$ groups in a DEPT-HMQC spectrum. Generally, the GlcNAc cross peaks were observed already in TOCSY spectra with short mixing times because of the relatively large H^1-H^2 coupling constants. In contrast, mannose residues have small H^1-H^2 and H^2-H^3 coupling constants, and thus the cross peaks in these residues were primarily observed in spectra with longer mixing times in which many cross peaks of the polypeptide were no longer observed due to faster transverse relaxation.

Connections between the spin systems identified with the TOCSY experiments and thus sequential assignments were obtained by recording NOESY experiments in D$_2$O. In these NOESY spectra, cross peaks were observed between the H^1 and H^4 protons of the β1->4 glycosidic bonds (GlcNAc2- GlcNAc1; Man3- GlcNAc2), between the H^1 and H^6 protons of the two α1->6 glycosidic bonds (Man4'-Man3; ManB-Man4'),between the H^1 and H^3 protons of the two α1->3 glycosidic bonds (ManA-Man4'; Man4->Man3) and between the H^1 and H^2 protons of the α1->2 glycosidic bonds (ManD3-ManB; ManD2-ManA; ManC-Man4).

A major difficulty in assigning carbohydrate resonances in glycoproteins in general is represented by the heterogeneity of the glycans. In the case of hCD2, the heterogeneity of the glycan is clearly manifested in the NMR spectra. In particular, the ^{1}H-^{13}C HSQC spectra show different sets of cross peaks for the different glycoprotein forms. Fortunately, the heterogeneity of the glycan effects primarily the ends of the carbohydrate branches. In particular, the resonances of the penultimate mannose residues ManA and ManB are doubled representing the forms with and without the terminal ManD2 and ManD3 residues, respectively.

CONFORMATION OF THE HIGH-MANNOSE GLYCAN OF hCD2

To characterize the 3D structure of the carbohydrate in human CD2 we recorded highly resolved 2D NOESY spectra on a Varian UnityPlus750 spectrometer. We identified more than 30 NOEs between the first two carbohydrate residues (GlcNAc1 and GlcNAc2) and side chains of residues on the B, D and E β-strands of the protein. Numerous intra-glycan NOEs define the conformation of the branch ending in ManC. Interestingly, we observe several NOEs between ManA and GlcNAc2 indicating that the ManA branch of the glycan is folded back to the stem of the carbohydrate. However, this is only observed for those

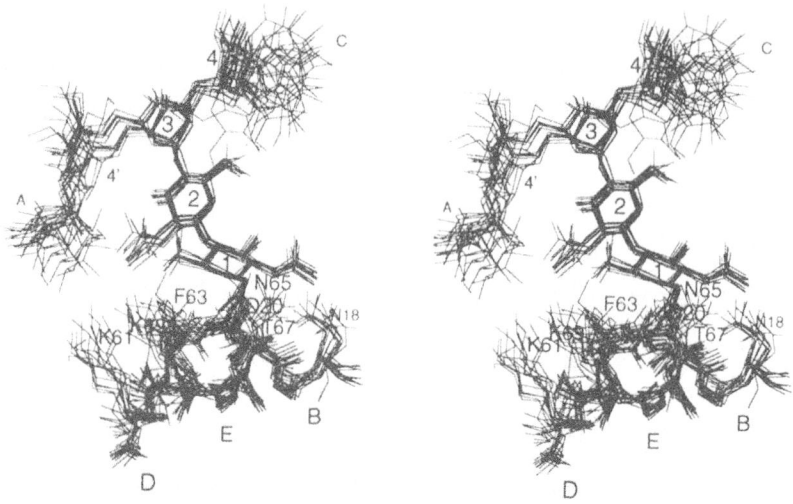

Figure 4. Stereodiagram of the carbohydrate core covering the BDE face of hCD2. Backbone atoms of strands D, E and B of the polypeptide are shown. Side chains of residues that have NOE contacts with the carbohydrate are also displayed.

glycan forms that do not have the ManD2 residue attached. Thus, the conformation of the A-branch truncated by one residue is better defined than that of the full-length branch. Fig. 4 shows an ensemble of distance geometry structures of the carbohydrate together with the B, E and D strands of the polypeptide. Details of the structure analysis of the carbohydrate and a discussion of its functional significance have been described elsewhere[22].

COMPARISON OF THE MOBILITY OF THE POLYPEPTIDE AND THE GLYCAN IN THE ADHESION DOMAIN OF hCD2

The relative mobility of the polypeptide and the glycan have been studied with measurements of ^{13}C line widths of CH groups in highly resolved ^{1}H-^{13}C HSQC spectra recorded on a Varian UnityPlus750 spectrometer. Since the glycoprotein has to be produced in CHO cells, isotope enrichment with ^{15}N and/or ^{13}C is very expensive. Thus relaxation experiments as they are now common for ^{15}N enriched proteins (for a review see for example ref[23]) are not possible. Precise measurement of the carbon line widths of methine CH groups can yield a reliable but qualitative measure of internal mobility. Here, large line widths are indicative of low mobility, small line widths represent higher mobility on a time scale faster than the overall motion of the protein. Thus, we expect that the rigid core of the protein will exhibit rather uniformly large line widths while more mobile parts (termini) should have sharper lines. Slow conformational exchange processes could be manifested in line widths that exceed the plateau values of the rigid core. Fig. 5 shows a comparison of the line widths of the C^{α} carbons of the polypeptide with those of side chain CH groups (Leu, Ile, Val) and of anomeric carbons of the glycan. The line widths of the C^{α} carbons show indeed a rather uniform plateau for the regions of regular secondary structure as indicated by the horizontal bars. The average line width for the regular secondary structures is 15.4 Hz. This is indicated as a reference in all three panels of Fig. 5. The lines are significantly sharper for the terminal regions and the loops connecting the β-strands. The average values for the loop regions and for the terminal regions are 14.5 Hz and 11.6 Hz, respectively. This correlation between structural features and C^{α} line widths assures that this method of analyzing internal

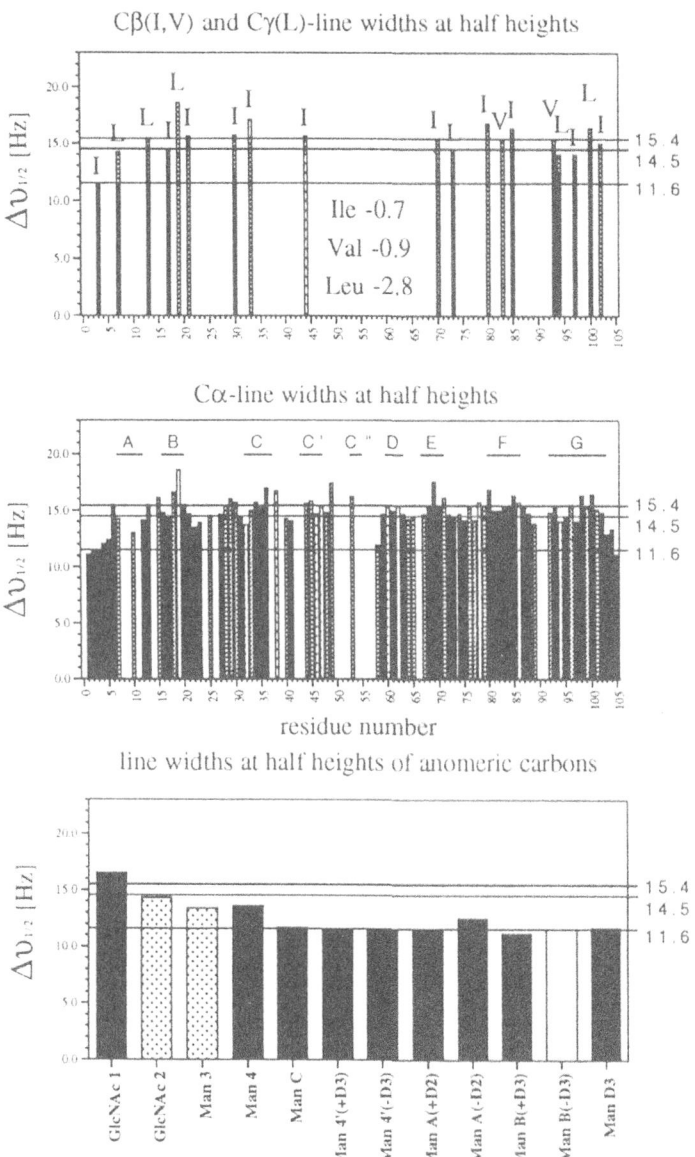

Figure 5. Carbon line widths as measured from highly resolved ^{1}H-^{13}C HSQC spectra (see text).

mobility is meaningful. There are several C^{α} resonances that have larger line widths than the average core line widths indicating that there might be slow conformational exchange. These include Q15, N18, L19, E36, T38, K49, T53, K69, K71, I80, I85, F98 and L100.

Line widths of CH groups of amino acid side chains of isoleucine, leucine and valine are shown in Fig. 5 as well and compared with the line widths of the backbone C^{α}. With the exception of I21 and I33 all side chain methine carbons have sharper lines than the C^{α} resonances indicating that the side chains are generally more mobile than the backbone. This is expected and again supports the feasibility of this method for characterizing internal mobility in proteins.

The lower panel of Fig. 5 shows the line width of the anomeric carbons of the carbohydrate. We observe that the line widths decrease from the stem towards the branches of the carbohydrate. This indicates that GlcNAc1, GlcNAc2, Man3 and Man 4 have similar mobility as the polypeptide. The outer branches of the carbohydrate have essentially a similar mobility as the terminal backbone residues of the polypeptide chain. Many side chain CH groups of leucine, valine and isoleucine residues, including many residues from the protein core have line width comparable to those of the outer branches of the glycan, in particular those of L7, L13, I30, L73 and L94.

Of particular interest is the line width of the anomeric carbon of ManA. This line width is significantly larger in the carbohydrate that lacks ManD2. This supports the observation mentioned above that loss of ManD2 leads to a fold back of the ManA branch while in the form of the carbohydrate that contains ManD2 the ManA branch is more mobile and doesn't seem to be folded back towards the core of the glycan.

FUNCTIONAL SIGNIFICANCE OF THE HIGH-MANNOSE GLYCAN IN hCD2

The adhesion domain of human CD2 requires the glycan attached to Asn65 to maintain adhesion function. In contrast, rat CD2 does not require such a carbohydrate for maintenance of its stable protein fold and adhesion function. The effect of removing the N-linked glycan has been studied with circular dichroism spectroscopy[22]. Treatment of the adhesion domain of hCD with PNGase F which removes the carbohydrate entirely results in a protein that has a CD spectrum of a random coil polypeptide. Treatment of the protein with endoglycosidase H (endo H) which removes the glycan except for the first GlcNAc destabilizes the protein without unfolding it. The CD spectrum is significantly changed but is still native-type. Thus, the carbohydrate seems to stabilize the protein conformation, and a single GlcNAc unit seems to be sufficient to keep it barely folded.

To explore what determines the necessity of a glycan in human CD2 extensive mutations were carried out in full length protein of residues near the glycan (Table 1). The mutations were also guided by comparison of rat, human, mouse and horse CD2 in the region surrounding the N-linked glycosylation site[24]. Human CD2 differs from CD2 in all the other species at positions 61 (Lys vs. Glu) and 63 (Phe vs. Leu). The level of expression of CD2 on the cell surface was measured as the ratio of mean fluorescence intensity of M32B staining of mutant CD2 versus that of wt CD2. The structural integrity of the expressed protein was tested with three monoclonal antibodies that recognize different epitopes of the adhesion domain, 8B5, 35.1 and T11$_2$. The mutations can be divided in four classes. (i) The point mutations N65Q and T67A change the amino acid sequence on the DE loop so that no carbohydrate is attached cotranslationally. These mutants do not bind CD58 and the structural integrity is lost as indicated by the loss of mAb binding. (ii) The double mutants F63L/T67D, F63A/N65Q and K64A/N65Q remove the carbo-hydrate and change single residues that are in contact with the carbohydrate in the wild type protein to alanine or leucine. None of these is active and all of them appear to have an altered structure. (iii)The mutations N18A, D20A, K61A, T67S, K69A and F63A/T67S involve residues that show NOE contacts with the glycan. None of these mutations affect CD58 binding. This indicates that there is no single contact between an amino acid side chain and the glycan that is required for maintaining the functional conformation. (iv) The triple mutations K61E/F63L/T67D and K61E/F63L/T67A remove the carbohydrate and graft the homologous residues of rat CD2 in positions 61 and 63 onto the human protein. These mutants are fully active. The crucial mutation appears to be K61E. This leads to a consistent picture of the role of the carbohydrate.

Table 1. Effects on CD2 structure and function of human CD2 point mutations at glycan contact sites (for details see ref.[17,22]). The level of expression on the cell surface and recognition by three monoclonal antibodies (8B5, 35.1 and T11$_2$) is also given

Mutation	Level of CD2 expression	8B5	35.1	T11$_2$	CD58 binding	N-glycan attached
N65Q	0.49	–	–	–	0	–
T67A	0.60	–	–	–	0	–
F63L/T67D	1.73	±	±	–	0.22	–
F63A/N65Q	0.44	–	–	–	0	–
K64A/N65Q	0.92	–	–	–	0	–
N18A	1.22	+	+	+	1.03	+
D20A	0.98	+	+	+	1.06	+
K61A	1.30	+	+	–	1.35	+
T67S	1.02	+	+	+	0.88	+
K69A	1.14	+	+	+	1.09	+
F63A/T67S	1.37	+	+	–	0.75	+
K61E/F63L/T67D	1.42	+	±	–	0.95	–
K61E/F63L/T67A	1.12	+	±	–	1.37	–

The N-glycan at N65 in the adhesion domain of human CD2 stabilizes the polypeptide conformation and thus supports adhesion functions. Human CD2 contains an accumulation of five surface exposed lysines on the DEB face of the protein. K61 sits in the center of this patch surrounded by K69, K71, K64 and K55. This grouping of positive charges is unfavorable. The glycan can stabilize the protein conformation through hydrogen bonds, van der Waals contacts and possibly entropic contributions and thus, counterbalances the clustering of the positive charges. If the central K61 is replaced with a negatively charged glutamic acid, it may form a salt bridge with K69 analogous to an interaction of the corresponding side chains which occurs naturally in rat CD2[2,4] where the protein is stable even without the carbohydrate. It thus appears that the unfavorable charge at residue 61 of human CD2 is responsible for its glycan requirement. Placing a negatively charged side chain (glutamic acid) at the position 61 eliminates the need of the stabilizing carbohydrate for keeping the protein folded. These data further show the remarkable functional differences between highly structurally related CD2 molecules among different species resulting from sequence changes during evolution. Given the striking effects of point mutations on protein stability it is clear that at least certain Ig domains such as CD2 are not rigid protein scaffolds, but rather maintain their stable conformation as a result of dynamic interactions between the polypeptide and its attached glycan. Based on hydrogen exchange NMR measurements it has previously been observed that the presence of a carbohydrate in RNase B globally stabilizes the polypeptide compared to the glycan-free RNase A[25].

ACKNOWLEDGMENTS

This work was supported by NIH (grants GM 38608 and GM 47467 to GW and AI321226 to ELR) and by a grant from Sandoz.

REFERENCES

1. Williams, AF and Barclay, AN: The immunoglobulin superfamily domains for cell surface recognition. Annu. Rev. Immunol. 6, 381-406 (1988).

2. Driscoll, PC, Cyster, JG, Campbell, ID and Williams, AF: Structure of domain 1 of rat T lymphocyte CD2 antigen. Nature, 353, 762-765 (1991).

3. Withka, JM, Wyss, DF, Wagner, G, Arulanandam, ARN, Reinherz, EL and Recny, MA: Structure of the glycosylated adhesion domain of human T lymphocyte glycoprotein CD2. Structure, 1, 69-81 (1993).

4. Jones, EY, Davis, SJ, Williams, AF, Harlos, K and Stuart, DI: Crystal structure at 2.8Å resolution of a soluble form of the cell adhesion molecule CD2. Nature, 360, 232-239, (1992).

5. Bodian, DL, Jones, EY, Harlos K, Stuart, DI and Davis, SJ. Crystal structure of the extracellular region of the human cell adhesion molecule CD2 at 2.5 Å resolution. Structure 2, 755-766 (1994).

6. deVos, AM, Ultsch, M and Kossiakoff, AA: Human growth hormone and extracellular domain of its receptor: Crystal structure of the complex. Science 255, 306-312 (1992).

7. Wang, J, Yan, Garrett, TPJ, Liu, J, Rodgers, DW, Garlick, RL, Tarr, GE, Husain, Y, Reinherz, EL and Harrison, SC: Atomic structure of a fragment of human CD4 containing two immunoglobulin-like domains. Nature 348, 411-418 (1990).

8. Ryu, SE, Kwong, PD, Truneh, A, Porter, TG, Arthos, J, Rosenberg M, Dai, X, Xuong, N, Axel, R, Sweet, RW and Hendrickson, WA: Crystal structure of an HIV- binding recombinant fragment of human CD4. Nature, 348, 419-426 (1990).

9. Brady RL, Dodson EJ, Dodson GG, Lange G, Davis SJ, Williams AF and Barclay AN: Crystal Structure of Domains 3 and 4 of Rat CD4: Relation to the NH2-Terminal Domains. Science 260, 979-983(1993).

10. Lange G, Lewis SJ, Murshudov GN, Dodson GG, Moody PCE, Turkenburg JP, Barclay AN and Brady RL: Crystal Structure of an Extracellular Fragment of the Rat CD4 Receptor Containing Domains 3 and 4. Structure 2, 469-481 (1994).

11. Recny, MA, et al., & Reinherz, EL: N-Glycosylation is required for human CD2 immunoadhesion functions. J. Biol. Chem. 267, 22428-22434 (1992).

12. Davis SJ, Davis EA, Barclay AN, Daenke S, Bodian DL, Jones EY, Stuart DI, Butters TD, Dwek RA and van der Merwe PA. Ligand bindingby the immunoglobion superfamily recognition molecule CD2 is glycosylation-independent. J. Biol. Chem. 270, 369-375 (1995).

13. Wyss, DF, Withka, JM, Knoppers, MH, Sterne, KA, Recny, MA & Wagner, G: ^{1}H resonance assignments and secondary structure of the 13.6 kDa glycosylated adhesion domain of human CD2. Biochemistry 32, 10995-11006 (1993).

14. Arulanandam, ARN, Withka, JM, Wyss, DF, Wagner, G, Kister, A, Pallai, P, Recny, MA and Reinherz, EL.: The CD58 (LFA3) binding site is a localized and highly charged surface area on the AGFCC'C"face of the human CD2. Proc. Natl. Acad. Sci. USA. 90, 11613-11617 (1993).

15. Karlsson GB, Butters TD, Dwek RA and Platt FM. Effects of the imino sugar N-butyldeoxynojirimycin on N-glycosylation of recombinant gp120. J. Biol. Chem. 268, 570-576 (1993).

16. Leahy, DJ, Axel, R and Hendrickson, WA: Crystal structure of a soluble form of the human T cell coreceptor CD8 at 2.6Å resolution. Cell 68, 1145-1162 (1992).

17. Arulanandam, ARN, Kister, A, McGregor, MJ, Wyss, DF, Wagner, G and Reinherz, EL.: Interaction between CD2 and CD58 involves the major β-sheet surface of each of their respective adhesion domains. J. Exp. Med., 180, 1861-1871 (1994).

18. Chothia, C, Novotny, Bruccoleri, R and Karplus, M: Domain association in immunoglobulin molecules. The packing of variable domains. J. Mol. Biol. 186, 651-663 (1985).

19. Homans SW: Oligosaccharide conformations: Application of NMR and energy calculations. Progr. NMR Spectrosc. 22, 55-81 (1990).

20. Wyss, DF, Choi JS, andWagner G: Composition and sequence specific resonance assignments of te heterogeneous N-linked glycan in the 13.6 kDa adhesion domain of human CD2 as determined by NMR on the intact glycoprotein. Biochemistry, 34, 1622-1634 (1995a).

21. Wormald, MR, Wooten, EW, Bazzo, R, Edge, CJ, Feinstein, A, Rademacher, TW, and Dwek, RA: The conformational effects of N-glycosylation on the tailpiece from serum IgM, Eur. J. Biochem. 198, 131-139 (1991).

22. Wyss, DF, Choi JS, Li, JL, Knoppers, MH, Willis, KJ, Arulanandam, ARN., Smolyar, A, Reinherz, EL andWagner G: Conformation and function of the N-linked Glycan in the Adhesion Domain of Human CD2. Science, 269, 1273-1278 (1995b).

23. Wagner, G: NMR relaxation and protein mobility. Current Opinion in Structural Biology, 3, 748-754 (1993).

24. Travenor, AS, Kydd, JM, Bodian, DL, Jones, EY, Stuart, DI, Davis, SI, Butcher, GW: Expression cloning of an equine T-lymphocyte glycoprotein CD2 cDNA, structure-based analysis of conserved sequence elements. Eur. J. Biochem. 219, 969-976 (1994).

25. Joao, HC, Scragg, IG, Dwek, RA: Effects of glycosylation on protein conformation and amide proton exchange rates in RNase B. FEBS Lett. 307, 343-346 (1992).

ABSTRACTS

1. PROTEIN STRUCTURES ABSTRACTS

1.1. Solution Structure of the ETS-Domain from Murine Ets-1: a Winged-Helix-Turn-Helix Motif

Logan W. Donaldson,[1] Barbara J. Graves,[2] and Lawrence P. McIntosh[1]

[1]Department of Biochemistry, University of British Columbia, Vancouver, British Columbia V6T 1Z3, Canada and [2]Department of Cellular, Viral and Molecular Biology, University of Utah School of Medicine, Salt Lake City, UT 84132

Several diverse transcription factor families employ the winged helix-turn-helix (wHTH) motif to accomplish sequence-specific DNA-binding. Situated on an anti-parallel β−sheet platform, a helix-turn-helix motif facilitates core contacts while loop-elements, termed "wings," facilitate additional contacts. We have determined the structure of a 110 residue protein fragment from murine Ets-1 containing the 85 residue ETS-domain by NMR methods and show that the ETS-domain is a new member of the wHTH class. Believed to play a regulatory role, the remaining 25 residues extending from the ETS-domain to the native C-terminus of Ets-1 adopt a helical conformation, packed anti-parallel to the first helix of the ETS-domain. In particular, the v-Ets oncoprotein sequence diverges from Ets-1 in this C-terminal region.

1.2. Assessment of Spatial Hydrophobic and Electrostatic Properties of Protein Molecules

A. P. Golovanov, R. G. Efremov and A. S. Arseniev

Shemyakin and Ovchinnikov Institute of Bioorganic Chemistry, Russian Academy of Sciences, Ul. Miklukho-Maklaya, 16/10, 117871 Moscow, V-437, Russia

A new method is proposed for analyzing NMR-derived spatial structures from the point of view of their hydrophobic and electrostatic properties. This approach enables us to find parts of a protein molecule with an "untypical" environment, which can form active sites and antigenic determinants. The role of amino acid residues in maintaining spatial structure can be displayed by analyzing two-dimensional maps of hydrophobic or hydrophilic contacts together with the information about the conservativity of amino acid residues among the homologous proteins. The method was tested on neurotoxin II (NTII) from the venom of *Naja naja oxiana* and on some other closely related neurotoxins.

The solution structure of NTII was determined by two-dimensional ^{1}H-NMR techniques. The cross-peak volumes in NOESY spectra, spin-spin coupling constants of vicinal protons and the observation of slow deuterium exchange of amide protons along with computational analysis were used to obtain a set of 19 energy minimized structures of NTII and to propose the mode of its dimerization. Environmental hydrophobicities of residues in NTII and its dimer were characterized in terms of 3D molecular hydrophobicity potential (MHP). 2D MHP maps and experimental data were used to identify residues taking part in dimerization of NTII.

The proposed approach can be used to study packing, hydrophobic and electrostatic properties of both NMR-derived and crystal protein models and to map residues important for the maintenance of 3D structure and/or binding with another molecules.

1.3. NMR Structure Determination of Bovine Angiogenin, a Ribonuclease Involved in Neovascularization

Olivier Lequin, Christine Reisdorf-Albaret, François Bontems and Jean-Yves Lallemand

Laboratoire de RMN, DCSO, Ecole Polytechnique, F-91128 Palaiseau, France

Angiogenins are 14-kDa proteins able to induce blood vessel growth in various preparations and are thought to be involved in the development of solid tumors. Sequence comparison shows significant similarities with pancreatic ribonuclease A. Indeed, angiogenins possess a ribonucleolytic activity which is critical for their angiogenic activity. In addition, these proteins are able to induce second-messenger pathways, enter endothelial cells and translocate to the nucleus. In order to better define structure-function relationships in angiogenins, it seems important to get structural data.

We have studied the solution structure of bovine angiogenin. At the same time, the crystal structure of human angiogenin has been determined (Acharya et al., 1994). Human and bovine angiogenins show 70% sequence similarity. However, the putative cell receptor binding site in bovine angiogenin possesses an RGD motif (characteristic of the recognition of proteic membrane receptors) absent in its human counterpart. This could signify that these molecules may recognize cells differently.

Since bovine angiogenin was directly purified from cow milk, it was not possible to enrich our protein with ^{15}N or ^{13}C isotopes. Extensive use of two—dimensional and three-dimensional proton NMR experiments enabled us to assign most of the proton resonances (Reisdorf et al., 1994) and to identify about 1000 NOEs. Structure calculations were performed using a procedure combining minimization in the dihedral space with the DIANA program followed by simulated annealing with the X-PLOR program. The resulting structure consists of three α–helices and an antiparallel twisted β-sheet structure with three strands on the N-terminal side and four strands on the C-terminal side.

The overall structure is very similar to that of ribonuclease A, but differs in the putative receptor binding site and in the ribonucleolytic active site. Preliminary results show that the side chain of Glu-118 is orientated toward the active site and could obstruct it, which could account for the low activity with conventional ribonuclease substrates. These results tend to confirm those of Acharya et al. reported for crystal structure of human angiogenin.

This solution structure determination opens the way to the study of the interaction of angiogenin with substrates and inhibitors. It will contribute to the analysis of divergence in the pancreatic RNase superfamily and between angiogenins.

Acharya, K. R., Shapiro, R., Allen, S. C., Riordan, J. F. and Vallee, B. L. (1994) *Proc. Natl. Acad. Sci. USA 91*, 2915-2919.

Reisdorf, C., Abergel, D., Bontems, F., Lallemand, J. Y., Decottignies, J. P. and Spik, G. (1994) *Eur. J. Biochem. 224*, 811-822.

1.4. Structural Information of M13 Major Coat Protein in Both Sodium Dodecyl Sulfate and Dodecyl Phosphocholine Micelles

C. H. M. Papavoine, R. N. H. Konings, C. W. Hilbers and F. J. M. van de Ven

NSR Center, Laboratory of Biophysical Chemistry, University of Nijmegen, Toernooiveld, 6525 Ed Nijmegen, The Netherlands

We are studying DNA binding proteins of Coliphage M13 as model systems to learn about virus infection and reproduction. The major coat protein, gene VIII protein (gVIIIp) of this phage has two conformations during its life cycle. It is a coat protein in the virus particle and it is a membrane protein. Upon phage penetration, gVIIIp is inserted into the membrane while the DNA is released onto the cytoplasm. During phage production, newly synthesized coat protein accumulates within the membrane and "picks up" DNA when the phage leaves the cell.

A combined use of solid state and high resolution NMR revealed a structure with a long hydrophobic helix spanning the membrane and a shorter amphipathic helix in the plane of the bilayer (McDonnell et al., 1993). Although considerable progress has been made during the last year, a complete three-dimensional structure has not yet been determined. In our group we have studied gVIIIp using high-resolution NMR. Small deuterated micelles are used a model systems to mimic the membrane-bound form of the coat protein. Here we present structural information on gVIIIp in both Sodium Dodecyl Sulfate (SDS) and Dodecyl PhosphoCholine (DPC) micelles. The secondary structure of gVIIIp in SDS has been determined to consist of two α-helices (van de Ven et al., 1993). Line-width measurements showed a large difference in flexibility between the two helices. The peaks from the hydrophobic residues in the first helix are broad, whereas the resonances of the other helix have much stronger peaks. We address this difference in the location of the protein relative to the micelle.

We have nearly completed the assignment of the ^{1}H, ^{15}N and ^{13}C resonances of gVIIIp in DPC micelles, which more closely resemble the membrane surface. The NMR data for gVIIIp in DPC micelles reveal a structure with two α-helices too, with the same difference in peak intensity. This is an indication that the structures of gVIIIp in SDS micelles and gVIIIp in DPC micelles are very similar. The ^{1}H$_\alpha$ and ^{13}C$_\alpha$ chemical shift deviations from random coil values of gVIIIp in both SDS and DPC also denote the close similarity between the two systems. This indicates that detergents are very good mimetics of the lipid bilayer for gVIIIp.

More detailed information on the location of gVIIIp in the micelles is gained with the use of two fatty acid spin labels, 5-doxylstearate and 16-doxylstearate, with the radical located near the headgroup and at the end of the carbon chain, respectively. The effects of both spin labels on the resonances of SDS and gVIIIp are consistent with the earlier described structure of the two helices. The spin label results also indicate that the SDS micelles are distorted around the DNA binding site of the protein (Papavoine et al., 1994). The results of gVIIIp in DPC suggest that the DPC micelles occur both as distorted and "normal" elliptical micelles.

McDonnell, P. A., Shon, K., Kim, Y. and Opella, S. J. (1993) *J. Mol. Biol. 233*, 447-463.

Papavoine, C. H. M., Konings, R. N. H., Hilbers, C. W. and van de Ven, F. J. M.(1994) *Biochemistry 33*, 12990-12997.
van de Ven, F. J. M., van Os, J. W. M., Aelen, J. M. A., Wijmenga, S. S., Remerowski, M. L., Konings, R. N. H. and Hilbers, C. W. (1993) *Biochemistry 32*, 8322-8328.

1.5. Structure and Stability of Titin Modules

S. Improta, C. Joseph, A. Pastore, M. Pfuhl and A. Politou

EMBL, Meyerhofstr. 1, D-69012 Heidelberg, Germany

Titin is a giant muscle protein of ca. 3000 kDa (i.e. the largest known protein) located in the thin and in the thick filament of striated muscles where it is thought to contribute to assembly and maintenance of the sarcomere (Maruyama et al., 1981; Trinick et al., 1984; Wang, 1985; Fuerst et al., 1988). The putative functions of titin range from interactions with myosin or other muscle proteins to providing an elastic component which maintains the thick filament in register and resists overstretching of the sarcomere (Horowits et al., 1989; Granzier and Wang, 1993; Maruyama et al., 1994).

The recent determination of the titin sequence shows a complex modular architecture (Labeit et al., manuscript in preparation). Most of the sequence is assembled by ca. 100 amino acid repeat modules, type I and type II, which have been predicted to belong to the fibronectin type III and to the immunoglobulin super-families respectively (Labeit et al., 1990). These two motifs are assembled in super-repeats in the center of the titin sequence (located in the A-band). A protein kinase is present in the C-terminus of the protein, near the M-line in the sarcomere. It is flanked by type II modules and unique insertions which do not find homology in the database (Gautel et al., 1993). The type I domains disappear also around the I-band, the region thought to account for the elastic behavior of the whole protein.

Because of this modularity, we have started a long-term plan to study the molecular basis of the properties of titin by dissecting it into units small enough to be structurally characterized in full detail. We determined the structure of two type II modules using multi-dimensional Nuclear Magnetic Resonance techniques. One is in the C-terminus of the titin molecule located around the M-line, the other is toward the N-terminus approximately located in the I-band. The two modules were selected as representative of two sequence sub-families obtained from the complete multiple alignment of type II modules. Our data provide the first structure determination of titin modules. Both domains belong to the I fold but with differences mainly noticeable in some of the loops. The high sequence similarity between titin modules will allow them to be used as molecular templates to model most if not all of the other immunoglobulin-like modules.

Starting from these structures, the stability of several other type II modules was investigated by measuring key thermodynamic parameters for thermal and chemical denaturation and by monitoring amide proton exchange as a function of time (Politou et al., 1994). Despite the overall structural similarity, the stability of modules from different regions of the titin molecule varies considerably. This might derive from distinct mechanical properties of domains located in different regions of the sarcomere or might simply reflect the influence of function-specific sequence consensus corresponding to the specific sarcomere origin. As structural analysis of more titin domains and their interactions with the surrounding is proceeding, these and related observations are expected to establish clear-cut structure/function relationships and to unveil the exact cellular role of this protein.

Fuerst, D. O., Osborn, M., Nave, R. and Weber, K. (1988) *J. Cell Biol. 106*, 1563-1572.
Gautel, M., Leonard, K. and Labeit, S. (1993) *EMBO* J. *12:10*, 3827-3834.
Granzier, H. L. M. and Wang, K. (1993) *Biophys J. 65*, 2141-2159.

Horowits, R., Maruyama, K. and Podolsky, R. J. (1989) *J. Cell Biol. 109*, 2169-2176.

Labeit, S., Barlow, D. P., Gautel, M., Gibson, T., Holt, J., Hsieh, C.-L., Francke, U., Leonard, K., Wardale, J., Whiting, A. and Trinick, J. (1990) *Nature 345*, 273-276.

Maruyama, K., Kimura, S., Ohashi, K. and Kuwano, Y. (1981) *J. Biochem. 89*, 701-709.

Maruyama, K. et al. (1994) *Biophys. Chem. 50*, 73.

Politou, A. S., Gautel, M., Pfuhl, M., Labeit, S. and Pastore, A. (1994) *Biochemistry 33*, 4730-4737.

Trinick, J., Knight, P. and Whiting, A. (1984) *J. Mol. Biol. 180*, 331-356.

Wang, K. (1985) *Cell and Muscle Mobility 6*, 315.

1.6. NMR Order Parameters Calculated by Normal Mode Analysis and Molecular Dynamics Simulation

Shinji Sunada and Nobuhiro Gō

Department of Chemistry, Faculty of Science, Kyoto University, Kyoto 606-01, Japan

We have developed the method of calculating a generalized order parameter (S^2) by normal mode analysis (NMA) in dihedral angle space (DAS). When an S^2 that includes effects from inter-nuclear distance fluctuations is calculated by NMA, both in DAS and in Cartesian coordinate space (CCS), it is necessary to expand the distance (r) in a Taylor series in the displacement (Δ) of the vector connecting two nuclei. To obtain a second order expression of S^2, Henry and Szabo (1985) expanded r up to the first order (not second!) in Δ, because they recognized that a proper expansion up to the second order causes unphysical results in the limiting case where the distance does not change. They used NMA in CCS, in which motions of each atom in normal modes are linear in space and it causes bond length changes (this is an unphysical result). However, we show that the formula of S^2 obtained by the second order expansion of r works in NMA in DAS, where curved motions of each atom in space are attained without leading to the unphysical results (Sunada and Go, 1995).

Second, to study how internal motions of a protein affect the S^2 value, molecular dynamics simulation (MD) and NMA of eglin-c (a proteinase inhibitor) were carried out. To define collective motions in water, principal component analysis was applied to the MD trajectory (Kitao and Go, 1990). Interpretations of S^2 values for N-H and H-H spin pairs in terms of normal modes and principal components will be presented.

Henry, E. R. and Szabo, A. (1985) *J. Chem. Phys. 82*, 4753.

Kitao, A. and Go, N. (1990) *Chem. Phys. 158*, 447.

Sunada, S. and Go, N. (1995) *J. Comp. Chem.*, in press.

1.7. Investigation of the Motions of Two Very Different Proteins by ^{15}N NMR

Carine van Heijenoort and Eric Guittet

CNRS, Institut de Chimie des Substances Naturelles, Laboratoire de RMN, 19 190 Gif sur Yvette, France

The analysis of ^{15}N spin relaxation enables a mapping of fast scale motions of the backbone of proteins in the picosecond range. Knowledge of these motions allows the correlation and interpretation of the root mean square deviations observed in the reconstructed 3D structures and brings information about local flexibilities along the sequence.

The information about motions theoretically available from ^{15}N NMR is the spectral density $J(\omega)$ at the five frequencies of the NH spin system: 0, ω_N, ω_H, and $\omega_H \pm \omega_N$ (Abragam,

1961). It has been shown that the evaluation of these five spectral densities requires the measurement of six relaxation parameters (Peng and Wagner, 1992). The main problem is that only three of these parameters (^{15}N T1, T2 and steady state heteronuclear NOE) can be measured precisely enough to be reliable. Approximations have then to be done on spectral densities to be able to use NMR data.

From now on, almost studies were based on the direct analysis of the three ^{15}N NMR relaxation parameters using the Lipari and Szabo formalism (Lipari and Szabo, 1982). This approach, however, doesn't take into account the potential existence of slow motions (microsecond range) in proteins. A less reducing approach is to calculate the three spectral densities $J(0)$, $J(\omega_N)$ and $J(\omega_H)$ from the three available NMR relaxation parameters, as proposed by Wagner and coworkers (Peng and Wagner, 1992), and then interpret these spectral densities in terms of motion.

The results of these approaches will be discussed for two very different proteins: capsicein, a 98 residue α–elicitin (Bouaziz et al., 1994) and fruR, a 60 residue DNA binding domain fragment of *Eschericia coli* fru transcriptional regulator (Scarabel et al., 1995). FruR shows a dynamical behavior compatible with the Lipari and Szabo model. On the contrary, this model is unable to account for the NMR data obtained on capsicein. The analysis of the spectral densities of this protein actually indicates the presence of slow time scale motions along the whole protein (van Heijenoort et al., 1994).

Abragam, A. (1961) *The Principles of Nuclear Magnetism*, Clarendon Press, Oxford.

Bouaziz, S., Huet, J-C. and Pernollet, J-C. (1994) *Biochemistry 33*, 8188-8197.

Lipari, G. and Szabo, A.(1982) *J. Am. Chem. Soc. 104*, 4546.

Peng, J. and Wagner, G. (1992) *J. Magn. Reson. 98*, 308.

Scarabel, M., Penin, F., Bonod-Bidaud, C., Negre, D., Cozzone, A. J. and Cortay, J.-F. (1995) *Gene*, in press.

van Heijenoort, C., Bouaziz, S. and Guittet, E. (1994) *J. Chim. Phys. 91*, 776-784.

2. PROTEIN INTERACTIONS ABSTRACTS

2.1. Designing Mutant Hemoglobins

Marcela Madrid,[1] H. Kim,[2] and Chien Ho[2]

[1]Pittsburgh Supercomputing Center, Pittsburgh, PA 15213, USA and [2]Department of Biological Sciences, Carnegie Mellon University, Pittsburgh, PA 15213

We have used molecular dynamics simulations to design hemoglobin molecules with specific oxygen binding properties. The molecules were then prepared by means of site directed mutagenesis techniques. Structural information, obtained by NMR, was correlated with the observed functions.

We will describe two designed molecules: In the first case, normal function was partially restored to an abnormal hemoglobin. In the second case, the hemoglobin molecule was re-designed to have low oxygen binding properties, an important step towards the use of hemoglobin as a blood substitute, replacing donated blood.

Our studies show that molecular dynamics and NMR can be used to design molecules of desired properties and to gain insight into the structure-function relationship. Properties of the novel molecules will be described, with important potential applications in the treatment of abnormal hemoglobins and in the design of blood substitutes.

2.2. Thermodynamic and Crystallographic Analysis of the Binding of Ligands within an Internal Cavity of T4 Lysozyme

Andrew Morton and Brian W. Matthews

Institute of Molecular Biology, Eugene, OR 97403-1229

To better understand the role of shape complementarity and hydrophobicity in protein-ligand interactions, we have assayed the binding of 91 different compounds to a cavity created in the core of the T4 lysozyme by site directed mutagenesis. The cavity is able to discriminate between different ligands on the basis of shape, polarity and volume. The binding energetics of 16 ligands were determined by titration calorimetry and analyzed by dividing the reaction into three processes: desolvation, immobilization and packing. While all three processes contribute significantly to binding specificity, a subset of ligands was found that interact with the protein in a manner very similar to the way in which they interact with nonpolar liquids. Crystallographic analysis of 10 of the protein-ligand complexes revealed the structural basis for the observed binding specificity. The most important aspect of the protein's response to binding was seen to be the dynamic behavior of a particular helix which forms part of the internal binding site. The flexibility of this helix, which appears to be induced upon ligand binding, allows ligands which have a particular type of shape to bind, while ligands of different shapes are not able to take advantage of this particular flexibility and are unable to bind. The implications of this behavior for ligand design and core packing algorithms will be discussed.

2.3. The Leucine Zippers of the HLH-LZ Proteins Max and c-Myc Preferentially form Hetero-Dimers

C. Muhle-Goll, M. Nilges, and A. Pastore

European Molecular Biology Laboratory, 69012 Heidelberg, Germany

c-Myc and Max are members of a sub-family of the basic-helix-loop-helix family. Their function is mediated by switches in the dimerization partners: c-Myc does not homo-dimerize *in vivo*, but competes with Mad, another member of the subfamily, to form hetero-dimers with Max leading to either activation or repression of transcription. Max is able to also form homo-dimers. In an attempt to identify which regions of the proteins carry the information to determine specific recognition of the dimerization partner, we have investigated the dimerization properties of synthetic peptides corresponding to the leucine zipper sequence of Max and c-Myc using circular dichroism and nuclear magnetic resonance techniques. These leucine zippers (denoted as Max-LZ and Myc-LZ) differ from those of the basic-leucine-zipper family in two important respects. First, the leucine zipper constitutes a second dimerization domain in addition to the helix-loop-helix motif, second, more of the positions typically occupied by hydrophobic residues are taken by polar or charged amino acids. We show that the hetero-dimer Max-LZ/Myc-LZ is obtained easily by simply mixing the peptides, and that it is more stable than the homo-dimer of the Max leucine zipper at neutral pH. We have shown in a recent paper (Muhle-Goll et al., 1994) that the leucine zipper of c-Myc does not form stable homo dimers under these conditions. We show further that hetero-dimerization induces a conformational change in the peptides. Thus, the leucine zipper regions of these two proteins by themselves display the same behavior as the entire protein. However, even the hetero-dimer is less stable than dimers of leucine zippers of the basic-leucine-zipper family, such as GCN_4 and fos-Jun. The specificity of the interaction between different monomers can be explained by polar interactions. We investigate the structural role of the polar and charged residues in the hydrophobic interface by molecular modeling studies.

Muhle-Goll, C., Gibson, T., Schuck, P., Schubert, D., Nalis, D., Nilges, M. and Pastore, A. (1994) *Biochemistry* 33, 11296-11306.

2.4. Protein:Protein Interactions Studied by NMR: Does Cytochrome *c* Bind to Plastocyanin on its Acidic Patch?

M. Ubbink,[a] X.-S. Gong,[b] J. C. Gray,[b] and D. S. Bendall[a]

[a]Department of Biochemistry and [b]Department of Plant Sciences, Cambridge University, Tennis Court Rd., Cambridge CB2 1QW, United Kingdom

Plastocyanin is a type I copper protein that shuttles electrons from cytochrome *f* to photosystem I in chloroplasts. Two pathways have been suggested for the electron transfer from the cytochrome *f* haem to the plastocyanin copper ion: via copper ligand His87, located in the 'hydrophobic patch' or via Tyr83 and ligand Cys84 at the 'acidic patch' of plastocyanin. Kinetic studies with wild type and mutant plastocyanin (e.g. He et al., 1991; Modi et al., 1992) have suggested that electron transfer proceeds via the acidic patch route, not only in the reaction with the natural redox partner cytochrome *f*, but also with the non-physiological partner, cytochrome *c*.

We are performing ^{1}H-NMR studies in order to obtain direct evidence for the site of binding of cytochromes *c* and *f* on the plastocyanin surface. In this work Cd-plastocyanin is used as a substitute of Cu(II)-plastocyanin because the latter is paramagnetic. The resulting Cd-plastocyanin is redox inactive so the effects of binding of both oxidized and reduced cytochromes can be studied. Proton NMR assignments for nearly all protons of Cu(I)- as well as Cd-plastocyanin have been obtained (D. S. Bendall, S. Modi and M. Ubbink, unpublished results).

As a first approach, binding is studied by analyzing the effects of cytochrome *c* on the chemical shifts of plastocyanin protons. Using the proton assignments, circa 500 peaks were identified in a NOESY spectrum of pea Cd-plastocyanin. Then, NOESY spectra of cytochrome *c* and of an equimolar mixture of Cd-plastocyanin and cytochrome *c* were obtained. This was done both with oxidized and with reduced cytochrome *c*. The effects of complex formation on the chemical shifts of the plastocyanin protons are being analyzed. The combination of the software package AZARA for NMR data processing and the program ANSIG (Kraulis, 1989) proves to be very useful for this purpose. Subtraction of 2D spectra is straightforward and overlaying of spectra can be done accurately; therefore even small shifts (0.01 ppm) can be detected. The results of this chemical shift analysis will be presented. Other approaches to characterize the properties of the cytochrome:plastocyanin complexes may include relaxation and amide exchange studies.

He, S., Modi, S., Bendall, D. S. and Gray, J. C. (1991) *EMBO J. 10*, 4011.
Kraulis, P. J. (1989) *J. Magn. Reson. 84*, 627
Modi, S., He, S., Gray, J. C. and Bendall, D. S. (1992) *Biochim. Biophys. Acta 1101*, 64.

3. NUCLEIC ACIDS AND NUCLEIC ACID-PROTEIN INTERACTIONS ABSTRACTS

3.1. A Study on the Dynamics of a DNA Binding Protein

L. M. Horstink, R. N. H. Konings, C. W. Hilbers, and F. J. M. van de Ven

NSR Center, Laboratory of Biophysical Chemistry, University of Nijmegen, Toernooiveld 1, 6525 ED Nijmegen, The Netherlands

We use the gene 5 protein of bacteriophage M13 as a model system to study the relation between protein dynamics and DNA binding. This gene 5 protein binds aspecifically to 4 consecutive bases of single-stranded DNA. The protein occurs as a dimer of two equivalent subunits of 87 amino acid residues which both contain a DNA binding loop. The 3D structure of this protein has been elucidated previously in our laboratory. These studies suggested an enhanced flexibility of the DNA binding wing of the protein.

Here we will present deuterium exchange and relaxation data for the free protein to further substantiate these results. Furthermore, the influence of DNA binding on these properties has been investigated. The relaxation experiments have been carried out using pulsed field gradients; the water magnetization was put back to its equilibrium position prior to acquisition. A comparison of the results obtained with and without DNA binding is used to obtain insight into the role of the flexibility of the DNA binding wing.

Dayie, K. T. and Wagner, G. (1994) *J. Magn. Reson. 111A*, 121-126.
Folkers, P. J. M., Nilges, M., Folmer, R. H. A., Konings, R. N. H. and Hilbers, C. W. (1994) *J. Mol. Biol. 236*, 229-246.
Kay, L. E., Nicholson, L. K., Delaglio, F., Bax, A. and Torchia, D. A. (1992) *J. Magn. Reson. 97*, 359-375.
Stonehouse, J., Shaw, G. L., Keeler, J. and Laue, E. D. (1994) *J. Magn. Reson. 107A*, 178-184.

3.2. What's a Polar Residue Doing at a Hydrophobic Dimer Interface? Asn in the Jun-Jun Leucine Zipper

Joel P. Mackay,[#] F. Keith Junius,[#] Sean I. O'Donoghue,[†] and Glenn F. King[#]

[#]Department of Biochemistry, University of Sydney, NSW 2006, Australia and [†]EMBL, Heidelburg, Germany

Jun and Fos are oncoproteins (*i.e.*, cancer-causing proteins when expressed at a different level or in an altered form). They function as nuclear transcription factors—both the Jun–Jun homodimer and the Jun–Fos heterodimer bind to specific DNA sequences, initiating the transcription of different genes. During tumorogenesis, elevated levels of Jun and Fos are found in the affected cells. We have been seeking to elucidate the factors which govern dimerisation affinity and specificity with a view to the eventual design of novel and specific anti-tumor agents. We have cloned and overexpressed the dimerisation domain of Jun (JunLZ, a 46-residue leucine zipper sequence which forms a coiled coil when dimerised), and determined the solution structure of the 92-residue coiled coil Jun homodimer using high-resolution NMR.

^{1}H-^{15}N HSQC spectra of the Jun homodimer recorded at different temperatures reveal an unusual feature—a pair of Asn residues (one from each half of the dimer) at the otherwise hydrophobic dimer interface, which appear to be involved in a motional averaging process on the millisecond timescale. It appears that this interaction destabilizes the dimer, as mutagenesis of the Asn to Val markedly increases the thermal stability of the dimer. Since this Asn residue is not present in the Fos leucine zipper domain, we have proposed that it may serve to mediate homo- vs. heterodimerisation affinities.

In an attempt to characterize this Asn–Asn interaction in more detail, we have measured ^{15}N NMR relaxation parameters (T_1, T_2, {^{1}H}–^{15}N NOE) for the backbone amide nitrogens of the uniformly ^{15}N-labeled JunLZ homodimer. No indication is seen, however, of different motional behavior at or near the interfacial Asn residue, compared to the bulk of the protein. The exchange process therefore appears to be localized to the Asn sidechain. A model is proposed which accounts for the observed averaging process, and is consistent with asymmetry previously observed in the crystal structure of the homologous leucine zipper homodimer, GCN4 (O'shea et al., 1991).

O'shea, E. K., Klemm, J. D., Kim, P. S. and Alber, T. (1991) *Science 254*, 539.

3.3. Hetero-TOCSY-Based Experiments for Measuring Heteronuclear Relaxation in Nucleic Acids and Proteins

Barry I. Schweitzer, Kevin H. Gardner, and Gregory Tucker-Kellogg

Walt Disney Memorial Cancer Institute, Florida Hospital, Orlando, FL 32826

While both ^{31}P and ^{113}Cd are present at locations of interest in many different macromolecular systems, heteronuclear-detected relaxation measurements on these nuclei have been restrained by limitations in either resolution or signal-to-noise. In the present study, we have developed heteroTOCSY-based methods to overcome both of these problems and thus to facilitate the study of heteronuclear relaxation rates on systems containing either of these nuclei. 2-D versions of these experiments were utilized to measure ^{31}P T_1 and T_2 values in DNA oligonucleotides; the additional resolution offered by a second dimension allowed determination of these values for most of the ^{31}P resonances in a DNA dodecamer. We used this data to ascertain that the terminal portions of the helix have higher mobility than internal portions, in agreement with previous studies using other methods. ^{31}P relaxation measurements also suggested that incorporation of the nucleoside analog cytosine arabinoside into a DNA dodecamer causes a reduction in backbone mobility. One-dimensional, frequency-selective versions of these experiments were also developed for use on systems containing a smaller number of heteronuclear spins. These methods were applied to investigate the heteronuclear relaxation properties of ^{113}Cd in ^{113}Cd(2)LAC9(61), a Cys(6)Zn(2) DNA binding domain. Data from these experiments confirms biochemical evidence that there are more significant differences in the metal-protein interactions between the two metal-binding sites than has been previously identified for proteins containing this motif. These demonstrations of the utility of heteroTOCSY-based relaxation measurements suggest that these methods may be more generally applicable to other systems.

4. POSTER ABSTRACTS

4.1. NMR Cross-Relaxation in Ubiquitin Investigated by Molecular Dynamics Simulation

Roger Abseher, Susanna Lüdemann, Hellfried Schreiber, and Othmar Steinhauser

Institute for Theoretical Chemistry, University of Vienna, Währingerstrasse 17, A-1090 Wien, Austria

In the theory of spin-lattice relaxation the dipolar correlation function of the NOE distance vector plays a key role. In many cases this correlation function can be separated with good approximation into two parts accounting for elongation and reorientation, respectively (Brüschweiler et al., 1992; Abseher et al., 1994; R. Abseher, S. Lüdemann, H. Schreiber and O. Steinhauser, *manuscript submitted*). This separation allows a decomposition of the cross-relaxation rate constant σ into contributions from overall rotation, distance fluctuations and internal rotation. These "partial cross-relaxation rate constants" allow a direct assessment of the validity of the "rigid-body assumption" for individual NOEs. Using the correlation function data extracted from a 1 nanosecond molecular dynamics simulations of solvated ubiquitin, a globular protein with 76 residues, partial cross-relaxation rate constants have been determined for more than 150 NOEs. The rigid-body assumption cannot be verified in general. The implications for the calibration of NOE distances have been investigated.

Abseher, R., Lüdemann, S., Schreiber, H. and Steinhauser, O. (1994) *J. Am. Chem. Soc. 116*, 4006-4018.
Brüschweiler, R., Roux, B., Blackledge, M., Griesinger, C., Karplus, M. and Ernst, R. R. (1992) *J. Am. Chem. Soc. 114*, 2289-2302.

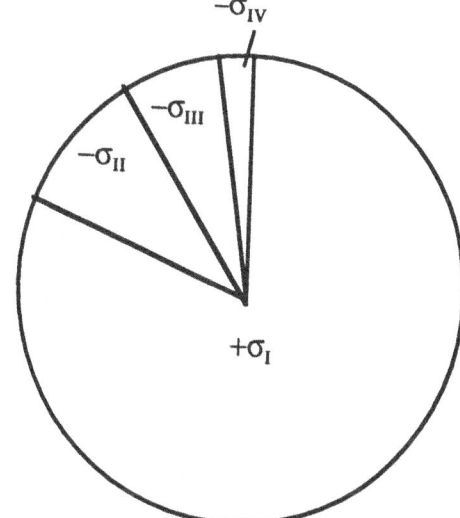

Figure 1. Partial cross-relaxation rate constants for $d_{\gamma\gamma'}^{VV}(17, 26)$ in ubiquitin. σ_I originates from overall rotation scales by the order parameters of radial and angular motion. σ_{II} collects correlations times of internal rotation, σ_{III} collects correlation times of distance fluctuation and σ_{IV} is a "second order" correction term.

4.2. Some Aspects of Molecular Modeling–NMR Interplay

M. Baginski and E. Borowski

Department of Pharmaceutical Technology and Biochemistry, Technical University of Gdansk, Narutowicza St 11/12, 80-952 Gdansk, Poland

Molecular modeling and experimental NMR measurements can give new quality of information in conformational analysis. Amphotericin B (AmB) is a well-known polyene macrolide antibiotic widely used in the treatment of systemic fungal infection. The mode of action of this antibiotic is still under study, but until now it has been accepted that the drug binds to the membrane sterols and causes impairment of their barrier function and loss of cell constituents through membrane ionic channels.

In order to elucidate the molecular level of action of AmB, theoretical studies were undertaken in addition to experimental efforts. These studies concerned the molecular structure of AmB. The crucial role for the overall shape of AmB depends on the mutual position of two semirigid fragments: macronolidic and glycosidic. Previously the structure of AmB was studied by x-ray and NMR measurements.

Our present work is devoted to an advanced conformational analysis of AmB performed by molecular mechanics methods. The results of this analysis (i.e. conformational 2D or population maps) were compared with NMR data, especially of NOE effects. We constructed similar 2D maps taking into account distances between hydrogen atoms from different semirigid fragments of molecule, for which NOE effects were observed by NMR. We have found that by comparing these 2D maps (from molecular modeling and those based on NMR data) it is possible to find more precisely which conformers are populated, i.e. are observed by NMR. Thus, the linkage of two approaches gave us extra knowledge.

4.3. Structure and Folding-Kinetics of Bovine α-Lactalbumin, Determined with NMR-Spectroscopy

J. Balbach, V. Forge, and C. M. Dobson

New Chemistry Laboratory, University of Oxford, South Parks Road, Oxford OX1 3QT, UK

Bovine α-Lactalbumin (BLA) is a 'modifier' protein in mammary glands, which promotes the synthesis of milk lactose through modification of the substrate specificity of galactosyl transferase. NMR assignments and regions of secondary structure were determined with a ^{15}N labeled sample at 35°C and pH 5 using standard techniques. The study of BLA folding is interesting, because it forms well-defined equilibrium intermediates under different conditions. The timescale of the refolding of BLA is also highly dependent on the Ca^{2+} concentration. Therefore it has been possible to probe the refolding of different domains of BLA with NMR. The results have been related to the measured kinetic parameter from optical techniques including CD- and fluorescence spectroscopy.

4.4. Conformational Studies of Saccharides by Selective ¹H NMR Measurements at Different Magnetic Fields and Molecular Dynamics Calculations

J. Bella,[1] M. Bosco,[1] R. Toffanin,[1] S. Miertu,[1,2] and S. Paoletti[1,3]

[1]POLY-bios Research Center, Padriciano 99, I-34012 Trieste, Italy; [2]International Institute for Chemistry, ICS, UNIDO, Area Science Park, Padriciano 99, I-34012 Trieste, Italy and [3]Istituto di Chimica Biologica, Università di Sassari, I-07100 Sassari, Italy

Solution conformation and internal motion of saccharides have been the subject of considerable interest in recent years. Several different approaches have been followed to obtain time-averaged geometrical parameters such as internuclear distances (r_{ij}), dihedral angles (ϕ/ψ) and correlation times (τ_c). ¹H and ¹³C relaxation experiments and measurements of long range J_{H-C} coupling constants are the most useful methods for obtaining the geometrical parameters. Usually these data are combined with molecular mechanics (MM) and molecular dynamics (MD) calculations.

In the present study we describe an approach based on selective 1D NOESY, 1D ROESY and proton selective T1 experiments carried out at different magnetic fields to derive internuclear distances and correlation times without any reference distances. This approach is compared to the conventional approach that uses 1D NOESY at only one field and needs reference distances. The experimental data for neutral and pyruvilated neoagaro-oligosaccharides are compared with those obtained by different molecular mechanics (MM) and molecular dynamics (MD) calculations to get information on the most probable minimum energy conformations and on the flexibility/rigidity of the glycosidic linkages. Moreover, the data from the quenched molecular dynamics (QMD) are used to calculate time-averaged interglycosidic distances and long range J(H-C) coupling constants to find a model describing the behavior of glycosidic linkages.

4.5. Structural Studies of Ras p21 and its Interactions

L. Ben-Tovim,[a] B. Smith,[a] F. Luh,[a] E. D. Laue,[a] Y. Ito,[b] T. Terada,[b] S. Yokoyama,[b] P. N. Lowe,[c] E. Jacquet,[d] and A. Parmeggiani[d]

[a]Department of Biochemistry, University of Cambridge, Tennis Court Road, Cambridge CB4 2QW, UK; [b]Biodesign Research Group, The Institute of Physical and Chemical Research, 2-1 Hirosawa, Wako-shi, Saitama 351-01, Japan; [c]Department of Cell Biology, Wellcome Research Labs, Langley Court, Beckenham BR3 3BS, Kent, UK and [d]Laboratoire de Biochemie, Ecole Polytechnique, 91128 Palaiseau Cedex, France

Ras p21 is a key regulator of intra-cellular signaling. It is a member of a ubiquitous eukaryotic gene family, which encode small GTPase proteins which are highly conserved in evolution. In the last few years, remarkable progress has been made, such that the pathway that Ras controls can now be mapped from changes that occur at the cell surface to alterations in gene expression in the nucleus. Ras couples these events by acting as a molecular switch. It cycles between the biologically active GTP form and the inactive GDP form; it has an intrinsic GTPase activity. Two classes of molecule act directly on Ras:

i. GTPase activating proteins (GAPs) - which promote the hydrolysis of GTP and increase the rate of inactivation

ii. guanine nucleotide exchange factors (GEFs) - which promote dissociation of GDP, allowing GTP to bind and thereby activate it.

When activated, Ras interacts, for example, with the protein kinase Raf-1 which associates with MAP kinase kinase eventually leading to phosphorylation of transcription factors in the nucleus.

At present there is no structural information available as to how Ras interacts with proteins modulating its activity (GAPs, GEFs) or with its downstream effectors such as Raf-1. In addition despite extensive crystallographic studies the mechanism of GTP hydrolysis and the accompanying structural changes are still unclear. The Ras on/off switch occurs via changes in conformation between finely balanced states, which are likely to be affected by crystal packing forces. Therefore detailed studies in solution using NMR spectroscopy would be useful for investigating the structure and dynamics of Ras, for providing insight into the mechanism of GTP hydrolysis, and for studying its interactions. So far we have the structure and dynamics of the GDP form of Ras p21 obtained by three and four dimensional NMR spectroscopy. These studies showed two flexible regions in solution, which may be altered on binding GTP. It is these regions which I hope to see in the Ras complexes as they become frozen in a particular conformation. The work involves:

1. Studies of the interactions of GEFs with Rasp21.GDP. We have the assignments and solution structure of the GDP form.
2. Assignment of 3D NMR spectra of $^{13}C/^{15}N$ labeled Ras p21 bound to the slowly hydrolyzed GTP analogue GMPPNP. This is the first step towards studies of GAP and Raf-1 interactions since they associate with Ras.GTP. We also have the corresponding spectra of the Ras p21.GMPPNP.Raf-1 complex.
3. Once assigned, studies of the structure and dynamics of Ras p21.GMPPNP, and its complexes with other proteins, in order to understand how this key intra-cellular regulator is able to recognize and bind specifically to such a diverse range of proteins.
4. Development of new techniques for studies of large (by NMR standards) molecules.

4.6. NMR Studies of the Citotoxic Ribonuclease α-Sarcin

R. Campos-Olivas, J. Santoro, M. Bruix, J. Lacadena, A. Martinez del Pozo, J. G. Gavilanes, and M. Rico

Instituto de Estructura de la Materia, CSIC, Calle Serrano 119, E-28006 Madrid, Spain

The ribonuclease α-sarcin from *Aspergillas giganteus* is an attractive system for studying the bases of biomacromolecular recognition. The citotoxicity of α-sarcin lies in its ability to cross cell membranes to inactivate the ribosome by hydrolyzing a strictly conserved sequence in the largest rRNA. Therefore, it interacts with membrane lipids and hydrolyzes RNA with high specificity. Interestingly, α-sarcin (150 amino acids) is larger than the rest of the known extracellular ribonucleases and shares a limited sequence identity with them. We have undertaken the determination of the solution structure and the study of the dynamics of α-sarcin by multidimensional heteronuclear NMR. Our goal is to establish the differences observed at the 3D level and to interpret its function in structural and dynamical terms. It is our working hypothesis that the larger size and higher mobility of α-sarcin would account for its interaction with lipids, not shown by other, shorter ribonucleases.

We recently succeeded in expressing and purifying uniformly ^{15}N-labeled α-sarcine. High quality homo- and heteronuclear bidimensional spectra have been recorded, and we are currently in the process of assigning the ^{1}H and ^{15}N resonances. Given the size of this protein and its a.a. composition (13 Pro, 13 Gly, ...), together with the high degeneracy of aromatic resonances observed, the assignment of α-sarcin with conventional bidimensional techniques as well as alternative, specific approaches to simplify the problem are now being applied. A few stretches of residues have already been assigned and we are hopeful that we can present most of the assignment and the secondary structure of α-sarcin in this communication.

4.7. Determination of Deoxyribose Conformations in Large DNA Duplexes: Combined NOE and Coupling Constant Analysis

Maria R. Conte,[1] Christopher J. Bauer,[2] and Andrew N. Lane[1]

[1]Laboratory of Molecular Structure and [2]MRC Biomedical NMR Centre, National Institute for Medical Research, London, UK

The conformations of the sugars in nucleic acids can in principle be determined from the spin-spin couplings (van Wijk et al., 1992). However, in larger DNA fragments where the line-widths are comparable to the coupling constants and there is severe spectral overlap, insufficient accurate values are available for a complete analysis. To obtain accurate coupling constants requires simulations of the spin system using the minimal number of approximations and assumptions. In addition, coupling constants can be supplemented with NOEs for the H4'-H1' and H4'-H2" vectors.

We have analyzed data from DQF-COSY, P.E. COSY, NOESY and TOCSY for the 16 bp ATF-2 recognition site d(CATGTGACGTCACATG)$_2$, acquired at different temperatures to examine the effects of correlation time and linewidths on the determination of coupling constants. In addition, we have determined the intrinsic relaxation rates by measuring the rotating frame T_1 values (equivalent to T_2) using a novel two-dimensional experiment (T1RHOSY), which revealed substantial differences in relaxation rates for different nucleotides. These values have been used as starting points for simulation of cross-peak shapes in DQF-COSY and P.E. COSY experiments in the program Gamma (Smith et al., 1994).

Detailed analysis of the apparent $\Sigma_{1'}$ and $\Sigma_{2'}$ values extracted from cross-sections under different conditions and about the ^{3}J couplings from P.E. COSY spectra, highlights the strong effects of the sugar proton linewidth on the coupling measurements. At 25°C, coupling constants were severely underestimated, and highly variable between different kinds of experiments. Consistency was obtained only for data acquired at higher temperatures (40 to 50°C). The simulations, in which the linewidths were also adjusted, showed that improved results can be obtained and that simultaneous fitting of NOE and coupling data by extensive searching yields improved and more reliable estimates of the sugar conformations.

Smith, S. A., Levante, T. O., Meier, B. H. and Ernst, R. R. (1994) *J. Magn. Reson. 106*, 75-105.
van Wijk, J., Huckriede, B. D., Ippel, J. H. and Altona, C. (1992) *Methods Enzymol. 211*, 286-306.

4.8. Conformational Analysis of Cyclic Hexapeptides by NMR and Molecular Dynamics: Ensemble Averaging and Peptide Flexibility

Philippe Cuniasse and Vincent Dive

C. E. A., Département d'Etude et d'Ingénierie des Proteines, C. E. Saclay, 91191, Gif sur Yvettes Cedex, France

The 3D NMR solution structure of biomolecules is based on the determination of the cross relaxation rate (σnOe/rOe) related to interproton distances and coupling constants (J) related to torsion angles. Most of the strategies used to define the structure of these compounds rely on the basic hypothesis that the molecule is rigid. However, these data are collected as time- and ensemble-averaged quantities. In the case of flexible molecules, the rigid model may lead to meaningless or virtual structures.

In the context of the development of zinc metalloprotease inhibitors, the conformational analysis of a series of cyclic pseudohexapeptides have been undertaken by analysis of the cross relaxation rates in rotating frame nuclear Overhauser experiments (σnOe). The quantitative analysis of these data suggests that several conformations may interconvert on the NMR time-scale. Thus we have developed a simulated annealing protocol which uses the NMR data as ensemble-averages restraints.

Under the hypothesis that the time of interconversion between the different conformations involved is slow as compared to the overall correlation time of the molecule, the σnOe between protons i and j depends on the ensemble averaged quantity r_{ij}^{-6}®. In that protocol, both distance and anti-distance restraints can be taken into account. Further, the observed coupling constants are treated as ensemble-averaged values J_i® of the individual conformations. The J_i ($J_{NH-H}\alpha$ coupling constants related to the φ angles and $J_{H\alpha-H\beta}$ related to the χ_1 angles) are calculated from Karplus type relations.

To evaluate this protocol, it has been applied to a cyclic hexapeptide (Gly-Pro-Phe-Gly-Pro-Ile). The resulting ensemble gives a better agreement with the experimental data (rOe and J) as compared to the standard approach. Further, the analysis of the different conformations obtained by the ensemble-averaging approach suggests that the conformational space accessible to this compound in solution is larger than the one obtained by the standard approach. In particular, several conformations are obtained for the reverse turns. We have investigated the possibility of determining homonuclear spectral densities to validate this protocol.

4.9. Structure Determination of the Native and Deglycosylated α-Subunit of Human Chorionic Gonadotropin

Tonny de Beer,[1] Carol W. E. M. van Zuylen,[1] Kalle Hård,[2] Bas Leeflang,[1,2] Rolf Boelens,[2] Robert Kaptein,[2] Johannis P. Kamerling,[1] and Johannes F. G. Vliegenthart[1]

Departments of [1]Bio-Organic Chemistry and [2]NMR Spectroscopy, Bijvoet Center for Bimolecular Research, Utrecht University, Padualaan 8, NL-3584 CH Utrecht, The Netherlands

The glycoprotein hormone human Chorionic Gonadotropin (hCG) consists of two non-identical, non-covalently linked subunits (α and β). The α–subunit is N-glycosylated at both Asn-52 and Asn-78, and particularly glycosylation at Asn-52 seems to be essential for the hormonal function; although deglycosylation at this site does not affect receptor binding of the dimer, subsequent stimulation of the cAMP-cycle is abolished (Matzuk et al.,

1989). This suggests that the carbohydrate chains themselves interact with the hCG receptor and/or that the presence of the N-glycans affects the conformation of the α–subunit. This conformational change might affect the mode of binding of the α–subunit to the β–subunit or of the hormone to the receptor, resulting in the inability to stimulate the cAMP-cycle. In order to understand the functional role of the N-linked carbohydrate chains, we are investigating conformational differences between α–hCG that is fully deglycosylated (van Zuylen et al., 1995) and fully glycosylated α–hCG. Using homonuclear (^{1}H) 2D COSY, NOESY, TOCSY, 3D TOCSY-NOESY and gradient-enhanced natural abundance 2D ^{1}H-^{13}C correlation spectroscopy, an almost complete assignment of the ^{1}H chemical shifts was achieved for both protein and carbohydrate resonances. Preliminary structure calculations based on our NOE data show partial homology with the X-ray structure of the α–subunit in the αβ–dimer (Lapthorn et al., 1994). Furthermore, deglycosylation does not seem to have a dramatic effect on protein conformation. However, profound differences in chemical shifts, linewidths and NOE patterns are observed for the two proximal GlnNAc residues at the glycosylation sites Asn-52 and Asn-78.

Lapthorn, A. J., Harris, D. C., Littlejohn, A., Lustbader, J. W., Canfield, R. E., Machin, K. J., Morgan, F. J. and Isaacs, N. W. (1994) *Nature 369*, 455-461.

Matzuk, M. M., Keene, J. L. and Boime, I. (1989) *J. Biol. Chem. 264*, 2409-2414.

van Zuylen, C. W. E. M., et al. (1995) *Eur. J. Biochem.*, in press.

4.10. Use of Heteronuclear NMR to Determine the 3D Structure of *E. coli* Initiation Factor 3

C. Garcia,[a] P. L. Fortier,[a] J. Y. Lallemand,[a] F. Dardel,[b] and S. Blanquet[b]

[a]Laboratoire de Synthèse Organique and [b]Laboratoire de Biochemie, Ecole Polytechnique, 91128 Palaiseau Cedex, France

Initiation factor 3 (IF3) is a 180 residue basic protein which plays an essential role in the initiation of protein translation in *E. coli* by binding to the 30S ribosomal subunit and proofreading the interaction between initiator tRNA[fmet] anticodon stem and loop and the AUG initiation codon.

In order to study the three dimensional structure of this large protein, we have produced doubly enriched (^{13}C-^{15}N) IF3. The concentration of the protein sample was not possible to increase above 0.8 mM. For this reason, 3D heteronuclear NMR experiments (e.g. HNCA, HOHAHA-HMQC, NOESY-HSQC) were time consuming and difficult to realize. Observation of the 2D NOESY spectrum of IF3 led us to explore another line of research. IN fact, the severe cross peak overlaps observed in some regions of the spectrum led us to expect the existence of a non-structured part in the protein.

A limited proteolysis cleavage of IF3 revealed a highly segmented structure with two independent domains of roughly equal size (IF3 N- and C-terminal fragments) which retained their spatial folding (Fortier et al., 1994). The isolation of both these domains has opened the way to their structural study by NMR. These two domains were overproduced in a recombinant system and uniformly ^{15}N enriched to overcome spectral overlap of the proton resonances. A combination of 3D heteronuclear NMR experiments allowed the identification of spin systems with standard ^{1}H sequential assignment methods and led to the determination of the secondary structure. We have determined from the NMR data the tertiary structure of these two domains of IF3 using X-PLOR software. We showed that the N-terminal domain contains a five-strand antiparallel β sheet with two type II turns and two α-helices (Garcia et al., 1995). The C-terminal domain is defined by a two-strand parallel and two-strand antiparallel β sheet with one type II turn and two parallel α-helices.

The structural data obtained for these two domains will be very useful to determine the spatial structure of the native IF3 and could give insights into how this protein specifically recognizes domains of the 16S ribosomal RNA.

Fortier, P. L., Garcia, C. and Dardel, F. (1994) *Biochimie 76*, 376-383.
Garcia, C., Fortier, P. L., Blanquet, S., Lallemand, J. Y. and Dardel, F. (1995) *Eur. J. Biochem. 228*, 395-402.

4.11. Structure, Dynamics and Engineering of a Double Strand β–sheet. An NMR and Molecular Dynamics Study

M. Guenneugues and B. Gilquin

Département d'Ingénierie et d'Etudes des Protéines, CE-Saclay, 91191 Gif-sur-Yvette Cedex, France

Toxin α is a 61 amino acid residue curaremimetic protein. Its overall folding consists of three major loops arising from a region cross-linked by four disulfide bridges. The central loop, loop II -folded in a β–sheet structure, contains on one face 7 of the 11 residues that have been shown to be essential for the binding to the nicotinic acetylcholine receptor (nAChR). Moreover, loop II, when isolated from the toxin, binds to the nAChR with a low affinity. Constraining this peptide to adopt a β–sheet structure should improve its affinity. A chimera was thus synthesized by introduction of the residues of loop II in the β–sheet of charybdotoxin - a small protein that comprises a β–sheet coupled to an α helix through three disulfide bridges. The solution structure of the chimera exhibits a β–sheet backbone close to the loop II in toxin α (rms 1.36Å on C,N,Ca atoms) and the well defined side-chains adopt conformations close to those in toxin α. However, this chimera binds to the nAChR with a very low affinity.

The structure determination protocol used was based on distances derived from NOEs interpreted in a rigid isotropic model. As it was unsatisfactory to explain the different biological activity of this β–sheet, we decided to characterize its dynamical properties in the two structural contexts: flanked by a β–strand on both sides in the toxin α or coupled to an α helix in the chimera.

In certain conditions, off-resonance ROESY experiments permit measurement of the values of the spectral densities at various frequencies for protons submitted to dipolar correlation. These values contain both structural and dynamical information. They were measured for toxin α and analyzed. These data indicate that the conformational space accessible to some of the side-chains is larger that the one obtained by a crude analysis of the set of 3D structures determined previously. The spectral densities were calculated from a trajectory and compared to the experimental values.

This work was partly supported by a fellowship from the Institut de Formation Superieure Biomedicale.

4.12. Calcium Binding of EGF Modules

Karin Julenius,[#] Maria Sunnerhagen,[#] Goran Carlstrom,[#] Johan Stenflo,[*] and Torbjorn Drakenberg[#]

[#]Department of Physical Chemistry 2, University of Lund, Chemical Center, P O B 124, S-221 00 Lund, Sweden and *Department of Clinical Chemistry, University of Lund, Malmo General Hospital, S-214 01 Malmo, Sweden.

Many of the mosaic proteins involved in blood coagulation, anticoagulation and fibrinolysis contain EGF-like modules, of which many are Ca^{2+}-binding. In many instances impaired calcium binding of these modules leads to bleeding disorders. Although a consensus sequence for the calcium binding is known, the reasons for the large differences in calcium affinities are not well understood. The three-dimensional structure of the calcium form of the N-terminal EGF module of coagulation factor X has previously been determined using NMR spectroscopy (Selander-Sunnerhagen et al., 1992). Whereas protein S, an anticoagulant protein, binds calcium with nM affinities, EGF modules from factor IX and X bind with only mM affinities. A statistical analysis of the residue occurrence in the calcium binding sites has shown a correlation between a hydrophobic residue in position 47 and a negatively charged one in position 49. While protein S shows this feature in three of its calcium sites, factor IX and X do not. We have made mutations in the calcium site of factor X in order to stepwise change the sequence to more resemble that of protein S. Residues Gly47 and Gln49 were mutated to Ile and Glu respectively. Single mutations did not significantly affect the calcium binding, while the double mutant showed a tenfold increase in calcium affinity. A detailed study of the double mutant is being pursued using high resolution NMR in order to determine the effects of the mutations on the structure.

Selander-Sunnerhagen, M., Ullner, M., Persson, E., Teleman, O., Stenflo, J. and Drakenberg, T. (1992) *J. Biol. Chem. 267*, 19642-19649.

4.13. Hydrogen-Deuterium Exchange in Native Human Carbonic Anhydrase I

Annika Kjellsson,[1] Bengt-Harald Jonsson,[1] and Ingmar Sethson[2]

[1]Department of Biochemistry and [2]Organic Chemistry, Umeå University, S-901 85 Umeå, Sweden

Hydrogen-deuterium exchange has been used to measure backbone amide proton exchange rates in carbonic anhydrase (HCA I). We have produced uniformly [15]N- and ^{13}C-labeled HCA I as well as HCA I enzyme-labeled with various isotopically enriched amino acid residues. 3D NMR experiments where the amide resonances (^{1}H/^{15}N) are correlated with C_α- and C_β-resonances of both the same and preceeding residue have been recorded (Muhandiram and Kay, 1994). With the aid of these experiments, the work with assignments is presently proceeding rapidly and we have successfully assigned 95% of all NH resonances. Using these assignments, we have performed a series of NH⊃ND exchange experiments at pH 7.6, 8.3, 8.8 and 9.5. Analysis of exchange rates in terms of the free energy required for transient opening of hydrogen bonds at NH will be presented.

Muhandiram, D. S. and Kay, L. E. (1994) *J. Magn. Reson., Ser. B 103*, 203.

4.14. Vicinal Coupling Constants and Motion: Experiment and Theory

George J. Maalouf

Rowland Institute, 100 Edwin H. Land Boulevard, Cambridge, MA 02142

Insights to motion are essential for an understanding of life at the molecular level. The well-known Karplus equation relates vicinal (three bond) coupling constants to torsion angles. These coupling constants have been used to determine dihedral angles by means of a static picture. However, vicinal coupling constants also contain information about the distribution of dihedral angles. Assuming a Gaussian model for the distribution, the theory of averaged coupling constants is presented. This theory provides a simple method for estimating both the central dihedral angle and the range of motion.

First, this theory is applied to synthetic data obtained from molecular dynamics simulations and the results compared with known quantities from the synthetic system. Next, very precise vicinal coupling constants are obtained for small experimental systems by making use of an extension to the Hellmann-Feynman theorem for the simulation and fitting of NMR spectra. The theory of averaged vicinal coupling constants is then applied to these experimentally determined coupling constants and the width of the distributions is estimated.

4.15. Solution Structures of Parathyroid Hormone Fragments

U. C. Marx,[1] P. Bayer,[1] S. Austermann,[2] K. Adermann,[2] W.-G. Forssmann,[2] and P. Rösch[1]

[1]Lehrstuhl für Struktur und Chemie der Biopolymere und Bayreuther Institut für Makro-molekülforschung, Universität Bayreuth, 95440 Bayreuth, Germany and [2]Niedersächsisches Institut für Peptidforschung, 30625 Hanover, Germany

Human parathyroid hormone (hPTH) is known to contain several functionally distinct domains. Two functional domains, a receptor domain and a cAMP-cyclase activating domain, are required to maintain normocalcemia. A third domain is responsible for the cAMP-independent signal transduction pathway for stimulation of DNA synthesis in chondrocytes (Schlüter et al., 1989) and osteoblasts (Sömjen et al., 1991). These functionally active domains of the 84 amino acid hormone are located in the NH_2-terminal part of the protein (Sömjen et al., 1991; Potts et al., 1982).

We carried out structural analysis of PTH(1-37) which was found to be the naturally occurring form extractable from human blood. It shows higher cAMP-generation activity than shorter fragments (Forssmann et al., 1993). Furthermore, we studied the structures of hPTH(1-34) and bovine PTH(1-37).

The structures of these hPTH fragments were investigated in aqueous buffer solution under near physiological conditions - pH 6.0 (50 mM potassium phosphate buffer) and salt concentration 270 mM - by NMR spectroscopy followed by molecular dynamics calculations. We found partial helical structures in all cases, even without addition of helix stabilizing agents such as trifluoroethanol (Marx et al., 1995).

Forssmann, W.-G., Schulz-Knappe, P., Meyer, M., Adermann, K., Forssmann, K., Hock, D. and Aoki, A. (1993) In: *Peptide Chemistry 1992, Proceedings of the 2nd Japan Symposium on Peptide Chemistry* (N. Yanaihara, ed.), ESCOM, Leiden, pp. 553-557.

Marx, U. C., Austermann, S., Bayer, P., Adermann, K., Ejchart, A., Sticht, H., Walter, S., Schmid, F.-X., Jaenicke, R., Forssmann, W.-G. and Rösch, P. (1995) *J. Biol. Chem.*, manuscript submitted.

Potts, J. T. Jr., Kronenberg, H. M. and Rosenblatt, M. (1982) *Adv. Prot. Chem. 35*, 323-396.

Schlüter, K.-D., Hellstern, H., Wingender, E. and Mayer, H. (1989) *J. Biol. Chem. 264*, 11087-11092.

Sömjen, D., Schlüter, K.-D., Wingender, E., Mayer, H. and Kaye, A. M. (1991) *Biochem. J. 277*, 863-868.

4.16. Determination of the 3D Structure of the (12-53)NCP7:DNA Double Strand Complex by ¹H NMR in Solution and Molecular Modeling

S. Mellac, H. de Rocquigny, N. Morellet, C. Auclair, and B. P. Roques

Laboratoire de Pharmacochimie Moléculaire et Structurale, INSERM U266, CNRS URA D 1500, UFR de Sciences Pharmaceutiques et Biologiques, Faculté de Pharmacie, 4, rue de l'Observatoire, 75270 Paris Cedex 06, France

Mature NCp7 human immunodeficiency virus type 1 (HIV-1) is a 72 amino acids basic protein which contains two CCHC retroviral zinc finger domains characterized by a high binding affinity. *In vitro*, NCp7 was shown to promote oligonucleotide dimerization and annealing activities, two critical processes for viral infectivity.

The tridimensional structure of NCp7 has recently been elucidated by 600 MHz ¹H NMR and distance geometry calculations. It is characterized by a kink at the L-P31 residue in the interfinger sequence RAPRKKG, inducing the spatial proximity of F16 and W37 of each zinc finger. Moreover, the interactions of (12-53)NCp7 with oligonucleotide single strands, TAATTAAT, ACGCC and AUUUUU have been studied. They imply the interaction of amino acid W37 and the hydrophobic residues V13, F16, T24, A25, Q45 and M46.

We are currently determining the structure of the NCp7:DNA double strand complex, which occurred during integration of the viral DNA. This self-complementary oligonucleotide ACTGCACT adopts a B-DNA conformation. Fluorescence experiments have shown that about two molecules of (12-53)NCp7 interact with an affinity of c.a. 10^7 with the duplex. Moreover, we have also done UV experiments indicating that the T_m of the oligonucleotide decreased in the complex compared to the isolated duplex. It seems that the NCp7 destabilized the double helical structure. These results are analyzed in detail by 600 MHz ¹H NMR: the chemical shifts of F16 and W37 residues show strong shielding ($0.2 < \Delta\delta < 0.5$ppm). Correlations are observed between protons of these residues and protons of the duplex.

The NMR experiments which allow us to determine the accurate structure of the complex by molecular modeling are in progress.

4.17. The KH Module Folds into an αβ Structure

Giovanna Musco, Maria Antonietta Castiglione Morelli,* Gunter Stier, Toby Gibson, Catherine Joseph, Gilles Travé, and Annalisa Pastore

EMBL, Meyerhofstr. 1, W-69012 Heidelberg, Germany and *Universitá della Basilicata, Potenza, Italy

The recently described KH domain (for hnRNP-K homologous domain) is one of four motifs (RNP, dsRBD, RGG/RGY and KH) which frequently recur in proteins associated with RNA, often in multiple copies (for a review see Burd and Dreyfuss, 1994). FMR1, which contains two KH domains (Gibson et al., 1993; Siomi et al., 1993) is of particular interest since mutations of the gene causes fragile X syndrome, the most frequent cause of mental retardation in humans, affecting 1 in 1250 males and 30% of female carriers (Gustavson et al., 1986; Webb et al., 1986; Mandel and Heitz, 1993). The KH domain is present in many proteins in large numbers: 8 in yeast scp160 and 14 in vigilin, the highest number so far (Schmidt et al., 1992; Wintersberger et al., 1993). The only common property shared by proteins belonging to the KH family is that they are all involved with RNA. Several are also known to bind to RNA *in vitro* or were isolated from RNP complexes. The definitive role of the KH domain is yet to be defined, although the distribution strongly suggests that it will bind RNA itself. In the case of FMR1, there is now evidence that the KH domains do indeed modulate RNA-binding affinity (Siomi et al., 1992).

The determination of the 3D structure of the KH motif would greatly benefit biochemical studies on whether and how it binds to RNA and inspire experiments to elucidate the functional role of the proteins containing this domain. Structural information may also give new insight into the molecular details of fragile X. However, no direct experimental evidence has been reported so far about the structure of the KH. As a first step toward the elucidation of the 3D structure of the KH module, we have produced genetically engineered fragments containing the KH motif and examined them by NMR spectroscopy. The sequence was selected from vigilin, which has a particularly clear subunit structure making it especially suitable for the selection of individual domains and for studying the precise definition of the domain boundaries. NMR spectroscopy was used to determine the secondary structure. Almost complete assignments were obtained for the ^{1}H and ^{15}N resonances using uniform ^{15}N-labeling of the protein combined with homonuclear 2D ^{1}H NMR and 3D ^{15}N correlated ^{1}H NMR. On the basis of NOE patterns, secondary chemical shifts and amide solvent exposure, the secondary structure consists of an antiparallel three stranded β sheet connected by two helical regions. This domain may also be stabilized by an appended C-terminal helix which is common to many, but not all, members of the KH family. From preliminary calculations on the tertiary structure, we have observed that the first helix is approximately perpendicular to the β–sheet and the C-terminal helix folds back and packs into the KH.

Burd, C. G. and Dreyfuss, G. (1994) *Science 265*, 615-621.
Gibson, T. J., Rice, P. M., Thompson, J. D. and Heringa, J. (1993) *Trends Biochem. Sci. 18*, 331-333.
Gustavson, K. H., Blomquist, H. and Holmgren, G. (1986) *Am. J. Med. Genet. 23*, 581-588.
Mandel, J. L. and Heitz, D. (1992) *Curr. Opin. Genet. Dev. 2*, 422-430.
Schmidt, C., Henkel, B., Poeschl, E., Zorbas, H., Puschke, W. E., Gloe, T. R. and Mueller, P. K. (1992) *Eur. J. Biochem. 206*, 625-634.
Siomi, H., Choi, M., Siomi, M. C., Nussbaum, R. L. and Dreyfuss, G. (1994) *Cell 77*, 33-39.
Siomi, H., Siomi, M. C., Nussbaum, R. L. and Dreyfuss, G. (1993) *Cell 74*, 291-298.
Webb, T. P., Bundey, S. E., Thake, A. I. and Todd, J. (1986) *Am. J. Med. Genet. 23*, 573-580.
Wintersberger, U., Kuchne, C. and Karwan, A. (1993) EMBL Database Acc. No. X65645.

4.18. Three-Dimensional Structure of Ectatomin in Aqueous Solution Determined from NMR Data

Dmitry E. Nolde, Alexander G. Sobol, Kirill A. Pluzhnikov, and Alexander S. Arseniev

Shemyakin and Ovchinnikov Institute of Bioorganic Chemistry, Russian Academy of Sciences, Ul. Miklukho-Maklaya 16/10, Moscow 117871, Russia

Two-dimensional ^{1}H-NMR techniques were used to determine the spatial structure of ectatomin, a toxic principle of the venom of the ant *Ectatomma tuberculatum*. Nearly complete proton resonance assignments for two chains of ectatomin (37 and 34 amino acid residues, respectively) were obtained using TOCSY, DQF-COSY and NOESY experiments. The spatial structure of the ectatomin in aqueous solution was determined by a distance geometry approach based on accurately quantified distance constraints obtained using the MARDIGRAS algorithm. Three disulfide bonds were located by analysis of the global fold of ectatomin calculated using NMR data. Spatial structures of two chains of the ectatomin are similar. Each chain consists of two antiparallel α-helices connected by a hinge region of

4 residues, and its hairpin structures are stabilized by disulfide bridges. Hinge regions of two chains are connected by a third disulfide bridge. Thus, ectatomin forms four α-helical bundle structure.

It has been demonstrated that ectatomin can form a membrane pore or ion channel. The presence of four amphipathic α-helices in the ectatomin structure suggests that ectatomin may be able to incorporate into a membrane bilayer without an appreciable change to its secondary structure. Several models for the ectatomin folding into a membrane bilayer were proposed based on the spatial structure of the ectatomin in aqueous solution.

4.19. NMR Studies of the Tetrameric Mnt Repressor, a β-Sheet DNA Binding Protein

I. Nooren,[1] M. Burgering[1], K. Knight,[2] R. Kaptein,[1] and R. Boelens[1]

[1]Bijvoet Center for Biomolecular Research, Padualaan 8, Utrecht University 3584 CH Utrecht, The Netherlands and [2]Department of Biology, Massachusetts Institute of Technology, Cambridge, MA 02139, USA

The Mnt repressor of *Salmonella* bacteriophage P22 forms, together with P22 Arc repressor and the *E. coli* MetJ repressor, a new family of DNA binding protein, which use an anti-parallel β–sheet as the DNA recognition element (Raumann et al., 1994).

The Mnt repressor is a tetrameric protein of 82 residues per monomer with a total molecular weight of 37 kDa. By deletion of the last 6 residues, the truncated Mnt (1-76) forms a dimer in solution. A well refined structure of Mnt (1-76) was determined using distance geometry and restrained simulated annealing calculations (Burgering et al., 1994). The N-terminal part of Mnt (residues 1-44), which shows a 40% sequence homology with the Arc repressor, has a similar secondary and tertiary structure, a DNA binding β–sheet and two helices. Mnt (1-76) continues with a loop region of irregular structure, a third α–helix and a random coil peptide. It was found that the carboxy-terminal third helix is less stable than the remainder of the protein. This C-terminal part probably becomes stabilized in the tetrameric Mnt wild-type.

By triple resonance experiments, ^{1}H, ^{15}N and ^{13}C NMR assignments have been obtained for about 70% of the Mnt wild-type. A comparison of the assignments with those of Mnt (1-76) shows that the largest differences occur in or near the third helix. Also, it was found that a number of NMR resonances of atoms in this region of the sequence and in the β–sheet are doubled. While the dimeric Mnt (1-76) is fully symmetric, the symmetry of the Mnt wild-type seems to be lower. This is consistent with a model of the structure of the tetrameric Mnt repressor that was built by superimposing the first 44 residues of two Mnt (1-76) dimers on the structure of the Arc repressor-operator complex, where two Arc dimers are bound in two successive major grooves. The observed doubling of resonances can be explained by a similar dimer of dimers arranged as in the Arc complex and resulting in two different environments for the third helix where one of them approaches the β–sheet. The later implies a role of the C-terminal part in the DNA-binding, which was also found in biochemical studies (Knight and Sauer, 1988).

NMR studies on a peptide which consists of the last 36 residues (47-82) of Mnt (this includes the third helix) shows similar doubling of resonances. According to CD studies, the peptide packs as tetrameric helices at high concentrations and probably forms a good model for the protein-protein interactions in the Mnt repressor. Also, similar NOE-contacts can be identified in the NOESY spectra of this peptide and those of the Mnt wild-type. Furthermore, the NMR spectra of the peptide and the Mnt wild-type revealed an important role of Tyr 78, which is involved in a strong hydrogen-bond and might be an essential determinant of tetramer formation by Mnt repressor (Knight and Sauer, 1988).

Burgering, M. J. M., Boelens, R., Gilbert, D. E., Breg, J. N., Knight, K. L., Sauer, R. T. and Kaptein, R. (1994) *Biochemistry 33*, 15036-5045.
Knight, K. L. and Sauer, R. T. (1988) *Biochemistry 27*, 2088-2094.
Raumann, B. E., Brown, B. M. and Sauer, R. T. (1994) *Curr. Opin. Struct. Biol. 4*, 36-43.

4.20. Preparation of Isotope Labeled Bradykinin and NMR Studies of Peptide-Fab-Fragment Interactions

H. Ottleben,[a] M. Haasemann,[b] M. Görlach,[a] Oliver Ohlenschläger,[a] W. Müller-Esterl,[c] and L. R. Brown[a]

[a]Institut für Molekulare Biotechnologie, Postfach 100813, 07708 Jena, Germany; [b]Institut Jacques Monod, 2 Place Jussieu, 75251 Paris Cedex, France and [c]Johannes Gutenberg Universität, 550099 Mainz, Germany

The linear nonapeptide bradykinin (BK, Arg-Pro-Pro-Gly-Phe-Ser-Pro-Phe-Arg) displays a variety of functions in a number of physiological and pathophysiological processes, e.g., pain, hyperanalgesia and asthma.

Information about the structure of BK bound to its receptor is of substantial theoretical and practical importance. Spectroscopic data provide evidence that BK possesses a high degree of conformational freedom in solution and forms a β-turn at the C-terminus in DMSO and SDS- and LPC-micelles (Young and Hicks, 1994). Since it has been shown that the conformation and peptides bound to their receptors can be altered compared to the unbound peptide (Wüthrich et al., 1991), we are involved in determining the structure of BK bound to the 55 kD MAb F_{ab}-Fragment of an antibody which acts as a surrogate receptor (Haasemann et al., 1991). This approach was found to be successful in cases where the native receptor is not amenable to structural studies (Garcia et al., 1992).

We are carrying out NMR experiments to determine the conformation of receptor bound BK. Our approach relies on recording NMR data for bradykinin, without observing NMR signals from the F_{ab}-fragment. Therefore, we produced completely isotope labeled BK. We have expressed part of an artificial kininogen gene in a fusion-protein-system and can release BK specifically cleaved by the action of kallikrein, the plasma protease which liberates bradykinin from its kininogen precursor in humans. By this means we facilitate purification and prevent degradation of the recombinant peptide. Initial NMR results will be presented.

Garcia, K. C., Ronco, P. M., Veroust, P. J., Brünger, A. T. and Amzel, L. M. (1992) *Science 257*, 502-507.
Haasemann, M., Buschko, J., Faussner, A., Roscher, A. A., Hoebecke, J., Burch, R. M. and Müller-Esterl, W. (1991) *J. Immunol. 147*, 3882-3892.
Wüthrich, K., Freyberg, B. v., Weber, C., Wider, G., Traber, R., Widmer, H. and Braun, W. (1991) *Science 254*, 953-955.
Young, J. K. and Hicks, R. P. (1994) *Biopolymers 34*, 6111-623.

4.21. Modeling Solvation Contributions to Conformational Free Energy Changes of Biomolecules Using a Potential of Mean Force Expansion

Matteo Pellegrini[1] and Sebastian Doniach[1,2]

[1]Departments of Physics and [2]Applied Physics, Stanford University, Stanford, CA 94305-4090

The standard Free Energy Perturbation (FEP) techniques for the calculation of conformational free energy changes of a solvated biomolecule involve long Molecular Dynamics (MD) simulations. We have developed a method for performing the same calculations many orders of magnitude faster. We model the average solvent density around a solute as the product of the relevant solute-solvent correlation functions (CF), following the work of A. Garcia and Hummer and Soumpasis (1994a & 1994b). We calculate the CF's by running Monte Carlo simulations of a single solute atom in a box of explicit water molecules and also angular dependent CF's for selected pairs of solute atoms. We then build the water shell around a larger solute (e.g. alanine dipeptide) by taking the product of the appropriate CF's. Using FEP techniques we are able to calculate free energy changes as we rotate the dihedral angles of the alanine dipeptide and we find they are in close agreement with the MD results. We also compute the potential of mean force as a function of distance between two solvated methanes and calculate the contribution of the solvent to the free energy change that results from rotating n-butane about its dihedral angle.

Hummer, G. and Soumpasis, D. M. (1994a) *Phys. Rev. E 49*, 591-596.
Hummer, G. and Soumpasis, D. M. (1994b) In: *Structural Biology: The State of the Art*, (eds., H. S. Ramaswamy and M. H. Sarma), Adenine Press, p 273-278.

4.22. Approaches to the Structure of Mitochondrial Porins by Means of CD-Spectroscopy and Functional Analysis of Deletion Mutants

B. Popp,[1] S. Gebauer,[1] D. Court,[2] R. Lill,[2] and R. Benz[1]

[1]Lehrstuhl für Biotechnologie Theodor-Boveri-Institut (Biozentrum) der Universität Würzburg, Am Hubland, D-97074 Würzburg, Germany and [2]Institut für Physiologische Chemie, Physikalische Biochemie und Zellbiologie der Universität München, Goethestraße 33, D-80336 München, Germany

The mitochondrial porin (also called VDAC, voltage-dependent anion-selective channel) is a large diffusion pore in the outer mitochondrial membrane. (For a recent review see Benz, 1994).

To get more insight into the structure and the relation between structure and function of the mitochondrial porin, we used several biophysical methods. The secondary structure of isolated and membrane inserted porin is investigated by means of CD-spectroscopy. Current models of the mitochondrial porin were supported by our CD-data. They propose a cylinder made from amphipathic β–barrels that forms the channel. Only an about 20 amino acid stretch at the N-terminus of this relatively hydrophilic membrane protein is presumably α–helical.

In addition, we investigated several deletion mutants of the *Neurospora crassa* mitochondrial porin in the black lipid bilayer system. For this purpose we developed a method to reconstitute overexpressed mitochondrial porin from inclusion-bodies. In the form of these inclusion bodies, the mitochondrial porin and its mutants can be obtained easily and in high amounts by lac-controlled expression in *E. coli* of the porin gene fused to a His-tag. We succeeded in solubilizing and purifying the porin from the inclusion bodies. The isolated porin could, after the addition of sterol, be used for the investigation in the black lipid bilayer system. With this system, the influence of the deletions on pore properties like single channel conductance and voltage dependence could be measured under different conditions.

Benz, R. (1994) *Biochim. Biophys. Acta 1197*, 167-196.

4.23. Sequence-Specific Assignments of the Backbone ^{1}H, ^{13}C and ^{15}N Resonances and Secondary Structure of *Fusarium solani pisi* Cutinase by Heteronuclear Multidimensional NMR

J. J. Prompers,* J. Vergeer,* A. Groenewegen,[§] H. A. M. Pepermans,[§] and C. W. Hilbers*

*NSR Center, University of Nijmegen, Toernooiveld, 6525 ED Nijmegen and [§]Unilever Research Laboratory, Olivier van Noortlaan 120, 3133 AT Vlaardingen, The Netherlands

Essentially complete sequence-specific assignments were made for the backbone ^{1}H, ^{13}C and ^{15}N resonances of *Fusarium solani pisi* cutinase, produced as a 214-residue heterologous protein in *E. coli*. Three-dimensional spectra of a doubly uniformly labeled sample in H_2O correlated the peptide ^{1}H-^{15}N with backbone nuclei for the residue itself and the preceding residue, namely with ^{13}C', ^{13}C$_\alpha$ and ^{1}H$_\alpha$. Concerted analysis of all these spectra using interactive graphics allowed to transverse the protein backbone in both directions, yielding continuous stretches of assigned residues terminating at Pro residues and at residues with undetectable or ambiguous resonances. Some ambiguities were solved by checking whether all previously identified sets of ^{13}C', ^{13}C$_\alpha$ and ^{1}H$_\alpha$ signals do indeed come from the same residue in a H_2O version of the COCAH experiment. Stretches of residues were positioned in the sequence by identification of the amino acid types using the combination of C_α, C_β and H_β chemical shifts. The latter were obtained from 3D ^{1}H-^{15}N-^{13}C$_\beta$ and ^{1}H-^{15}N-^{1}H$_\beta$ correlation experiments, respectively. The backbone assignment was performed using a single sample containing 2mM cutinase in H_2O.

Secondary structure elements were identified by the typical secondary chemical shifts and NOEs as observed in a ^{15}N-edited NOESY. This allows a first comparison of the structure in solution with the crystal structure from X-ray diffraction.

4.24. NMR Investigations of Human Angiogenin

M. Robin,[1] H. Thuring,[2] J. Y. Lallemand,[2] and M. Vuilhorgne[1]

[1]Rhône Poulenc Rorer, 13 quai Jules Guesde, 94400 Vitry sur Seine and [2]Ecole Polytechnique, 91128 Palaiseau, France

Human angiogenin was first isolated from the medium of human colon adenocarcinoma cells and is involved in the induction of blood vessel growth. Later it was also found in human plasma [0.11-0.38 mg/ml] as well as bovine milk [1.5-4.5 mg/ml]. mRNAs coding for angiogenin are found in several other tissues.

The angiogenins (~14 kD) with known sequence have 70% sequence homology. They also share a 35% primary sequence homology with several pancreatic RNases. Despite catalytically essential groups of RNase A being conserved, only minimal activity with low molecular weight RNA is observed. However, angiogenin exhibits ribonucleolytic activity with high molecular weight RNA such as 16S and 24S RNA.

To understand the different functionality of angiogenin, we have started NMR investigations of human angiogenin. A coding gene was synthesized and cloned in *E. coli*. Isotopic labeling was achieved by growing the bacteria on minimum medium with ^{15}NH4Cl as sole nitrogen source. After expression of the gene, angiogenin [Met+1] is obtained in an insoluble form consisting of up to 15% of total protein. The protein was denatured in 7 mM Guanidinium chloride, renatured, and purified.

In the ^{1}H NMR spectra, multiple resonances were observed for the indole NH of W89. NMR and Circular Dichroism spectra were then recorded at different temperatures and conditions. The analysis of these spectra show that this multiplicity is due to the existence of several conformers in slow exchange rather than to differently folded isoforms. The cis/trans isomerization of a proline residue close to W89 could be responsible for this phenomenon. Homonuclear 2D ^{1}H and heteronuclear 2D and 3D ^{1}H-^{15}N NMR spectroscopy was then used to obtain the secondary structure of human angiogenin. Further experiments are under way to determine its 3D structure in solution.

4.25. Comparison of Wild-Type C70A and C70S *Azotobacter vinelandii* Flavodoxin - Backbone Assignments and Secondary Structure of Oxidized C70A Flavodoxin as Determined by Three-Dimensional Heteronuclear NMR Spectroscopy

Elles Steensma, Dirk Heering, Fred Hagen, and Carlo van Mierlo

Department of Biochemistry, Wageningen Agricultural University, Dreijenlaan 3, 6703 HA Wageningen, The Netherlands

We have chosen *Azotobacter vinelandii* flavodoxin as a model system to study protein folding since it adopts one of the nine protein domain superfolds: the a/b doubly wound superfold which consists of four α–helices and a β–sheet (Orengo et al., 1994). Flavodoxins are able to function as electron carriers between redox proteins because they contain a non-covalently bound FMN molecule. *A. vinelandii* flavodoxin is thought to be involved in electron transport to nitrogenase (Klugkist et al., 1986; Mayhew and Tollin, 1992).

A. vinelandii flavodoxin is a 180-residue protein which has been cloned and overexpressed in *E. coli* (Van Mierlo et al., 1995). Since the intermolecular disulfide bonds are formed between individual wild-type flavodoxin molecules, the single cysteine at position 70 was replaced by an alanine as well as a serine residue. The resulting C70A and C70S mutants had conformations indistinguishable from that of wild-type *A. vinelandii* flavodoxin, as shown by NMR spectroscopy. Besides, EPR-monitored titrations and cyclic voltametry measurements showed no significant difference in midpoint potentials between wild-type, C70A and C70S flavodoxin. Guanidinium hydrochloride induced unfolding/refolding of wild-type, C70A and C70S flavodoxin has been monitored by fluorescence and indicates that the three proteins have comparable stability's (Van Mierlo et al., 1995). Since C70A flavodoxin has the smallest difference in stability as compared to the intermolecular disulfide bond forming wild-type flavodoxin, we decided to continue our studies on this mutant.

Three-dimensional heteronuclear NMR studies were performed using both uniformly enriched ^{15}N and ^{15}N/^{13}C labeled C70A flavodoxin. Assignment of the backbone resonances and determination of the secondary structure of oxidized C70A flavodoxin using data obtained from 3D-NOESY-HMQC, 3D-TOCSY-HMQC, CT-HNCO, CT-HNCA, CT-HN(CO)CA and CT-HN(CA)CO experiments is currently in progress.

Future research will be directed towards the determination of the three-dimensional structures and dynamical properties of both oxidized and reduced C70A flavodoxin and the elucidation of the folding pathway of C70A flavodoxin with use of NMR techniques. In addition, the interaction between *A vinelandii* C70A flavodoxin and nitrogenase, its putative electron acceptor protein, will be studied using NMR.

Klugkist, J., Voorberg, J., Haaker, H. and Veeger, C. (1986) *Eur. J. Biochem. 155*, 33-40.

Mayhew, S. G. and Tollin, G. (1992) in: *Chemistry and Biochemistry of Flavoenzymes*, (ed., F. Muller), CRC Press, Boca Raton, Florida, p. 384.

Orengo, C. A., Jones, D. T. and Thornton, J. M. (1994) *Nature 372*, 631-634.

Van Mierlo, C. P. M., Van den Berg, W. A. M., Van Berkel, W. J. H., Van Dongen, W. M. A. M. and Steensma, E. (1995) *manuscript in preparation.*

4.26. Solution Structure of the Spectrin EF-hands in the Absence and in the Presence of Calcium: Anatomy of an Ion-Induced Conformational Change

G. Travé,* M. Saraste, and A. Pastore[#]

*CNRS-UPR9003, ESBS, Université Louis Pasteur, 67400 Illkirch Graffenstaden, France and [#]EMBL, Meyerhofstr. 1, W-69012 Heidelberg, Germany

Background. The EF-hands are calcium-binding helix-loop-helix motifs shared by all the members of the calmodulin superfamily. EF-hands always occur in pairs tightly connected to each other by helix-helix interactions. Each loop binds one calcium ion. A current model solely based on the comparison of different domains, suggests that EF-hands involved in signal transduction undergo a major conformational change ("closed" towards "open" transition) allowing protein-protein interaction. The major effect of the transconformation is a change in the relative orientation of the two helices in each motif.

Results. We have determined the solution structure of the EF-hand pair from the multi-domain protein α-spectrin in the absence and in the presence of calcium. The domain opens as predicted by the model . We have analyzed the structural basis of the movement. The first EF-hand is the motor, whereas the second EF-hand is a transistor which tunes the affinity of the whole domain for calcium. Cooperativity between the two sites is transmitted via a small β-sheet connecting the two calcium-binding loops.

Conclusions. For the first time, a pair of EF-hand motives is proven to undergo the closed-to-open transition. This result has allowed us to extract the rules of the conformational change. Based on the high sequence conservation of EF-hands, we expect these rules to be extendible to other members of the family, including calmodulin and troponin C.

4.27. Conformational Studies of Microcystin-LR using NMR Spectroscopy and Molecular Dynamic Calculations

Gull-Britt Trogen,[1] Arto Annila,[2] Janusz Zdunek,[3] Ingmar Sethson,[1] and Ulf Edlund[1]

[1]Department of Organic Chemistry and [3]Department of Medical Biochemistry and Biophysics, Umeå University, S-90187, Umeå, Sweden and [2]VTT, Chemical Technology, POB 1401 FIN-02044 Espoo, Finland

Some genera of blue-green algae produce hepatotoxic cyclic heptapeptides, microcystins (Rinehard et al., 1994). The general structure of microcystins is described as cyclo(-D-Ala-L-X-erythro-β-methyl-D-Asp-L-Z-Adda-D-Glu-N-methyl-dehydro-Ala), where X and Z are variable amino acids and Adda refers to a β-amino acid (2S, 3S, 8S, 9S)-3-amino-9-methoxy-2,6,8-trimethyl-10-phenyldeca-4(E),6(E)-dienoic acid.

Three-dimensional structures of microcystin-LR were determined in aqueous and dimethyl sulfoxide/water solutions. The conformations of this cyanobacterial toxin were studied using NMR spectroscopy and a simulated annealing (SA) protocol followed by refined SA calculations *in vacuo* (Nilges et al., 1988) using the program XPLOR. In order to gain an insight into dynamic processes of microcystin-LR in solution, unrestrained MD simulations (Brooks et al., 1983) in water were performed within the program CHARMm.

The peptide ring was found to have a saddle-shaped form, essentially the same in both solvent systems. The structural difference between the two solution structures was most significant for the part consisting of Mdha, Ala and Leu. The Arg side chain is very flexible, while the side chain of Adda, essential for activity, is constrained in the vicinity of the backbone ring, but flexible in the more remote part.

Brooks, B. R., Bruccoleri, R. E., Olafson, B. D., States, D. J., Swaminathan, S. and Karplus, M. (1983) *J. Comp. Chem. 4*, 187-217.

Nilges, M., Gronenborn, A. M., Brünger, A. T. and Clore, G. M. (1988) *Protein Eng. 2*, 27-38.

Rinehard, K. L., Namikoshi, M. and Choi, B. W. (1994) *J. Appl. Phycol. 6*, 157-176.

4.28. Structural Analysis of *cis/trans* Proline Isomerization in Staphylococcal Nuclease

Dagmar M. Truckses,[1] John R. Somoza,[2] Kenneth E. Prehoda,[1] and John L. Markley[1]
[1]Department of Biochemistry, College of Agricultural and Life Sciences, University of Wisconsin-Madison, 420 Henry Mall, Madison, WI 53706, USA and [2]Graduate Group in Biophysics, University of California at Berkeley, Berkeley, CA 94720, USA

We have refined the crystal structures of H124L staphylococcal nuclease (recombinant protein whose sequence is identical to the nuclease produced by the V8 strain of *Staphylococcus aureus*) and the point mutants H124L+P117G and H124L+P47G+P117G. The H124L structure is identical to the wild type (WT) structure. The increased stability of H124L (approximately 1.2 kcal/mol at 20°C, pH 7.0 (Alexandrescu et al., 1990)) compared to WT staphylococcal nuclease (nuclease) is explained in terms of greater stability of the helix containing residue 124. We hypothesize that this increased helical stability strengthens the C-terminal anchoring of the loop containing the Lys[116]-Pro[117] peptide bond (Hodel et al., 1993) and thus causes the 10% change in the *cis/trans* equilibrium of this bond as observed by NMR spectroscopy (Shortle, 1986). The H124L+P117G structure is identical to the WT P117G structure (Hynes et al., 1994). Residues 115-118 adopt a type I' β-turn conformation, shifting the loop containing this residue (residues 112-118) closer to the adjacent loop around residue 80. This rearrangement makes residues Tyr[115] and Lys[116] point into the active site. A water molecule appears to stabilize this conformation by hydrogen bonding to residues on both of the adjacent loops. The WT and H124L nuclease populations with a *trans* Lys[116]-Pro[117] peptide bond (approximately 6% in WT and 13% in H124L nuclease) probably have a local conformation similar to that observed in the H124L+P117G structure. Local structural changes caused by the P47G mutation could not be analyzed in detail because of the weak electron density for this flexible part of the structure. However, this mutation did not cause any global structural changes.

Alexandrescu, A. T., Hinck, A. P. and Markley, J. L. (1990) *Biochemistry 29*, 4516-4525.

Hodel, A., Kautz, R. A., Jacobs, M. D. and Fox, R. O. (1993) *Protein Science 2*, 838-850.

Hynes, T. R., Hodel, A. and Fox, R. O. (1994) *Biochemistry 33*, 5021-5030.

Shortle, D. (1986) *J. Cell. Biochem. 30*, 281.

4.29. NMR Structure Determination of the Wheat HMW DX5 Domains

Eric van Swieten,[1] Alard A. van Dijk,[1] Tjibbe Bosma,[1] Christin Choma,[1] Ruud M. Scheek,[2] and George T. Robillard[1]

[1]Department of Biochemistry and [2]Department of Biophysical Chemistry, Groningen Biomolecular Sciences and Biotechnology Institute, University of Groningen, Nijenborgh 4, 9747 AG Groningen, The Netherlands

4.29.1. Introduction. Wheat gluten proteins play a fundamental role in determining the breadmaking quality of wheat. An important portion of these proteins consists of the high molecular weight (HMW) subunits. It is known that the HMW subunits in particular show a strong relation with the breadmaking performance in varieties of wheat cultivars. To elucidate the role of the HMW subunits, fundamental structural research at atomic resolution is required. Our project comprises the subcloning , expression, global characterization (CD, DSC) and NMR structure determination of the HMW DX5 subunit. The HMW DX5 subunit has three distinctive domains. The A and C domains are thought to be globular and contain 110 and 42 amino acids, respectively. Predictive analysis based on amino acid composition gives α-helical like structure for both domains. The central repetitive B-domain contains 680 amino acids (consensus sequence: PGQGQQ) and possesses the unusual β-helix substructure, which is also found for elastin.

4.29.2. A-Domain. The expression and purification of the A-domain is still in progress. Attempts to isolate the protein in a soluble form under non-denaturing conditions have failed until now. After ample expression of the A-domain in *E. coli*, it appeared that the A-domain is formed in inclusion bodies. This was demonstrated by EM immunogold labeling.

4.29.3. B-Domain. The HMW DX5 B-domain is a large domain composed of 680 amino acid residues. It has a repetitive sequence with several consensus peptides of different sizes. One of them is the frequently occurring hexapeptide PGQGQQ. For the B-domain, NMR structural determination would be impossible by its size. The subcloned B-domain, which is split into four equal parts would fulfill the size criterion, but then extensive resonance overlap would cause too many problems in the spectral analysis. To overcome these limitations, a cyclic dodecapeptide was synthesized which contains two times the consensus hexapeptide sequence. The cyclic form was chosen to prevent the peptide from being too flexible. Structural research was initiated by looking to the secondary structure by means of circular dichroism (CD) measurements. From the CD spectrum a predominantly β-turn structure was calculated (CONTIN). To start structural determination on this peptide, several 2D NMR experiments were performed. The analysis of ROESY, TOCSY and ^{13}C-HSQC spectra have lead to the complete assignment of ^{1}H and ^{13}C resonances. Further ROESY spectra were taken to examine flexibilities of specific protons by variation in mixing time and temperature. More experiments combined with a thorough analysis are needed to get more insight in the structure and dynamic processes of the B-peptide.

4.29.4. C-Domain. To alleviate the time-consuming process of expression and purification for the C domain, this protein was prepared synthetically. For the initial characterization we used CD, DSC and NMR (1D, 2D). It appeared that the C-domain adopts a random coil structure in water. Variation of pH, temperature and salt concentration had no effect on the overall structure. To overcome this situation, trifluorethanol (TFE) was added to favor the α-helical propensity, which was found by GOR secondary structure prediction. The addition of TFE effected α-helical structure, which was maximal at 40% TFE (v/v) and higher TFE concentrations. The standard concentration for further NMR experiments was therefore kept at 40% TFE. Two-dimensional NMR experiments were performed to solve the three-dimensional structure. Spectra were taken on a VARIAN UNITY 500 spectrometer. The sample is not isotopically enriched, so assignment via heteronuclear experiments is not possible. Thus far, 38 out of 42 spin systems have been identified from TOCSY spectra. 18 spin systems from this set were assigned to specific residues in the sequence. The most distinctive spin systems were used in the sequential assignment, consisting of the single arginine, lysines (2), histidines (2), valines (3) and glycines (2). A main problem in the assignment is the relatively high peak overlap for this protein. In the near future we hope to obtain spectra at a higher resolution to solve this problem.

4.30. Detection and Classification of Hyperfine-Shifted ^{1}H, ^{2}H, and ^{15}N Resonances of *Clostridium pasteurianum* Rubredoxin

Bin Xia,[1,2] William M. Westler,[2] Hong Cheng,[2] Jean-Marc Moulis,[3] and John L. Markley[1,2]

[1]Graduate Biophysics Program and [2]Department of Biochemistry, University of Wisconsin at Madison, Madison, WI 53706-1569, USA and [3]CEA, DBMS-Métalloprotéines, CENG 17 Rue des Martyrs, 38054 Grenoble, France

Rubredoxins belong to the simplest class of iron-sulfur proteins. They contain a single iron coordinated by four cysteinate sulfurs. The rubredoxin from *Clostridium pateurianum* was overproduced in *Escherichia coli*, and the metal was incorporated into the apoprotein by *in vitro* reconstitution. Protein samples were prepared at natural isotopic abundance, labeled uniformly with ^{15}N, and labeled specifically with [^{2}H$^\alpha$]cysteine, [^{2}H$^{\beta2,\beta3}$]cysteine, and [^{15}N]cysteine. One-dimensional ^{2}H, ^{2}H, and ^{15}N nuclear magnetic spectroscopy was used to study the electron-nuclear interactions. Previously unreported hyperfine-shifted resonance signals were observed in the ^{1}H and ^{2}H NMR spectra of rubredoxin samples in both the oxidized and reduced states. Signals from the α- and β-hydrogens of the four cysteines were identified unambiguously from ^{1}H and ^{2}H NMR spectra of samples labeled selectively with deuterium. The cysteine hydrogen signals are resolved more clearly by ^{2}H (lower magnetogyric ratio) than by ^{1}H (higher magnetogyric ratio) NMR spectroscopy. In the oxidized state, signals from two of the four α-hydrogens are located downfield in the 150 to 175 ppm range; the other two are found upfield at about -10 ppm. Signals from all eight β-hydrogens were detected downfield in the 300 ppm to 900 ppm region. Upon reduction, the ^{1}H NMR signals from all eight β-hydrogens lie downfield between 150 and 240 ppm; signals from two of the four α-hydrogens lie upfield near 0 ppm, and those from the other two are downfield between 10 ppm and 20 ppm. Thirteen hyperfine-shifted signals were resolved in one-dimensional ^{15}N NMR spectra of the sample labeled uniformly with ^{15}N. The two signals located farthest upfield and two signals in the downfield region were assigned to the cysteines that ligate the iron on the basis of selective labeling with [^{15}N]cysteine. Other ^{15}N hyperfine-shifted signals were also classified by amino acid types through selectively isotopic labeling.

4.31. Structural Investigation of PsaD Using Multi-Dimensional NMR

Z. Xia,[a] R. W. Broadhurst,[a] E. D. Laue,[a] D. A. Bryant,[b] J. H. Golbeck,[c] and D. S. Bendall[a]

[a]Department of Biochemistry, University of Cambridge, Tennis Court Road, Cambridge CB2 1QW, UK; [b]Department of Molecular and Cell Biology, Pennsylvania State University, University Park, PA 16802 and [c]Department of Biochemistry, University of Nebraska, Lincoln NE 68583, USA

PsaD is located at the stromal side of the PSI reaction center which catalyses the electron transfer from plastocyanin (or cytochrome c-552) to ferredoxin, and is believed to play a crucial role in transferring electrons from PsaC - the terminal electron acceptor in PSI - to ferredoxin (Golbeck, 1993). We are determining the solution structure of this protein using multidimensional NMR techniques.

The PsaD of Nostc sp. PC 8009 was expressed in *E. coli* grown on a minimal medium, labeled with 15-Nitrogen and purified as previously described (Li et al., 1991).

^{15}N-^{1}H HSQC spectra of PsaD indicate that a large part of the protein is not well structured and that there are flexible regions. From a 2D ^{15}N-^{1}H HSQC-COSY spectrum we roughly estimate, using chemical shifts as an index (Wishart et al., 1991), that the helical part accounts for 19% at most, β–strand around 13% and random coil at least 54%.

At least three domains may exist in this protein as indicated by 15-Nitrogen relaxation measurements. The ^{15}N-^{1}H HMQC spectra at several different pH values from 5.7 to 8.5 suggest that there is a conformational equilibrium in slow exchange with a pK_a of about 6.3 and that the structured part of the acid form is less stable. The functional relevance of this conformational change is being investigated.

Golbeck, J. H. (1993) *Curr. Opin. Struct. Biol. 3*, 508.
Li, N., Zhao, J., Warren, P. V., Warden, J. T., Bryant, D. A. and Golbeck, J. H. (1991) *Biochemistry 30*, 7863.
Wishart, D. S., Richards, F. M. and Sykes, B. D. (1991) *FEBS Lett. 293*, 72.

4.32. Backbone Dynamics of *trp* Repressor

Zhiwen Zheng,* Jerzy Czaplicki,[†] and Oleg Jardetzky*

*Stanford Magnetic Resonance Laboratory, Stanford University, Stanford, California 94305-5055 and [†]Centre de Recherche de Biochimie et Génétique Cellulaires, CNRS, 118, route de Narbonne, 31062 Toulouse Cedex, FRANCE

Backbone dynamics of *trp* repressor, a 25 kDa DNA binding protein, have been studied using ^{15}N relaxation data measured by proton-detected two-dimensional ^{1}H-^{15}N NMR spectroscopy. ^{15}N spin-lattice relaxation time (T_1), spin-spin relaxation time (T_2), and heteronuclear NOEs were determined for all visible backbone amide ^{15}N nuclei. Monte Carlo simulations of the amplitudes of backbone motions led to the conclusion that a wobbling in a cone model with consideration of the anisotropic reorientation of the molecule was appropriate to describe the underlying motions, allowing us to derive semi-angle of the cone (α) and the effective correlation time for internal motions (τ_e) for each N-H bond vector. The final optimized rotational diffusion coefficients parallel ($D_{\parallel}$) and perpendicular ($D_\wedge$) to the unique axis of the molecule were found to be $1.48 \pm 0.06 \infty 10^7$ and $1.15 \pm 0.05 \infty 10^7$ s^{-1}, respectively. The average semi-angle of the cone (α) describing the amplitude of NH vector motions on the picosecond time scale was found to be $20.9° \pm 5.7°$. Large amplitude motions on the picosecond time scale are found at both the N and C termini, but are restricted both in the hydrophobic core and DNA-binding regions. The ^{15}N experiments were also performed for the apo repressor. Similar T_1, T_2, and NOE profiles were obtained. No significant differences in the segmental motions on the picosecond time scale were found between the apo and holo repressor.

4.33. NMR Studies of the Combining Site of an Anti-GP120 HIV Neutralizing Antibody with its Peptide Antigen

Anat Zvi, Irina Kustanovich, Miriam Eisenstein, and Jacob Anglister

Department of Structural Biology, Weizmann Institute of Science, Rehovot 76100, ISRAEL

The 24 amino acid peptide RP135 (NNTRKSIRIQRGPGRAFVTIGKIG) corresponds in its amino acid sequence to the principal neutralizing determinant (PND) of the HIV-1$_{IIIB}$ isolate (residues 308-331 of the envelope glycoprotein gp120). In order to map the antigenic determinant recognized by 0.5β, the complex of RP135 with an anti-gp120 HIV neutralizing antibody, 0.5β, which cross reacts with the peptide was studied by 2D NMR spectroscopy. A combination of HOHAHA and ROESY spectra of the Fab/peptide complex measured in H$_2$O was used to eliminate the resonance of the Fab and the tightly bound peptide residues and to obtain sequential assignments for those part of the peptide which retain considerable mobility upon binding. In this manner, a total of 14 residues (Ser^6-Thr^{19}) were shown to be part of the antigenic determinant recognized by the antibody 0.5β. To study the interactions with the PND of HIV-1$_{IIIB}$ we used the peptide RP135a (RKSIRIQRGPGRAFVT), which contains the epitope recognized by the antibody. The NOESY difference spectra measured using specifically deuterated derivatives of RP135a show exclusively the interactions of the deuterated residues both within the bound peptide and with the Fab fragment of the antibody. These measurements reveal within the bound peptide hydrophobic interactions that for a 12-residue loop with the conserved GPGR sequence at its bottom.

PARTICIPANTS PHOTOS

Figure 1. Faculty and students-first week of the second course. "Dynamics and the Problems of Recognition in Biological Macromolecules," at the International School of Biological Magnetic Resonance on May 19–30, 1995, at EMCSC, Erice, Italy.

Figure 2. Faculty and students-second week of the second course, "Dynamics and the Problems of Recognition in Biological Macromolecules," at the International School of Biological Magnetic Resonance on May 19–30, 1995, at EMCSC, Erice, Italy.

AUTHOR INDEX

SUBJECT INDEX

The manufacturer's authorised representative in the EU is Springer
Nature Customer Service Centre GmbH, Europaplatz 3, 69115 Heidelberg,
Germany. If you have any concerns regarding our products, please
contact ProductSafety@springernature.com

Printed and bound by CPI Group (UK) Ltd, Croydon, CR0 4YY
23/04/2026
02095585-0020